Raptors in Human Landscapes

This book is dedicated to the memory of Richard "Butch" Olendorff (1943–1994) whose neverending optimism and amazing foresight allowed him to conceive the theme of this book over twenty years ago when he wrote these words in his now classic work on raptors, *Golden Eagle Country*:

"... birds of prey are exploiting the potential of living in concert with men. Given half a chance, they will even breed in spite of us."

Raptors in Human Landscapes

Adaptations to built and cultivated environments

Edited by

David M. Bird
Avian Science and Conservation Centre,
McGill University, Montreal, Canada

Daniel E. Varland
Northwest Forest Resources,
Rayonier, Hoquiam, Washington, USA

Juan Jose Negro
Estacion Biologica de Donana,
Sevilla, Spain

Academic Press

Harcourt Brace & Company, Publishers

London   San Diego   New York   Boston   Sydney   Tokyo   Toronto

ACADEMIC PRESS LIMITED
24–28 Oval Road
LONDON NW1 7DX

US Edition Published by
ACADEMIC PRESS INC.
San Diego, CA 92101

This book is printed on acid-free paper

Copyright © 1996 ACADEMIC PRESS LIMITED
The chapter by C. J. Henny and J. L. Kaiser, pages 97 to 108, is a US Government work in the public domain
and not subject to copyright

All rights reserved

No part of this book may be reproduced or transmitted in any form
or by any means, electronic or mechanical including photocopying, recording,
or any information storage and retrieval system without permission
in writing from the publisher

A catalogue record for this book is available from the British Library

ISBN 0-12-100130-X

Typeset by Paston Press Ltd, Loddon, Norfolk
Printed and bound in Great Britain by WBC Book Manufacturers Ltd, Bridgend

Contents

Contributors	viii
Preface	xi
Acknowledgments	xix

Raptors in Urban Landscapes

1. Peregrine Falcons in Urban North America — 3
   Tom J. Cade, Mark Martell, Patrick Redig, Gregory A. Septon and Harrison B. Tordoff

2. Bridge Use by Peregrine Falcons in the San Francisco Bay Area — 15
   Douglas A. Bell, David P. Gregoire and Brian J. Walton

3. Eggshell Thickness and Contaminant Analysis of Reintroduced, Urban Nesting Peregrine Falcons in Wisconsin — 25
   Gregory A. Septon and Jim B. Marks

4. The Urban Buteo: Red-shouldered Hawks in Southern California — 31
   Peter H. Bloom and Michael D. McCrary

5. Urban Nesting Biology of Cooper's Hawks in Wisconsin — 41
   Robert N. Rosenfield, John Bielefeldt, Joelle L. Affeldt and David J. Beckmann

6. Urban Ecology of the Mississippi Kite — 45
   James W. Parker

7. Costs and Benefits of Urban Nesting in the Lesser Kestrel — 53
   Jose Luis Tella, Fernando Hiraldo, Jose Antonio Donázar-Sancho and Juan Jose Negro

8. Nesting Success of Western Burrowing Owls in Natural and Human-altered Environments — 61
   Eugene S. Botelho and Patricia C. Arrowood

9. Eastern Screech Owls in Suburbia: A Model of Raptor Urbanization — 69
   Frederick R. Gehlbach

Raptors and Artificial Nest Sites

10. Red-tailed Hawks Nesting on Human-made and Natural Structures in Southeast Wisconsin — 77
    William E. Stout, Raymond K. Anderson and Joseph M. Papp

11 Documentation of Raptor Nests on Electric Utility Facilities Through a Mail Survey 87
Roberta Blue

12 Osprey Population Increase along the Willamette River, Oregon, and the Role of Utility Structures, 1976–93 97
Charles J. Henny and James L. Kaiser

13 The Use of Artificial Nest Sites by an Increasing Population of Ospreys in the Canadian Great Lakes Basin 109
Peter J. Ewins

14 The Osprey in Germany: Its Adaptation to Environments Altered by Man 125
Bernd-Ulrich Meyburg, Otto Manowsky and Christiane Meyburg

15 Effectiveness of Artificial Nesting Structures for Ferruginous Hawks in Wyoming 137
James R. Tigner, Mayo W. Call and Michael N. Kochert

16 Peregrine Falcons: Power Plant Nest Structures and Shoreline Movements 145
Gregory A. Septon, John Bielefeldt, Tim Ellestad, Jim B. Marks and Robert N. Rosenfield

17 Competition for Nest Boxes Between American Kestrels and European Starlings in an Agricultural Area of Southern Idaho 155
Marc J. Bechard and Joseph M. Bechard

Raptors in Cultivated Landscapes

18 White-tailed Kite Movement and Nesting Patterns in an Agricultural Landscape 165
Andrea L. Erichsen, Shawn K. Smallwood, A. Marc Commandatore, Barry W. Wilson and Michael D. Fry

19 Association Analysis of Raptors on a Farming Landscape 177
Shawn K. Smallwood, Brenda J. Nakamoto and Shu Geng

20 Sparrowhawks in Conifer Plantations 191
I. Newton

21 Adaptations of Raptors to Man-made Spruce Forests in the Uplands of Britain 201
Steve J. Petty

22 Spotted Owls in Managed Forests of Western Oregon and Washington 215
Scott P. Horton

23 Goshawk Adaptation to Deforestation: Does Europe Differ From North America? 233
Robert E. Kenward

24	Rain Forest Raptor Communities in Sumatra: The Conservation Value of Traditional Agroforests *Jean-Marc Thiollay*	245
25	Diurnal Raptors in the Fragmented Rain Forest of the Sierra Imataca, Venezuela *Eduardo Alvarez, David H. Ellis, Dwight G. Smith and Charles T. Larue*	263
26	Value of Nest Site Protection in Ameliorating the Effects of Forestry Operations on Wedge-tailed Eagles in Tasmania *Nick J. Mooney and Robert J. Taylor*	275

Raptors in Industrial Landscapes

27	Use of Reservoirs and other Artificial Impoundments by Bald Eagles in South Carolina *A. Lawrence Bryan, Jr., Thomas M. Murphy, Keith L. Bildstein, I. Lehr Brisbin, Jr. and John J. Mayer*	285
28	Attraction of Bald Eagles to Habitats just below Dams in Piedmont North and South Carolina *Richard D. Brown*	299
29	Reclaimed Surface Mines: An Important Nesting Habitat for Northern Harriers in Pennsylvania *Ronald W. Rohrbaugh, Jr. and Richard H. Yahner*	307
30	Raptors Associated with Airports and Aircraft *S. M. Satheesan*	315

Raptors at Large

31	The Effect of Altered Environments on Vultures *David C. Houston*	327
32	The Impact of Man on Raptors in Zimbabwe *Ron R. Hartley, Kit Hustler and Peter J. Mundy*	337
33	Response of Common Black Hawks and Crested Caracaras to Human Activities in Mexico *Ricardo Rodríguez-Estrella*	355
34	Occurrence and Distribution of Diurnal Raptors in Relation to Human Activity and Other Factors at Rocky Mountain Arsenal, Colorado *Charles R. Preston and Ronald D. Beane*	365

Appendix: List of species mentioned in the text	375
Index	387

List of Contributors

Joelle Affeldt, W3832 Manitowoc Road, Kaukauna, WI 54130, USA.
Eduardo Alvarez, PO Box 15251, Gainesville, FL 32604, USA.
Raymond K. Anderson, University of Wisconsin, Stevens Point, WI 54481, USA.
Patricia C. Arrowood, Department of Biology, PO Box 30001/Department 3AF, New Mexico State University, Las Cruces, NM 88003, USA.
Ronald D. Beane, Department of Zoology, Denver Museum of Natural History, Denver, Colorado 80205, USA.
Joseph M. Bechard, Department of Biology, Boise State University, Boise, ID 83725, USA.
Marc J. Bechard, Department of Biology, Boise State University, Boise, ID 83725, USA.
David Beckmann, College of Natural Resources, University of Wisconsin, Stevens Point, WI 54481, USA.
Douglas A. Bell, Department of Ornithology & Mammalogy, California Academy of Sciences, Golden Gate Park, San Francisco, CA 94118, USA.
John Bielefeldt, 4940 Highway ZC, Dovsman, WI 53118, USA.
Keith L. Bildstein, Hawk Mountain Sanctuary, Kempton, PA 19529, USA.
Peter H. Bloom, Western Foundation of Vertebrate Zoology, 439 Calle San Pablo, Camarillo, CA 93010, USA.
Roberta Blue, Carolina Power & Light Company, 412 S. Wilmington Street, Raleigh, NC 27601, USA.
Eugene S. Botelho, Department of Biology, PO Box 30001/Department 3AF, New Mexico State University, Las Cruces, NM 88003, USA.
I. Lehr Brisbin, Jr., Savannah River Ecology Laboratory, Aiken, SC 29802, USA.
Richard D. Brown, Brunswick Community College, PO Box 30, Supply, NC 28462-0030, USA.
A. Lawrence Bryan, Jr., Savannah River Ecology Laboratory, PO Drawer E, Aiken, SC 29802, USA.
Tom J. Cade, The Peregrine Fund, Inc, 5666 West Flying Hawk Lane, Boise, ID 83709, USA.
Mayo W. Call, Consultant, Energy International Inc., 510 E. 2nd. Avenue, Afton, WY 83110, USA.
A. Marc Commandatore, 647 F Street, Davis, CA 95616, USA.
Jose Antonio Donazar-Sancho, AUDA Mª Luisa S.N. 41013 Sevilla, Spain.
Tim Ellestad, 2810 Mason Street, Madison, WI 53705, USA.
David Ellis, Peabody Coal Company, Kayenta, AZ 86033, USA.
Andrea L. Erichsen, Department of Avian Sciences, University of California, Davis, CA 95616, USA.

Peter J. Ewins, Canadian Wildlife Service, Environment Canada, 4905 Dufferin Street, Downsview, Ontario M3H 5T4, Canada.

Michael D. Fry, Department of Avian Sciences, University of California, Davis, CA 95616, USA.

Frederick R. Gehlbach, Professor Emeritus Department of Biology, Baylor University, Waco, TX 76798, USA.

Shu Geng, Department of Agronomy and Range Science, University of California, Davis, CA 95616-8515, USA.

David P. Gregoire, Santa Cruz Predatory Bird Research Group, Lower Quarry, University of California, Santa Cruz, CA 95064, USA.

Ron R. Hartley, Zimbabwe Falconers' Club, Falcon College, Esigodini, Zimbabwe.

Charles J. Henny, National Biological Service, Northwest Research Station, 3080 SE Clearwater Drive, Corvallis, OR 97333, USA.

Fernando Hiraldo, Estacion Biologica de Donana, CSIC, Apdo 1056, 41080 Sevilla, Spain.

Scott P. Horton, Washington State Department of National Resources, 411 Tillicum Lane, Forks, WA 98331, USA.

David C. Houston, Applied Ornithology Unit, Zoology Department, University of Glasgow, Glasgow G12 8QQ, Scotland.

Kit Hustler, PO Box 159, Victoria Falls, Zimbabwe.

James L. Kaiser, National Biological Service, Northwest Research Station, 3080 SE Clearwater Drive, Corvallis, OR 97333, USA.

Robert E. Kenward, Institute of Terrestrial Ecology, Furzebrook Research Station, Wareham, Dorset BH20 5AS, UK.

Michael Kochert, National Biological Survey, Raptor Research and Technical Assistance Center, 3948 Development Avenue, Boise, ID 83705, USA.

Charles T. Larue, Biology Department, Southern Connecticut University, 1501 Crescent St., New Haven, CT 06515, USA.

Michael D. McCrary, 3795 Calle Posadus, Newbury Park, CA 91320, USA.

Otto Manowsky, Schonebecker Str. 12, 16247 Joachimsthal, Germany.

Jim B. Marks, Wisconsin Peregrine Society, PO Box 1148, Milwaukee, WI 53201, USA.

Mark Martell, The Raptor Centre, 1920 Fitch Ave., St. Paul, MN 55108, USA.

John J. Mayer, Westinghouse Savannah River Co., Savannah River Site, Aiken, SC 29808, USA.

Christiane Meyburg, Wangenheimstr. 32, 14193 Berlin, Germany.

Bernd-Ulrich Meyburg, World Working Group on Birds of Prey, Wangenheimstr. 32, 14193 Berlin, Germany.

Nick J. Mooney, Parks & Wildlife Service, GPO Box 44A, Hobart, Tasmania 7001, Australia.

Peter J. Mundy, Department of National Parks and Wildlife Management, Box 2283, Bulawayo, Zimbabwe.

Thomas M. Murphy, Route 2, Box 167, Green Pond, SC 29446, USA.

Brenda J. Nakamoto, Department of Wildlife and Fisheries Biology, University of California, Davis, CA 95616-8515, USA.

Juan Jose Negro, Estacion Biologica de Donana, CSIC, Apdo 1056, 41080 Sevilla, Spain.
Ian Newton, Institute of Terrestrial Ecology, Monks Wood, Abbots Ripton, Huntingdon, Cambs PE17 2LS, UK.
Joseph M. Papp, Route 1, Box 158A, Drummond, WI 54832 USA.
James W. Parker, Aerie East, RR3, Box 3110, Holley Road, Farmington, ME 04938, USA.
Steve J. Petty, The Forestry Authority, Wildlife Ecology Branch, Ardentinny, Dunoon, Argyll PA23 8TS, Scotland
Charles R. Preston, Department of Zoology, Denver Museum of Natural History, Denver, Colorado 80205, USA.
Patrick Redig, The Raptor Centre, 1920 Fitch Ave., St. Paul, MN 55108, USA.
Ricardo Rodriguez-Estrella, Centro de Investigaciones Biologicas del Noroeste, Div. Biol. Terr., Apdo. Postal 128, La Paz 23000 B.C.S., Mexico.
Ronald W. Rohrbaugh, Jr., Sutton Avian Research Center, PO Box 2007, Bartlesville, OK 74005-2007, USA.
Robert N. Rosenfield, Department of Biology, University of Wisconsin, Stevens Point, WI 54481, USA.
S. M. Satheesan, A/303 Shantiniketan, Y.A.C. Nagar, Kondivitta Lane, Andheri (East), Bombay 400 059, India.
Gregory A. Septon, Milwaukee Public Museum, 800 West Wells St., Milwaukee, WI 53233, USA.
Shawn K. Smallwood, Institute for Sustainable Development, 516 Oeste Dr., Davis, CA 95616, USA.
Dwight G. Smith, Biology Department, Southern Connecticut State University, 1501 Crescent St., New Haven, CT 06515, USA.
William E. Stout, W2364 Heather Street, Oconomowoc, WI, 53066, USA.
Robert J. Taylor, Forestry Tasmania, 30 Patrick Street, Hobart, Tasmania 7000, Australia.
Jose Luis Tella, Estacion Biologica de Donana, CSIC, Apdo 1056, 41080 Sevilla, Spain.
Jean-Marc Thiollay, Laboratoire Ecologie, E.N.S. 46 rue d'Ulm 75230 Paris Cedex, France.
James R. Tigner, 104 East Kendrick St., Rawlins, WY 82301, USA.
Harrison B. Tordoff, 100 Ecology, 1987 Upper Buford Circle, St. Paul, MN 55108, USA.
Brian J. Walton, Santa Cruz Predatory Bird Research Group, Lower Quarry, University of California, Santa Cruz, CA 95064, USA.
Barry W. Wilson, Department of Avian Sciences, University of California, Davis, CA 95616, USA.
Richard H. Yahner, School of Forest Resources, Ferguson Building, The Pennsylvania State University, University Park, PA 16802, USA.

Preface

Everyone hears so much today about the destructive effects of human activity on raptor populations. Serious threats include loss and degradation of habitat, pesticides, industrial pollution, electrocution, and direct persecution by shooting, trapping and poisoning. As the human population continues to grow, the chances for the survival of many raptor species continues to diminish. The situation is not entirely bleak, however, as many birds of prey have found ways to survive and to breed successfully in increasingly human-altered environments.

The vast majority of the papers in *Raptors in Human Landscapes* provide evidence of successful coexistence between humans and raptors. The book is a direct result of a symposium, *Raptors Adapting to Human-Altered Environments*, held on 6 November 1993 in Charlotte, North Carolina in the United States in conjunction with the annual meeting of The Raptor Research Foundation, Inc. (RRF). The gathering brought together an audience of more than 175 interested participants to hear 24 papers from scientists originating from a dozen countries including the US, Canada, Scotland, England, France, Spain, Germany, Zimbabwe, Venezuela, Mexico, Czech Republic and Tasmania. Of the 23 manuscripts received, three were rejected through peer review. An additional 14 manuscripts were solicited from scientists known to be studying adaptations of raptors to human-altered landscapes.

The ultimate human-altered landscape must surely be urban centers and raptors are becoming quite familiar with this habitat the world over. Perhaps no better example of this phenomenon can be found than the urban peregrine falcon. It is certainly reason enough to choose as our lead paper the most recent update on this subject by Tom Cade, Mark Martell, Patrick Redig, Greg Septon and Harrison Tordoff, all of whom have had ample experience with peregrine falcons nesting in cities. In 1993 there were almost ninety pairs nesting in more than sixty urban areas in North America! All it takes is a safe nesting site because "sufficient food always seems to be available". Over a hundred different bird species are preyed upon by city peregrines. While buildings comprise the majority of nesting sites, bridges, overpasses and other tall structures are also selected. Bridge sites have not proven to be the best nesting sites as fledging success is rather poor; however, Douglas Bell, David Gregoire and Brian Walton studied bridge use by peregrines in the San Francisco area and they conclude that bridge nest sites must be enhanced to improve fledging success. Otherwise, the best alternative is to foster nestlings from bridge sites to natural eyries. Greg Septon and Jim Marks have done a remarkable job in Wisconsin in making midwestern cities more comfortable for peregrine falcons. And contrary to what the public might think about the cleanliness of urban environments for wildlife and humans alike, Septon and Marks found very low levels of organochlorines in midwestern city peregrine eggs.

In a population of red-shouldered hawks in California that Peter Bloom has studied for over a decade, he and Michael McCrary discovered that close to one-third of their 170 breeding territories were in urban environments or at least in situations frequently exposed to high human activity levels.

Even Cooper's hawks, well known for their secretivity, are adopting suburban lifestyles and breeding in highly fragmented forests. According to Robert Rosenfield, John Bielefeldt, Joelle Affeldt and David Beckmann, Cooper's hawks in Wisconsin suburbs are not only breeding in the highest densities ever recorded, but producing very healthy clutch sizes and raising good numbers of young to the banding stage.

James Parker, a self-employed wildlife consultant in Maine, claims that the Mississippi kite "may be the most abundant urban raptor in North America". While his studies have been restricted to only four states in the US, Parker has amassed a substantial body of information on this species, particularly known for its aggressive territorial defense against humans. Densities of urban-nesting Mississippi kites are quite high with roosting populations often exceeding 50 birds. Interestingly, the urban populations appear to be more successful than rural ones.

One of the reasons touted for better nesting success for urban-nesting raptors is lowered predation. This argument is backed by a long-term study of urban versus rural populations of lesser kestrels in Spain by Jose Luis Tella, Fernando Hiraldo, Jose Antonio Donazar and Juan Jose Negro of the Donana Biological Station based in Seville, Spain. At least nine predatory species were the primary cause of death for adults and nestlings in rural kestrels, whereas urban kestrels must contend with only two predatory species. On the other hand, food is not as plentiful in the latter environment and resulted in the loss of nestlings through starvation.

Urban-nesting behavior is not just a feature of diurnal raptors; the owls, especially great-horned owls, barred owls and screech owls, have long been attracted to protected greenspaces in cities and even to suburban landscapes with suitable tangle and large nesting trees.

Nesting below ground in burrows in open areas, the burrowing owl readily adapts to human landscapes. Eugene Botelho and Patricia Arrowood of New Mexico State University found that pairs nesting in human-altered areas fledged more young than those in natural areas. While they do attribute the lower nesting success in the latter population to higher predation pressure, they also suggest that close neighbor proximity in the natural population may have led to more disturbance of nesting activities. Contrary to the lesser kestrel study, Botelho and Arrowood proposed that the burrowing owls nesting in human-altered areas may have greater access to food resources, e.g. insects and bats attracted to lights and carcasses of birds in large passerine roosts on their university campus.

Perhaps no one has taken the study of urbanization in raptors as far as

Frederick Gelhbach of Baylor University in Waco, Texas. In a long-term study, he is using the suburban-nesting screech owl as a model of urbanization. He uses the term "ecologically plastic" to refer to species like the screech owl and he contends that greater survival of eggs and chicks, combined with enhanced urban resources and climatic stability are the driving forces behind the urbanization of the eastern screech owl.

As is commonly known by raptor biologists these days, the two limiting factors for raptor populations are availability of food and nesting sites. Raptors are proving to be equally opportunistic in the acquisition of either of these limited commodities.

A good example in North America is the red-tailed hawk. Not only is this species readily accepting power poles and high-voltage transmission towers as nesting substrates, William Stout, Raymond Anderson and Joseph Papp of Wisconsin have also documented them nesting on billboards in urban centers like Milwaukee! As a result, urban populations of red-tailed hawks are growing at a rapid pace.

The sponsorship of this symposium by a number of power utility companies was not serendipitous by any means. Throughout the world, raptors of all kinds, both cliff and tree nesters alike, have made use of power transmission structures for nesting purposes. While that behavior can lead to accidental death of raptors, it is in the best interests of both the utility industry and birds of prey to eradicate the problem of raptor electrocution. Thus, raptor biologists, cinematographers and utility industry personnel have worked long and hard to make power transmission structures safe nesting havens for raptors.

Roberta Blue of Carolina Power & Light Company based in North Carolina circulated a two-part questionnaire to solicit information on the number and species of raptors nesting on power line structures, as well as participation by electric utility companies in enhancing such structures for nesting raptors. Among the 12 species of raptors reported, ospreys and red-tailed hawks were the most common. Of 141 companies receiving questionnaires, 88 returned them. Of these, 58 were contributing to raptor enhancement projects either by installing artificial nesting platforms or by participating in hack release projects of raptors like peregrine falcons.

Without a doubt, the osprey provides the best example of a raptor adopting utility structures, e.g. poles, towers, for nesting substrates. While Charles Henny and James Kaiser of the National Biological Survey in Corvallis, Oregon attribute some of the success of the ospreys nesting on the Williamette River to the banning of DDT, improved water and fish conditions in the river, and notably to "a new enlightened attitude towards birds of prey which resulted in less shooting", the availability of utility structures in that region may now give the birds unlimited potential for nesting sites.

A similar situation exists in the Great Lakes drainage basin, according to Peter Ewins of the Canadian Wildlife Service. Prior to 1945, ospreys nested only in trees. Now the birds are using a wide range of artificial structures, including those installed specifically for them. On Lake Huron, for example, over four-fifths of artificial platforms were occupied within one year of installation. The birds are also showing a high degree of tolerance toward human activities near their nests.

The increase in osprey breeding populations is not just relegated to North America either. According to Bernd-Ulrich Meyburg, Otto Manowsky and Christiane Meyburg of Germany, a shortage of suitable nesting trees in their country was clearly a limiting factor in osprey numbers. As early as 1938 the ospreys began nesting on power pylons and now over three-quarters of the population raises their young on them. Moreover, nesting success is higher on the pylons than on tree sites.

Similar results were found by James Tigner, Mayo Call and Michael Kochert for ferruginous hawks in Wyoming. This species not only readily accepted artificial nesting platforms provided for them, they also enjoyed improved nesting success.

The peregrine falcon is even better known for its plasticity in nesting substrates, e.g. cliffs, ground, trees, buildings, etc. In the US midwest, this species is readily accepting nest boxes erected on man-made structures, particularly power plants, along the shores of Lake Michigan and the Mississippi River by an innovative raptor management team composed of Greg Septon, John Bielefeldt, Tim Ellestad, Jim Marks and Robert Rosenfield.

As a cavity-nester, the American kestrel readily accepts nestboxes of various designs and as a result, a number of nestbox colonies of kestrels have been established all over North America for research and management purposes. Marc Bechard and his son, Joseph, have set up such a colony in southern Idaho in farmland dominated by livestock grazing and hay production. One of the banes of kestrel nestbox colonies has been competition with European starlings, but the Bechards found that kestrels can successfully outcompete starlings and even exclude them from farmland areas.

Not all raptor species are embracing agricultural landscapes "with open wings". A geographical information system was used by Andrea Erichsen, Shawn Smallwood, Marc Commandatore, Barry Wilson and Michael Fry of Davis, California to examine white-tailed kite movement and nesting patterns in such habitat in the Sacramento Valley. While the kites chose areas of natural vegetation overall for nesting, the most successful nests were associated with more human development, but without human activity present, i.e. abandoned farms, vacant lots, a cemetery. They also preferred to hunt voles in sugarbeet and alfalfa fields, as well as in rice stubble fields in spring.

Shawn Smallwood also teamed up with Brenda Nakamoto and Shu Geng to survey and analyze raptor activity in general in the farming landscape of the Sacramento Valley, California. Fourteen species of raptors belonging to the Accipitridae and Falconidae "preferentially selected riparian, wetland and upland vegetation, alfalfa fields, and rice stubble and other crop debris, all of which occurred rarely". Human settlements, plowed fields, and row and grain crops were avoided. Areas with perches, e.g. utility poles, snags, were preferred by the Accipitridae. Smallwood and his colleagues hope to assist agriculturists in developing more effective strategies for increasing raptor populations in farmland.

Another form of agriculture is tree plantations, often consisting of a single species of desirable commercial value. Such monoculture has often been deemed by wildlife managers as undesirable for wildlife, including raptors. However, Ian Newton's long-term population study of sparrowhawks in Great Britain shows otherwise. Despite lower prey availability, these accipiters prefer to nest in conifer plantations, mostly in those where nesting success was the highest. Furthermore, the age of the plantation is also a factor. Sparrowhawks especially prefer those plantations that have been thinned for the first time, around 20 years of age. After roughly 10 years of good production of young, nesting success declines. While rotational management, i.e. felling and replanting, does tend to keep sparrrowhawk numbers stable, the birds will abandon some of the older plantations and occupy younger ones.

Steve Petty, a wildlife ecologist with Scotland's Forest Authority, takes a broader view of the impact of deforestation and reforestation on raptors in Britain. New raptor guilds and their associated prey have been created in the last 75 years with the creation of large spruce forests in the British uplands. Of the 16 raptor species inhabiting the uplands, most have benefited or been affected little by the successional changes in the forest. Petty suggests that forestry management practices can be improved to better enhance the habitat for raptors.

Scott Horton of the Washington State Department of Natural Resources could not agree more. In the last decade, perhaps no other species has better demonstrated the conflict between the forest industry and wildlife than the spotted owl of the Pacific Northwest. Horton reviewed its situation in the managed forests of western Oregon and Washington. While the spotted owl undoubtedly prefers old, unmanaged forest for nesting and foraging, Horton concludes that "forests can be managed to provide favorable conditions for the owls". He calls for integrated management plans for commercial forests to accommodate the spotted owl.

What works well for a species on one continent may not be applicable worldwide though. Robert Kenward's study of the goshawk in Europe is a case in point. Northern goshawks in North America nest mainly in areas of continuous woodland, but in Europe they do well in woodland/farmland mosaics. Kenward attributes the difference to winter diets, competition with *Bubo* and *Buteo* species, and forest management.

Unlike in Great Britain and the US, managed agroforests in western Indonesia, which are rapidly replacing natural forests there, do not appear to be as beneficial for raptor populations. Jean-Marc Thiollay examined the conservation value of agroforests in Sumatra and found that "both species richness and density in agroforests were more than twice as high as in cultivated areas, but they were twice as low as in primary forests." Thiollay concludes that agroforests conserve only a quarter of the original forest raptor community.

Whether any tropical rainforest raptor species can adapt to these severe habitat changes, if given the time, is not known, but there may be a glimmer of hope. Extensive deforestation to make way for agriculture has also occurred in eastern Venezuela. Moreover, mining and logging activities have resulted in the construction of access roads which have fragmented the forest habitat. During surveys of activity and habitat use by 42 raptor species in the rainforest of the Sierra Imataca, Eduardo Alvarez, David Ellis, Dwight Smith and Charles Larue noted that some raptors generally considered to be "forest interior species", as well as some of the open country species, were foraging and roosting in the man-made openings in the forest.

Nick Mooney, a biologist with the Parks and Wildlife Service in Tasmania, and Robert Taylor with Forestry Tasmania, suggest that some raptors can learn to adapt to disturbance caused by human alteration of landscapes, if given the proper circumstances. Their case in point is the endangered Tasmanian wedge-tailed eagle. If their nests are discovered before logging commences, protective measures such as retention of the nest tree, a buffer zone, and reduction of disturbance by the logging operation might not only maintain the eagles' numbers, but also help develop a population more tolerant of disturbance.

In some instances, the creation of new foraging habitat can result in associated increases in nesting territories. In 1977 there were only 12 active bald eagle nesting territories in South Carolina. Over a 22-year period, this number increased seven-fold! Part of this remarkable growth was due to the creation of water reservoirs by dams. According to Lawrence Bryan, Jr., Thomas Murphy, Keith Bildstein, Lehr Brisbin, Jr., and John Mayer, bald eagles not only rapidly locate and use these reservoirs, but enjoy better nesting success than those eagles in non-reservoir areas.

Even wintering bald eagles are finding the reservoirs attractive for foraging. Richard Brown of Charlotte, North Carolina conducted mid-winter surveys of bald eagles over roughly 10 years and found that 88 percent of the birds were seen just below dams. Some dams were more attractive than others, perhaps because of higher fish populations and/or greater availability of perch sites.

Northern harriers can also respond favorably to habitat changes, in particular to open grassland habitat created by the reclamation of surface mines in Pennsylvania. Ronald Rohrbaugh, Jr., and Richard Yahner of the Pennsylvania State University discovered that breeding harriers are seeking out this relatively new form of habitat in that state and that if managed properly, reclaimed grassland surface mines will not only provide important nesting habitat for northern harriers, but may help offset the loss of traditional nesting habitat elsewhere in their geographical range in North America.

Not always is a favorable adaptive response to human landscapes welcomed. Airports also constitute human-altered environments. The necessity of maintaining large expanses of short-cropped wild grasses and the occasional proximity of airports to cities, garbage dumps, marshes, and larger bodies of water often make them ideal habitat for microtines, insects of various kinds, open-country and water-dwelling bird species, and of course, the raptors that feed on them. Naturally, the economic losses, not to mention human fatalities, associated with collisions with aircraft cannot be ignored. Some airports are more attractive to raptors than others. S.M. Satheesan reviewed the situation in India and found that more than half of all aircraft strikes at airports between 1966 and 1993 involved raptors.

As highly adaptive scavengers capable of eating even human excrement, black kites are one of the most common and successful raptors in the world today. In North America, the turkey vulture continues to extend its range northward, especially in eastern Canada. Whether this is due to climatic warming trends or greater access to carcasses or both is not known. David Houston of the University of Glasgow, Scotland examines the effect of human activities and altered landscapes on vultures around the world. Some are adapting to change, with or without our help, while others are struggling. All animals die sooner or later and in most areas, do provide life to scavengers like the vultures. However, electrocution, chemical poisoning through the food chain, decreasing big game populations, and loss of breeding habitat are making existence difficult for some vulture species like the Cape Griffon.

Nowhere are these problems more evident than in Zimbabwe, according to Ron Hartley, Kit Hustler and Peter Mundy. While habitat changes have negatively affected 31 of 73 raptor species in that country, land clearance, dam construction and artificial nest sites have had positive influences on 18 other species.

In Mexico too, some raptor species are benefiting from increased human activity. Ricardo Rodriguez-Estrella reports that the damming of the Rio Yaqui in the mid 1930s has resulted in a 300% increase of riparian woodland favoring higher populations of the common black hawk. The crested caracaras in the Cape Region of Baja California Sur have responded favorably to changes in waste disposal practises there, i.e. greater carrion availability.

Finally, Charles Preston and Ronald Beane of the Denver Museum of Natural History recorded the occurrence and distribution of 12 raptor species in the Rocky Mountain Arsenal, a federal superfund site near Denver and recently designated to become a National Wildlife Refuge once extensive contamination cleanup is completed. They concluded that none of the species avoided areas of high human activity and that "attractive ecological features of the area may boost the tolerance of some species to human activity".

The last remark is crucial to the message of this book. While we were proceeding with organizing the symposium, a colleague remarked to DMB that maybe it was not such a good idea to hold such a symposium. Perhaps it would send out the wrong message to those more interested in economic matters than the environment and our wildlife heritage. In other words, raptors and perhaps other wildlife forms will simply adapt to any changes that humans inflict upon the earth for whatever reason.

Obviously, since you are reading these words, we did not agree with our learned colleague. First of all, the truth must be pursued and brought to the forefront. If raptors are capable of adapting, then the world must know about it. To do otherwise would be tantamount to retaining a species on an endangered species list just for its protection, even though it is no longer endangered. Second, progress, for better or worse, is here to stay. While some environmentalists might prefer it, we cannot step backward to the prehistoric days of cave-dwelling, clubs and loin cloths. Thus, it is important that we know that at least some wildlife species can adapt, even partially, to increasingly human-dominated landscapes. At the same time, this does not mean that we give up the fight to conserve natural habitat either. Raptor lovers would not be happy with a situation where peregrine falcons only nested in our cities and not on natural cliff sites. In fact, the best aspects of having peregrines and other raptors nest in urban environments are that their presence is accessible to children and other segments of society that may otherwise never have the opportunity to see them in wilderness situations, and second, that they provide a genetic pool or reservoir of birds capable of filling or refilling vacant territories in more natural environments.

Third, it is critical that those involved in raptor conservation and those more interested in economics keep an open dialogue with one another. The fact that several utility companies and some members of the forest industry supported the symposium is surely proof that they care about wildlife and that they wish to keep open the lines of communication between raptor lovers and their industries. By working together and not at cross purposes, we may be able to find some common ground and to develop techniques to make the world a safer place for raptors and thus, a better world for those that like to watch them.

Fourth, in these depressing times of skyrocketing human populations, massive changes to natural environments, and dwindling wildlife populations on a global scale, environmentalists desperately need a positive message. This book offers many examples of opportunistic raptors adapting to human landscapes. But they cannot do it alone. As Preston and Beane stated above, we must ensure that attractive ecological features still exist in the environment to help instill a tolerance in raptorial birds for our activities. Just as important, we must buy them time, time to adapt to change. Evolution does not happen overnight.

To our knowledge, *Raptors in Human Landscapes* is the first book to bring together papers on this subject, perhaps for wildlife of any kind. It is our fervent hope that this publication is of interest to readers who enjoy watching and learning about raptors, readers involved with conserving and managing raptors, readers seeking information on raptors on a worldwide basis, and especially readers looking for some good news about wildlife conservation.

David M. Bird
Daniel E. Varland
Juan Jose Negro

Acknowledgments

Generous financial support for symposium expenses as well as for the publication of these proceedings was provided by the National Audubon Society (Rocky Mountain Regional Office), by Carolina Power & Light Company, Central and South West Services, Inc., Duke Power Company, Edison Electric Institute, Florida Power and Light, Idaho Power and Light and Virginia Power of the utility industry, and by Rayonier and Weyerhaeuser Company of the forest products industry. The commercial companies were partly motivated to help largely because of the positive message put forth by the symposium and book. In these times of dwindling government budgets for support of wildlife conservation, it is becoming more and more important for groups with sometimes divergent interests, such as conservationists and industrialists, to work together to achieve the goals of species management and conservation.

The editors, particularly DEV and JJN, gratefully acknowledge the assistance of the following referees listed in alphabetical order: Victor Apanius, David Baker-Gabb, Thomas Balgooyen, Marc Bechard, James Bednarz, Steven Beissinger, Richard Bierregaard, Jr., Keith Bildstein, Peter Bloom, Petra Bohall-Wood, Lehr Brisbin Jr., Joe Buchanan, Javier Bustamante, Mitchell Byrd, Tom Cade, Grady Candler, Thomas Carpenter, Charles Collins, Jose Antonio Donazar, Eduardo Inigo-Elias, David Ellis, James Enderson, Peter Ewins, James Fraser, Mark Fuller, Frederick Gehlbach, Richard Gerhardt, Laurie Goodrich, James Grier, Edmund Henckel, Charles Henny, Fernando Hiraldo, Geoffrey Holroyd, David Houston, Grainger Hunt, Larry Irwin, Paul James, Patricia Kennedy, Robert Kenward, Steve Petty, Michael Kochert, Jeffrey Lincer, Bruce MacWhirter, Carl Marti, Mark Martell, Brian Millsap, Peter Mundy, Ian Newton, Michael Nicholls, Peter Nye, Penny Olsen, Jim Parker, Alan Poole, Charles Preston, Patrick Redig, Robert Rosenfield, William Russell, Ronald Ryder, Erran Seaman, Charles Schaadt, Josef Schmutz, Greg Septon, Robert Simmons, Dale Stahlecker, Mark Stalmaster, Jean-Marc Thiollay, Rodger Titman, Pablo Veiga, Andrew Village, Ian Warkentin, Clayton White, and Stanley Wiemeyer.

The Charlotte RRF conference was an excellent vehicle for holding the symposium. The editors appreciated the exemplary cooperation of Keith Bildstein and Laurie Goodrich as chairpersons of the General Scientific Program, as well as the assistance of Robert Gefaell and his Committee on Local Arrangements. Jim Fitzpatrick, the dedicated treasurer of RRF, is also thanked for his management of the finances.

Lastly and most important, *Raptors in Human Landscapes* would not have been possible at all without the concept in the first place. For that, the editors, particularly DMB, cannot take credit. This unique symposium was the brainchild of two men devoted to raptor conservation, Richard Olendorff (see dedication) and Richard Thorsell, formerly of the Edison Electric Institute and a long-time friend to The Raptor Research Foundation, Inc. We cannot thank them enough for involving us in what has turned out to be a highly worthwhile endeavor for all concerned, especially for raptors throughout the world.

Raptors in Urban Landscapes

1

Peregrine Falcons in Urban North America

Tom J. Cade, Mark Martell, Patrick Redig, Greg Septon and Harrison Tordoff

Abstract — In 1993, 88 territorial pairs of peregrine falcons were known from 60 urban areas in North America. The northeastern seaboard, midwestern states, and coastal Southern California had the densest concentrations of nesting pairs. Urban nesting sites are especially important in the Midwest and eastern United States (US) where urban birds represented 58% and 34% of the regional populations, respectively. The primary requirement for successful urban nesting is a safe nest site; sufficient food seems always to be available. Nest sites are provided in urban areas by buildings (61%), bridges and overpasses (30%), and other tall structures (9%). Nest boxes or trays, with washed pea-gravel on the bottom, are usually essential for the safety of eggs and young. A total of 104 species of avian prey was identified at 19 sites in the Midwest. Ten species were found at ≥10 sites; rock dove, northern flicker and blue jay were the most commonly found prey items. We conclude that urban nesting by peregrine falcons has been a significant factor in the recovery of the midwestern and eastern regional populations. It may be possible for peregrine populations to exceed their known historical abundance owing to the propensity of re-established falcons to use urban nest sites. The use of these sites will provide a unique, yet challenging opportunity for wildlife management in the future.

Key words: peregrine falcon; breeding; urban; North America; prey.

The peregrine falcon has been recorded nesting on human-built structures in cities and towns since the Middle Ages, and in the twentieth century reintroduced peregrines have adapted to tall buildings in urban areas of North America and Europe (Cade and Bird 1990). Cade and Bird (1990) reported 30–32 pairs of peregrine falcons nesting in 24 urban environments across the United States and Canada in 1988. Since then, peregrine falcon populations have increased across the continent due to natural reproduction and the continuing release of captive-bred birds. We report here on the concurrent increase in the number and distribution of urban nesting peregrines.

NUMBERS AND DISTRIBUTION

In 1993, we were able to document 88 pairs of peregrine falcons located in 61 urban areas in the US and Canada (Table 1, Fig. 1). The largest concentration of

Table 1. Numbers of urban nesting peregrine falcons compared to total numbers in North America, 1993.

	Number of urban areas with pairs	Number of urban pairs	Total pairs
Eastern US	20	32	95
Eastern Canada	3	3	22
Midwestern US	21	25	43
Middle Canada	4	5	12
Western US	11	19	566
Western Canada	2	4	15
Total	61	88	753

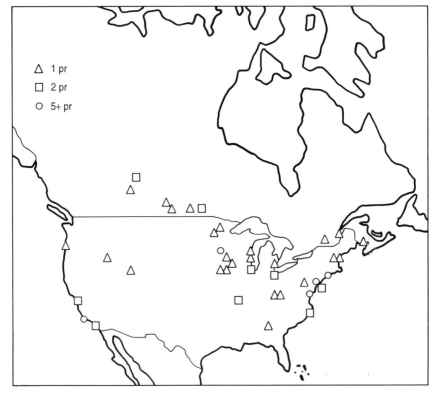

Figure 1. Urban peregrine falcon nest sites in the United States and Canada, 1993.

urban nesting peregrines occurred on the east coast and in the midwestern states of the US, as well as along coastal Southern California. Most urban areas had no more than one or two pairs of falcons. Notable exceptions include New York, New York (9 pr), Philadelphia, Pennsylvania (5 pr), Minneapolis-St. Paul, Minnesota (5 pr), San Francisco-Oakland, California (4 pr) and Long Beach-Los Angeles, California (8 pr; Table 2).

Urban nesting peregrine falcons were a significant proportion of the population in both the eastern and midwestern regions of the US. In the east, 32 of the estimated 95 nesting pairs (34%) were found in urban settings, while in the midwest 25 of 43 pairs (58%) were urban. In contrast, the western US population had only 19 of an estimated 566 nesting pairs in urban sites. In both the midwestern and eastern regions of the US, management of urban nesting peregrine falcons is critical to the maintenance of viable peregrine populations.

The situation in Canada is less easy to define as the total number of Canadian birds is not known. Also, coastal and tundra populations complicate the picture. However, with only 12 urban nesting pairs in Canada, their significance to the overall population is probably minimal. Urban nesters in Canada may be important to the recovery of populations south of the taiga region (e.g., Maritimes, southern Quebec, southern Manitoba, Saskatchewan and Alberta).

PREY SPECIES

In the summer of 1993, to elicit information on prey species being used by urban peregrines, we sent a questionnaire to 17 biologists familiar with nesting urban peregrine falcons in the midwestern US. The questionnaire requested information on the number of individuals of each species of prey found near the nest site between 1986 and 1993. Data were gathered at the site by one or more of the following methods: (1) retrieval of kill remains found around and below the nest site, (2) observations of kills by peregrines, (3) salvaging of prey remains from nest boxes during annual cleaning at the end of the nesting season, and (4) observation of prey brought to the nest.

Questionnaires on prey use were received from 10 respondents (59% response rate) representing 19 urban areas in the states of Illinois, Indiana, Iowa, Michigan, Minnesota, Ohio and Wisconsin. Appendix I lists the respondents and the sites.

From the 19 sites, 104 species of birds were identified (Table 3). The most commonly recorded species were rock dove, northern flicker, blue jay and mourning dove (Table 4). Five species of mammals were recorded: eastern cottontail rabbit, red squirrel, pocket gopher, little brown bat (Byre 1990), and hoary bat. The most unusual item recorded was a fish, a black crappie about 17 cm in length, found on a ledge 15 stories above the ground at the Colonnade Building in Minneapolis, Minnesota. The fish was not eaten.

Peregrine falcons nesting in urban settings seem to have a diverse prey base as indicated by the wide range of species taken in the nesting season. Studies

Table 2. Pairs of urban peregrine falcons in the US and Canada, 1993.

Location	Site
West	
Edmonton, Alberta	Alberta Government & Telephone Building
Edmonton, Alberta	Health Sciences Building
Edmonton, Alberta	Inland Cement Building
Calgary, Alberta	Building
Benicia, California	Florence Bridge
Long Beach, California	Vincent Thomas Bridge
Long Beach, California	City Hall
Long Beach, California	Grain Silo
Long Beach, California	Terminal Island Bridge
Los Angeles, California	California Federal Building
Los Angeles, California	Union Bank Building
Los Angeles, California	Building
Marina del Rey, California	Medical Center Building
Mission Valley, California	Bank of America
Oakland, California	Oakland Bridge
San Diego, California	Coronado Bridge
San Diego, California	Hitachi Crane
San Diego, California	Bank of America Building
San Francisco, California	Oakland Bridge West
Vallejo, California	Carquinez Bridge
Nampa, Idaho	Amalgamated Sugar Refinery
Portland, Oregon	Fremont Bridge
Salt Lake City, Utah	Cave in a rock quarry
Midwest	
Chicago, Illinois	125 S. Wacker Building
East Chicago, Indiana	Cline Avenue overpass
Gary, Indiana	US Steel Plant
Cedar Rapids, Iowa	Firstar Bank Building
Des Moines, Iowa	American Republic Building
Brandon, Manitoba	Building
Winnipeg, Manitoba	Delta Winnipeg Hotel
Winnipeg, Manitoba	Mary Speechly Hall, Univ. of Manitoba
Detroit, Michigan	Book Building
Minneapolis, Minnesota	Multifoods Building (nest box)
Minneapolis, Minnesota	Colonnade Building (nest box)
Minneapolis, Minnesota	NSP Black Dog Power Plant (nest box on smokestack)
St. Paul, Minnesota	Montgomery Wards Building (nest box)
St. Paul, Minnesota	North Central Life Building (nest box)
Bayport, Minnesota	NSP King Plant (nest box on smokestack)
Rochester, Minnesota	Mayo Medical Clinic (nest box)
Becker, Minnesota	NSP Sherco plant (nest box on stack)
Virginia, Minnesota	Nest box on cliff in iron mine
Cohasset, Minnesota	MPL Boswell Energy Center (nest box on smokestack)
St. Louis, Missouri	Southwestern Bell Building
St. Louis, Missouri	Park Plaza Building
Cincinnati, Ohio	Central Trust Tower

Continued

Table 2. *Continued.*

Location	Site
Cleveland, Ohio	Terminal Tower
Dayton, Ohio	Lazarus Building
Toledo, Ohio	Commodore Perry Motor Inn
Saskatoon, Saskatchewan	Building
Regina, Saskatchewan	Regina City Hall
LaCrosse, Wisconsin	Church Steeple
Milwaukee, Wisconsin	Firstar Building (nest box)
Sheboygan, Wisconsin	Edgewater Generating Station (nest box on building)

East

Location	Site
Washington, D.C.	National City Christian Church, Thomas Circle
Delaware Memorial Bridge, Delaware	Bridge
Atlanta, Georgia	Marriot Hotel
Annapolis, Maryland	Chesapeake Bay Bridge
Baltimore, Maryland	US F&G Building (nest box)
Baltimore, Maryland	F. Scott Key Bridge
Ocean City, Maryland	Building
Boston, Massachusetts	Old Customs Building (near hack site)
Springfield, Massachusetts	Building
St. John, New Brunswick	Bridge
Atlantic City, New Jersey	Golden Nugget/Ballys Casino (nest box)
Betsy Ross Bridge, New Jersey	Bridge
Tuckerton Fish Factory, New Jersey	Watertank tower (nest box)
New York City, New York	Hodges Memorial Bridge (nest box)
New York City, New York	Riverside Church (nest box)
New York City, New York	Cornell Medical College (nest box)
New York City, New York	Throgs Neck Bridge (nest box)
New York City, New York	Metropolitan Life (nest box)
New York City, New York	Verrazano Narrows Bridge (nest box)
New York City, New York	George Washington Bridge
New York City, New York	Building, (Wall Street) (nest box)
New York City, New York	Bridge, Goethals
Tappan Zee Bridge, New York	Bridge (nest box)
Pittsburgh, Pennsylvania	Building (nest box)
Philadelphia, Pennsylvania	Walt Whitman Bridge (nest box)
Philadelphia, Pennsylvania	Girard Point Bridge (nest box)
Philadelphia, Pennsylvania	Commodore Barry Bridge (nest box)
Philadelphia, Pennsylvania	Downtown Building (nest box)
Pennsylvania/New Jersey Turnpike, Pennsylvania	Bridge
Quebec City, Quebec	Bridge
Montreal, Quebec	Old Royal Bank Building (nest box)
Fort Eustis, Virginia	Ship (crows nest)
Newport News, Virginia	Bridge, James River (nest box)
Norfolk, Virginia	Bridge (Railroad drawbridge)
York River Bridge, Virginia	Bridge (nest box)

Table 3. Prey recorded at 19 urban peregrine falcon eyries in the midwestern US, 1986–93.

Species	Number of nest sites reporting (kills/yr)					Total
	1/Yr	2–5 Yr	6–10 Yr	10+/Yr	20+/Yr	
Pied-billed grebe	6	1	1	—	—	8
Horned grebe	1	1	—	—	—	2
Least bittern	2	1	—	—	—	3
Green-backed heron	4	—	—	—	—	4
Black-crowned night heron	1	—	—	—	—	1
Mallard	1	—	—	—	—	1
Gadwall	2	—	—	—	—	2
Green-winged teal	4	—	—	—	—	4
Blue-winged teal	4	—	—	—	—	4
Wood duck	3	—	—	—	—	3
Bufflehead	4	1	—	—	—	5
Ruddy duck	1	—	—	—	—	1
Hooded merganser	3	1	—	—	—	4
American kestrel	2	1	—	—	—	3
Virginia rail	3	—	—	—	—	3
Sora	5	—	—	2	—	7
Yellow rail	2	—	—	—	—	2
Common moorhen	1	—	—	—	—	1
American coot	6	—	—	—	—	6
Black-bellied plover	1	—	—	—	—	1
Killdeer	5	1	—	1	—	7
Lesser yellowlegs	4	—	—	—	—	4
Solitary sandpiper	3	—	—	—	—	3
Sandpiper (genus unknown)	1	—	—	—	—	1
Spotted sandpiper	3	—	—	—	—	3
Ruddy turnstone	1	—	—	—	—	1
Woodcock	9	3	1	—	—	13
Common snipe	5	1	—	—	—	6
Ring-billed gull	4	1	—	—	2	7
Bonaparte's gull	1	—	—	—	—	1
Caspian tern	1	—	—	—	—	1
Forster's tern	2	—	—	—	—	2
Common tern	1	1	—	—	—	2
Black tern	2	1	—	—	—	3
Rock dove	12	1	—	—	5	18
Mourning dove	8	—	1	1	2	12
Monk parakeet	1	—	—	—	—	1
Budgerigar	1	—	—	—	—	1
Cuckoo	1	—	—	—	—	1
Yellow-billed cuckoo	6	—	1	1	1	9
Black-billed cuckoo	8	2	1	—	1	12
Northern saw-whet owl	2	—	—	—	—	2
Whip-poor-will	1	—	—	—	—	1
Common nighthawk	3	1	—	—	1	5
Chimney swift	2	—	1	—	1	4
Belted kingfisher	2	1	—	—	—	3
Hairy woodpecker	1	—	—	—	—	1
Downy woodpecker	2	—	—	1	—	3

Continued

Table 3. *Continued.*

Species	1/Yr	2–5 Yr	6–10 Yr	10+/Yr	20+/Yr	Total
Yellow-bellied sapsucker	2	—	—	—	1	3
Red-headed woodpecker	2	3	—	1	—	6
Red-bellied woodpecker	2	—	1	—	—	3
Northern flicker	11	1	2	1	2	17
Eastern kingbird	—	2	—	—	—	2
Olive-sided flycatcher	—	1	—	—	—	1
Purple martin	2	—	—	—	—	2
Cliff swallow	2	—	—	—	—	2
Tree swallow	3	—	—	—	1	4
Blue jay	11	—	2	—	3	16
American crow	1	—	—	—	—	1
Black-capped chickadee	1	—	—	—	—	1
White-breasted nuthatch	1	—	—	—	—	1
Brown creeper	—	1	—	—	—	1
Golden-crowned kinglet	1	—	—	—	—	1
Eastern bluebird	3	—	—	—	—	3
Wood thrush	1	—	—	—	—	1
Gray-cheeked thrush	—	1	—	—	—	1
Swainson's thrush	3	1	—	—	—	4
American robin	9	2	—	1	—	12
Northern mockingbird	—	1	—	—	—	1
Gray catbird	3	1	1	—	—	5
Brown thrasher	3	2	1	—	—	6
Cedar waxwing	3	1	—	—	—	4
European starling	6	—	2	3	1	12
Red-eyed vireo	1	—	—	—	—	1
Yellow-rumped warbler	1	—	—	—	—	1
Bay-breasted warbler	1	—	—	—	—	1
Blackburnian warbler	1	—	—	—	—	1
Common yellowthroat	1	—	—	—	—	1
American redstart	3	—	—	—	—	3
Scarlet tanager	2	1	—	—	—	3
Dickcissel	1	—	—	—	—	1
Northern cardinal	5	—	—	1	—	6
Rose-breasted grosbeak	4	2	1	—	—	7
Rufous-sided towhee	1	1	—	—	—	2
Vesper sparrow	1	—	—	—	—	1
White-throated sparrow	2	1	—	—	—	3
Clay-colored sparrow	1	—	—	—	—	1
Dark-eyed junco	3	—	—	—	—	3
Song sparrow	1	—	—	—	—	1
Fox sparrow	1	—	—	—	—	1
Snow bunting	—	1	—	—	—	1
Bobolink	1	—	—	1	—	2
Brown-headed cowbird	5	1	1	—	—	7
Yellow-headed blackbird	3	—	—	—	—	3
Red-winged blackbird	5	3	2	—	—	10
Eastern meadowlark	1	3	1	—	—	5
Western meadowlark	1?	—	—	—	—	1

Continued

Table 3. *Continued.*

Species	\multicolumn{6}{c}{Number of nest sites reporting (kills/yr)}
	1/Yr	2–5 Yr	6–10 Yr	10+/Yr	20+/Yr	Total
Northern oriole	5	1	1	—	—	7
Common grackle	3	—	2	5	1	11
Evening grosbeak	—	1	—	—	—	1
Purple finch	1	—	—	—	—	1
House finch	—	—	—	1	—	1
Common redpoll	1	—	—	—	—	1
American goldfinch	1	—	1	—	—	2
Pine siskin	—	1	—	—	—	1
House sparrow	2	1	2	—	—	5

Table 4. Twelve most common prey items found at 19 urban nest sites in the midwestern US, 1986–93.

Species	Number of sites
Rock dove	18
Northern flicker	17
Blue jay	16
Woodcock	13
Mourning dove	12
Black-billed cuckoo	12
European starling	12
American robin	12
Common grackle	11
Red-winged blackbird	10
Yellow-billed cuckoo	9
Pied-billed grebe	8

conducted in Baltimore, Maryland during the 1980s had similar results (Barber and Barber 1983, 1988). There, rock doves were the most commonly found prey item of an unmated female as well as a mated pair. The pair also took a variety of prey and responded to seasonal abundance of migrants. Many respondents to our survey noted that prey use changed seasonally. During spring and fall migrations, for example, the prey taken reflect the local movement of migratory species (see also Barber and Barber 1983, 1988).

Large numbers of a single prey species have been observed at nest sites within a short period of time. For example, the remains of 23 common nighthawks were found at the Montgomery Ward Building in St. Paul, Minnesota in 1992. In 1987, urban-hacked peregrine falcons in Milwaukee, Wisconsin killed numerous cuckoos (*Coccyzus* spp.) late in the summer and were observed hunting at night (Wendt *et al.* 1991). From 8–16 September, the cached remains of 38 yellow-billed cuckoos and one black-billed cuckoo were found.

Midwestern peregrine falcons have also been observed striking birds which they did not eat. In most cases these appeared to be acts of territorial defense. Some of the species that were pursued and sometimes struck and killed include: red-tailed hawks, one killed and one injured, in Minneapolis, Minnesota; a great blue heron and a merlin in St. Paul, Minnesota; a turkey vulture (Septon 1991) and a snowy owl (B. Boldt pers. comm.) in Milwaukee, Wisconsin. We have also observed great horned owls being attacked in Milwaukee, Wisconsin and Bayport, Minnesota. Similarly, Barber (1989) reported 31 cases of interspecific aggression in Baltimore, Maryland which included attacks on red-tailed hawks, red-shouldered hawks, a broad-winged hawk and an osprey.

NEST SITES

In North America urban peregrine falcons nested on buildings, including power plants (61%), bridges and overpasses (30%), and other structures (9%; Table 2). Bridges were used more commonly in the eastern region, accounting for 56% (19/34) of the urban pairs. Seventy percent of all bridge nesting pairs in North America occurred in the eastern region. However, in the midwestern region, 93% (28/30) of the urban sites were on buildings.

Buildings used for nesting by peregrines were generally tall office buildings but also included industrial and other sites such as the US Steel Plant in Gary, Indiana, and the Golden Nugget/Ballys Casino in Atlantic City, New Jersey (Table 2). Power plant smokestacks were also used by peregrines, especially in the Midwest (see Septon *et al.* Chapter 16).

Eight sites were not typical buildings or bridges/overpasses but were still in urban areas. These sites included a loading crane (San Diego, California), a ship (Fort Eustis, Virginia), an elevated railroad bridge (Norfolk, Virginia), a quarry (Salt Lake City, Utah) an iron mine (Virginia, Minnesota), a water tank (New Jersey), a sugar silo (Nampa, Idaho) and a grain silo (Long Beach, California).

Most urban nesting sites have a nest box which is used by the birds. These boxes are extremely attractive to nesting peregrines. Pea-gravel placed in the bottom provides an excellent substrate for eggs. Typically, boxes have been placed on structures already being used by peregrines, although boxes have been installed in the hope of attracting a pair and have sometimes been used, particularly when placed on conspicuous structures.

Urban sites at which successful nesting has occurred are almost always reused. Abandonment has been recorded at only two sites where successful nesting occurred in previous years. Also, a Salt Lake City, Utah pair was forced to abandon a building site due to human disturbance, moving to a nearby quarry. Some sites where multiple years of occupancy resulted in no breeding success eventually became productive owing either to human improvement of the nest site or to replacement of the nesting pair by new birds.

Urban sites present unique hazards, which can lead to juvenile and adult mortality. Some of these hazards, such as collisions with airplanes, vehicles and windows, and falling down smokestacks (see Frank 1994) are part of the urban landscape and seem unavoidable. Hazards such as secondary poisoning from feral pigeon control efforts, shooting by vandals and pigeon fanciers, and disturbance associated with building maintenance can often be prevented or reduced. It will be interesting to see whether first-year survival of urban peregrines differs from that of rural peregrines. Perhaps mortality from collisions in cities merely replaces equal mortality from predators in the country. So far, our data are inadequate to answer this question.

The future success of urban nesting pairs, and the contribution they make to maintaining peregrine falcon populations, will depend on the cooperation of building owners, peregrine biologists, city and state planners, politicians and an interested public. The fact that the number of city-dwelling peregrines in North America has more than doubled in just six years is an encouraging indication that these human interests are working in the right direction.

ACKNOWLEDGMENTS

Our information is based on surveys we have done as well as information provided by many other individuals. Contributors included S. Frank, B. Walton, G. L. Holroyd, U. Banasch, and people from state and federal natural resource management agencies in many states. Thanks go to B. Gittings, K. Zunker and J. Marks for assistance in gathering and analyzing prey data. G. L. Holroyd and M. C. Byrd provided thoughtful reviews which improved the quality of this manuscript.

REFERENCES

BARBER, J. 1988. Prey of an urban peregrine falcon-Part II. *Maryland Birdlife* **44**: 37–39.
—. 1989. Interspecific aggression in urban peregrine falcons. *Maryland Birdlife* **45**: 134–135.
— AND M. BARBER. 1983. Prey of an urban peregrine falcon. *Maryland Birdlife* **39**: 108–110.
BYRE, V.J. 1990. A group of young peregrine falcons prey on migrating bats. *Wilson Bull.* **102**: 728–730.
CADE, T.J. AND D.M. BIRD. 1990. Peregrine falcons, *Falco peregrinus* nesting in an urban environment: a review. *Can. Field-Nat.* **104**: 209–218.
FRANK, S. 1994. *City peregrines: a ten year saga of New York City falcons.* Hancock House Publishers, Blaine, Washington USA.
SEPTON, G. 1991. Peregrine falcon strikes turkey vulture. *Passenger Pigeon* **53**: 192.
—, BIELEFELDT, J., T. ELLESTAD, J. MARKS AND R.N. ROSENFIELD. 1996. Peregrine falcons: Power plant nest structures and shoreline movements. Pp 145–153 in D.M. Bird, D.E. Varland and J.J. Negro, eds., *Raptors in human landscapes.* Academic Press, London.
WENDT, A., G. SEPTON AND J. MOLINE. 1991. Juvenile urban-hacked peregrine falcons (*Falco peregrinus*) hunt at night. *J. Raptor Res.* **25**: 94–95.

APPENDIX 1

Respondents to a questionnaire on prey species of urban peregrine falcons in the midwestern US.

RESPONDENT	LOCATION
John S. Castrale (Indiana DNR)	– US Steel Mill, Gary, Indiana
Dudley Edmondson (Raptor Works)	– Duluth Hotel/Central High School, Duluth, Minnesota – Palisade Head, Minnesota
Mary Hennen (Chicago Academy of Sciences)	– 125 S. Wacker Building, Chicago, – Hyde Park, Chicago, Illinois – Irving Park, Chicago, Illinois
Jim Marks (Wisconsin Peregrine Society)	– Edgewater Generating Station, Sheboygan, Wisconsin
Sara Jean Peters (Ohio DNR)	– Terminal Tower, Cleveland, Ohio – Metro Health Building, Cleveland, Ohio
Greg Septon (Milwaukee Public Museum)	– Firstar Center, Milwaukee, Wisconsin
Jim Surdick (Wisconsin DNR)	– State Capitol, Madison, Wisconsin
Harrison Tordoff (Univ. of Minnesota)	– Multifoods Tower, Minneapolis, Minnesota – Colonnade Building, Minneapolis, Minnesota – North Central Life, St. Paul, Minnesota – Montgomery Ward, St. Paul, Minnesota – Control Data., Bloomington, Minnesota
Mark Washburn (Iowa Falconers Association)	– Centennial Bridge, Davenport, Iowa
Judith Yerkey (Michigan DNR)	– Book Building, Detroit, Michigan – Fisher Building, Detroit, Michigan

2

Bridge Use by Peregrine Falcons in the San Francisco Bay Area

Douglas A. Bell, David P. Gregoire and
Brian J. Walton

Abstract – Peregrine falcons have been resident in the highly urbanized San Francisco Bay Area at least since the mid-1980s. Two pairs of falcons make annual nesting attempts on the Oakland-San Francisco Bay Bridge, while a third pair uses the Golden Gate Bridge as a hunting post. The last mentioned pair has shifted nest locations between near-urban sites and wild coastal sites, while the Bay Bridge pairs invariably use the bridge during the nesting season, but tend to move to downtown areas for the winter. Foraging habits and prey species differ between each pair, as do eggshell thinning and hatching success. In most years eggs were removed from bridge-nesting pairs; in one instance a pair was triple clutched. Three instances of chicks fledging from the Bay Bridge were noted. Fledging success from bridge sites is poor, and several factors appear to contribute to high mortality of young falcons at fledging. We conclude that bridge nest sites must be enhanced to improve fledging success. Fostering of nestling falcons from bridge sites to natural eyries may be the most effective option for enhancing productivity of bridge-nesting peregrines.

Key words: peregrine falcons; bridge use; nest success; San Francisco Bay.

With increasing frequency, peregrine falcons are inhabiting urban environments and using man-made structures for nesting (Tordoff and Redig 1988, Cade and Bird 1990, Frank 1992). The highly urbanized environment of the San Francisco Bay Area, California, currently harbors both wintering and resident peregrines. Traditionally, peregrines nested in many of the habitats surrounding the greater San Francisco Bay Area (Grinnell and Wythe 1927), yet only one confirmed nesting involved a man-made object – a barrell in the Redwood City marshes (Dawson 1923). Isolated reports from the 1950s suggest that peregrines may have frequented the San Francisco-Oakland Bay Bridge (S. Bunnell pers. comm.) as well as a hotel in downtown San Francisco (R. T. Orr pers. comm.). At best, urban use by falcons prior to their population crash in the 1950s and 60s was infrequent. Since the 1980s two pairs of peregrines have made regular nesting attempts on the San Francisco-Oakland Bay Bridge (SFOBB), while a third pair uses the Golden Gate Bridge (GGB) as a hunting post. Here we report on the habits and nesting success of these urban and near-urban resident peregrines and point out management concerns related to the urban environment.

STUDY AREA AND METHODS

The SFOBB is an 8.32 km long double-deck bridge that spans San Francisco Bay in an approximate east–west direction. The bridge consists of two major structures: a western suspension span between the City of San Francisco and Yerba Buena Island, and an eastern cantilever span between Yerba Buena Island and the City of Oakland. The San Francisco end of the SFOBB is characterized by a plethora of skyscrapers and a shoreline of cement breakwaters, piers and marinas. The Oakland end of the SFOBB is characterized by a shoreline to the south consisting of container shipyards, marinas and airfields, and to the north by a shoreline known as the Emeryville Crescent – a thin strip of mudflats and marshland backed by freeways, marinas and industrial lands. Yerba Buena Island, which bisects the SFOBB, is approx. 1 km^2 and consists of steep slopes with scrub-eucalyptus habitat. It is connected to the north with flat, man-made Treasure Island (approx. 2 km^2).

The GGB is a single deck suspension bridge approximately 2 km long that spans from north to south the "Golden Gate" – the entrance to San Francisco Bay. The southern shore of the Golden Gate consists of coastal bluffs, urban areas, and the parkland-like environs of the Presidio of San Francisco, while the northern end opens onto nearly 30 350 ha of the Golden Gate National Recreation Area, a region of coastal headlands, cliffs and scrub habitat. Urban areas on the north side of the Golden Gate are located on the bay side of the Marin penninsula.

Prior to 1986, most observations of peregrines in the San Francisco Bay Area were incidental in nature. Beginning in August 1986, one of us (DPG) began concerted, year-round monitoring of peregrines that has continued to the present. Observations were at first centered on a pair in downtown San Francisco. Monitoring efforts were expanded as peregrine sightings increased in both frequency and geographic extent. A network of volunteers and professionals, e.g. bridge employees, aided this effort. Sightings ranged in frequency from daily to incidental. Ancillary evidence, such as fresh excrement and prey remains, also indicated recent occupancy of breeding, roosting and hunting sites.

In the years 1988–92 most clutches were removed from the SFOBB pairs by the Santa Cruz Predatory Bird Research Group (SCPBRG) for incubation and subsequent release of chicks by fostering, cross-fostering and hacking. Chicks released by these three methods were banded with blue-anodized bands, while those banded in unmanipulated eyries received standard aluminium US Fish and Wildlife Service (USFWS) bands. In most cases, numbers on the blue and aluminum bands were not legible on free-living falcons. Addled eggs and eggshell fragments were retrieved from nest sites. In general, chicks were not returned to bridge eyries because of the risk of poor fledging success. It should be noted that the bridge eyrie sites chosen by the falcons were integral structures of the bridge. The only "improvement" to these sites was the addition of pea gravel to provide nesting substrate.

RESULTS

San Francisco-Oakland Bay Bridge–East Bay

A pair of blue-banded peregrines appeared to establish a territory here in 1983. The male had been fostered into an eyrie in Napa County, CA, while the female was from a prairie falcon cross-foster site in San Luis Obispo Co., CA. Unfortunately, the female was found shot near the bridge, and although bridge workers occasionally sighted peregrines here from 1987–89, occupancy of this territory was not confirmed until 1990, when a blue-banded adult female of undetermined origin was discovered with a three-egg clutch at the E2-N site on 18 May (Table 1). The female was paired with a blue-banded adult male, also of undetermined origin. By the beginning of 1992 an unbanded adult male had replaced this blue-banded male.

The peregrines on the eastern span of the SFOBB alternated between two eyrie sites on Tower E2 (Table 1). Tower E2 is located just a few meters east of Yerba Buena Island. Eyrie site E2-SE is on the inside of the south post of the E2 tower and faces east and north. It is located on a corner gusset plate (0.99 m × 1.45 m) that anchors a girder to the tower. The girder bisects the area available to the falcons into two triangles, but since the girder does not extend all the way to the tower, the falcons can move around the end of the girder to access both "triangular" ledges. A second gusset plate is positioned atop the bisecting I-beam and effectively "covers" the lower plate. Eyrie E2-SE is about 1.5 m beneath the lower deck of the bridge, and 55 m above the base of the tower. To enter the site, the falcons must approach from below and fly up to it, since all horizontal access and views are blocked by girders. The alternate nest site, E2-N,

Table 1. Summary of nesting history of San Francisco-Oakland Bay Bridge – East Bay peregrines. In 1991 and 1992 clutches were removed by the Santa Cruz Predatory Bird Research Group for captive propagation. A replacement male was present from 1992 onwards.

Year	Nest loc[a]	Clutch size	Laying date[b]	Egg fate[c]	No. fledged
1990	E2-N	3		3B	
1991	E2-N	4	4 March	2D, 2H	
	E2-SE	4	3 April	4H	
1992	E2-SE	4	7 March	3A, 1H	
	E2-SE	4	12 April	1A, 3H	
1993	E2-N	4	12 March	1B, 3H	1

[a] refers to bridge tower number (e.g. E2-) and exposure of eyrie site (e.g. −N = north).
[b] laying date of first egg.
[c] A = addled or infertile, B = broken in nest, D = large dead embryo, H = hatched.

is on the outside of the north post of the E2 tower in a rectangular recess in the tower. The floor of the recess measures 0.8×0.6 m and contains the remnants of what appears to be a common raven's nest. The eyrie faces north and is approximately 4 m below the lower deck of the bridge and 53 m above the waterline.

The year-round territory of the SFOBB-East Bay peregrines encompassed an area of about 39 km^2. It included a small group of skyscrapers in downtown Oakland (8.9 km southeast of E2) and several buildings at the Emeryville Crescent (6.5 km northeast of E2). From mid-January through June the falcons were observed most often at the bridge and E2 tower. The falcons spent considerable time perching, eating and roosting on girders beneath the lower deck of the bridge. Hunting forays were launched from any position on the bridge, but the most recent adult tiercel preferred to initiate hunts from the top of the cantilever bridge towers. This same tiercel took advantage of the prevailing northwest winds in the spring for exceptionally high (>500 m) aerial hunting flights over the bay. From July to October the adults tended to occupy the downtown Oakland area, and for the remainder of the year, October to January, they were often found at the Emeryville Crescent, where they appeared to prey on wintering shorebirds along the nearby mudflats.

In two instances the SFOBB-East Bay pair had chicks in eyrie sites on the bridge. In 1991 a 17 d chick was placed in the E2-SE site after the falcon's second clutch was pulled for incubation (Table 1). The chick was accepted by the adults and fed regularly for the next 10 d until it fell out of the eyrie. The chick was recovered unharmed from the shore, but it was not replaced in the eyrie. In the days preceding its fall from the site it was quite active, and could be observed running about the E2-SE ledge. It often appeared to crane for a better view of the adults approaching with food. Thus, the enclosed nature of the E2-SE ledge may have contributed to the chick's fall. In 1993 manipulation of peregrine eyries in California was curtailed. This presented an opportunity to observe "natural" breeding at the E2-N site (Table 1). Egg laying began on 12 March, incubation was initiated on 16 March, and hatching began on 18 April. Three of the four eggs hatched, and chick development proceeded normally. Three chicks fledged on 1–3 June (D. and L. Jesus pers. comm.). The first bird to fledge, a female, had apparently landed on the base of Tower E2. However, it was soon forced off the tower base by a nesting western gull. With several gulls in hot pursuit, the fledgling attempted to fly east at low altitude (1 m) over the water to the next tower, 0.42 km away. An adult peregrine followed the procession closely, but did not intervene. Unfortunately, the fledgling landed in the water after flying about 0.30 km. Immediately upon hitting the water the fledgling reversed its course, swimming west. Several gulls continued to harass the fledgling by either diving on it or landing in the water nearby. It did not appear that the gulls ever made physical contact with the fledgling. The fledgling swam strongly for 10 min, then weakly for another 5 min until its head went underwater. A rescue party recovered the drowned bird 5–10 minutes later. The second chick, a male, apparently disappeared during a wind storm the next day and was presumed drowned. Only the third chick, a female, successfuly fledged. Within a few days of fledging this chick exhibited defensive behavior towards harassing gulls – it would face a gull diving at it and jump up to present its feet to the oncoming gull. In addition, the adult female became very defensive and would drive harassing gulls away from the fledgling. As late as 25 August the fledgling was seen with both adults at the downtown Oakland buildings.

San Francisco-Oakland Bay Bridge–West Bay

A pair of blue-banded falcons, an adult male and immature female, appeared in the downtown San Francisco area late in the summer of 1986. Attempted nesting was not confirmed until 1988 (Table 2), when a five-week old chick was recovered from a pier on the waterfront below tower W-2. All subsequent clutches of this pair were removed by the SCPBRG for captive propagation purposes. The pair was triple-clutched in 1989 (Table 2). The female ceased laying eggs in 1992, but continued to occupy eyrie locations and make fresh scrapes in 1992 and 1993.

The SFOBB-West Bay pair favored site W4-E, known as the "central anchorage", but has alternated between six eyrie sites on three suspension towers and the central anchorage pillar (Table 2). In all cases, the same type of structure was

Table 2. Summary of nesting history of San Francisco-Oakland Bay Bridge – West Bay peregrines. Except for 1988, all clutches were removed by the Santa Cruz Predatory Bird Research Group for captive propagation.

Year	Nest loc[a]	Clutch size	Laying date[b]	Egg fate[c]
1988	W4-E	1	28 March	1A
	W2-W	2	19 May	1D, 1H
1989	W4-E	4	17 March	3D, 1H
	W2-E	4	20 April	1A, 1B, 1D, 1H
	W4-E	4	18 May	3A, 1H
1990	W4-E	4	23 March	4H
	W5-E	4	24 April	3A, 1B
1991	W4-E	3	22 March	1A, 1D, 1H
	W5-W	1	18 April	1A
	W3-W	1	15 May	1A
1992	W4-E	0		
	W5-E	0		
1993	W4-E	0		

[a] refers to bridge tower number (e.g. W4-) and exposure of eyrie site (e.g. -E. = east).
[b] laying date for first egg of clutch.
[c] A = addled or infertile, B = broken in nest, D = large dead embryo, H = hatched.

used: a large box (4 m × 0.9 m × 0.5 m) forced by upper and lower gusset plates that span central girders beneath the lower deck. This structure is found on all western and eastern faces of the suspension bridge towers. The boxes have two 0.5 m openings along the inner edges of the girders. There is essentially no structure in front of the box openings, a factor that probably contributed to the 1988 chick's fall. Distance from the box opening to the water below is 67 m, and the distance between towers is 0.35 km.

As a minimum estimate, the territory of the SFOBB-West Bay peregrines extended from an eastern boundary at Yerba Buena Island west to buildings at Van Ness Avenue and Fell Street (5.5 km) in San Francisco, and from Nob Hill in the north of the city south to the Islais Creek Channel (5.8 km), an area of about 32 km^2. In general, this pair of falcons remained on the bridge from January through July, and moved to the tall skyscrapers of the Financial District in San Francisco from August through December.

Golden Gate Bridge

In the spring of 1989 a USFWS banded adult male showed up on the North Tower of the Golden Gate Bridge. The following fall a blue-banded adult female joined the male. The pair remained on the bridge and in the area throughout the winter. In the spring of 1990 they took possession of a raven's old nest on a nearby seacliff. The first egg of a three-egg clutch was laid on 25 March and incubation began six days later on 30 March. Hatching took place exceptionally late, on days 38 and 39 (6–7 May). One chick was removed during banding on 21 May due to unusual behavior (it died within 24 hrs, and subsequent autopsy revealed a brain abscess). Rains the next day appeared to destroy part of the nest, and the two remaining chicks disappeared, probably by falling from the defunct nest, by 24 May. In the spring of 1991 the pair relocated to a natural ledge on a seacliff over 11.8 km away from the 1990 site. A clutch was initiated on 25 March, incubation of three eggs began on 29 March (a fourth egg was laid after beginning of incubation), and hatching took place 35 d later on 2 May. All four chicks developed normally and fledged between 12–16 June. In 1992 the pair was seen at various locations throughout their territory, but no breeding attempt was confirmed. In 1993 the pair again established on the same seacliff as in 1991. The first egg of a four-egg clutch was laid on 31 March, incubation began on 4 April, and hatching took place 35 d later. Unfortunately, all four downy chicks died at three weeks of age during exceptionally hard rains. The adult female disappeared during this storm; she was never seen again.

The GGB pair used the bridge primarily for hunting purposes. Favored lookout perches were on tower catwalks located 101 m and 155 m above the waterline, respectively. In most years this pair used the bridge primarily in the non-breeding season, from August to February. It appears that they preyed on the relatively steady stream of migrant birds crossing the Golden Gate. The territory of this pair was large; the adults were seen hunting over an area of at least 60 km^2.

Breeding Success

Mean eggshell thinning in the SFOBB-East Bay female was 15.2% (range 9.5–20.3%; SD = 0.033%; $N = 19$), whereas the SFOBB-West Bay female averaged 7.2% (range 0–21.4%; SD = 0.065%; $N = 28$). Although the SFOBB-East Bay female exhibited significantly greater eggshell thinning than the SFOBB-West Bay female (Mann-Whitney $U = 84.5$; $P = 0.0007$; two-tailed test), the former had a higher percentage of eggs hatch than the latter (53% versus 32%; Tables 1 and 2). Many of the unhatched eggs from the SFOBB-West Bay female were either infertile or exhibited the "slipped membrane" phenomenon (J. Linthicum pers. comm.). Fledging success from the SFOBB was poor: only one out of five chicks (20%) fledged successfully. Eggshell thinning in the GGB female averaged 11.4% ($N = 7$). Although all of her eggs hatched under natural incubation, only four of her 11 chicks survived long enough to fledge.

Prey Base

A qualitative list of identified prey items taken by each pair is provided in Appendix 1. All three peregrine pairs took rock doves, mourning doves, European starlings, varied thrushes and American robins. The SFOBB-East Bay pair seemed to take the widest spectrum of prey, including many different shorebirds. Escaped and feral psittacines have been taken by both the SFOBB-West Bay and GGB pairs.

DISCUSSION

Territory use by the SFOBB peregrines was typical for year-round residents (Ratcliffe 1980). During the breeding season they centered their activities at the eyrie site, in this case the bridge, but expanded their activities to include other areas of the territory during the non-breeding season, such as downtown areas. In most years, the GGB pair exhibited a pattern of bridge use opposite from that of the SFOBB pairs. They were drawn to the bridge in fall and winter, where the falcons were in an exceptional position to prey on birds migrating across the Golden Gate. The GGB peregrines may not have used the bridge for nesting simply because the Golden Gate Bridge, as opposed to the Oakland-San Francisco Bay Bridge, lacks suitable eyrie sites (see Cade and Bird 1990). In any case, bridge use by peregrines appears to be related to resource availability, be it a ready prey base or a nest site location, or both.

Although the SFOBB pairs are "urban" in their habits, the higher eggshell thinning in the SFOBB-East Bay female may be due to the fact that her territory includes mudflats where the opportunity to feed on shorebirds is greatest, a fact reflected in this pair's identified prey items. Shorebirds tend to exhibit higher contaminant levels of chlorinated hydrocarbons (Baril *et al.* 1990). Widely differing levels of eggshell thinning, even in neighboring pairs, seem typical of California peregrines and have been reported in eastern US peregrines as well (Gilroy and Barclay 1988). Similarly, the high intraclutch variation in eggshell thinning, observed here for the SFOBB pairs, has been observed in three species of falcons including peregrines (Monk *et al.* 1988). In general, peregrines in California exhibit high levels of eggshell thinning and contamination from a multitude of toxins (Peakall and Kiff 1988; Jarman *et al.* 1993). The urban peregrines of the San Francisco Bay Area reflect this pattern.

Bridges present unique problems for nesting peregrines. For instance, gull–peregrine fledgling interaction may be exacerbated by the reduced area available to both at tower bases. Frank (1992) reported that peregrines breeding on artificial ledges placed on bridges in the New York City area successfully fledged on average 1.1 chicks per brood versus 2.5 chicks per brood for falcons nesting on buildings. Peregrines nesting on the SFOBB, using unimproved ledges experienced very poor fledging success. The cramped ledges, lack of structure beneath the ledges, great distances between towers, and the extreme heights of the SFOBB probably contributed to high mortality of fledglings (see also Cade and Bird 1990). The state highway agency, Caltrans, is willing to enhance the peregrine nest sites on the SFOBB by installing gusset plates with borders to increase shelf area and possibly improve fledging success, but permission to do so has not been granted by the California Department of Fish & Game's Office of Endangered Species. The office is opposed to enhancing the SFOBB peregrine nest sites because of the potential adverse effect peregrines may have on a colony of endangered least terns. However, the threat posed by the SFOBB peregrines to the least tern colony appears to be minimal. Least terns have not been identified in the prey remains of these particular falcons (App. 1).

Enhancement of bridge nest ledges is one option for reducing mortality of falcons at fledging. Other management options could be undertaken, such as preventing gulls from nesting on tower bases to lessen gull-peregrine interactions at fledging, or encouraging the falcons to move off the bridge by installing nest ledges on city buildings. The peregrines are not likely to be discouraged from using the SFOBB since they have persisted in nesting on the bridge despite regular removal of their eggs for fostering purposes. Fostering of eggs or chicks from bridge sites to natural eyries may be the most effective option for enhancing productivity of bridge-nesting peregrines.

ACKNOWLEDGMENTS

We thank Lloyd Kiff and Sam Sumida at the Western Foundation of Vertebrate Zoology for providing eggshell thickness data. Janet Linthicum of the SCPBRG compiled data on egg fates. N. John Schmitt helped identify prey remains. David and Lynn Jesus permitted use of unpublished field notes. Many people volunteered their time to monitor these urban falcons. We especially thank Joe Didonato of the East Bay Regional Park District, Allen Fish and members of the Golden Gate Raptor Observatory, John Aiken and the San Francisco Zoological Society, and local Audubon societies. The cooperation and help of personnel from the Caltrans San Francisco-Oakland Bay Bridge District and the Golden Gate Bridge, Highway and Transportation District is greatly appreciated. Pat Redig and Greg Septon provided helpful comments on the manuscript.

REFERENCES

BARIL, A., J.E. ELLIOTT, J.D. SOMERS AND G. ERICKSON. 1990. Residue levels of environmental contaminants in prey species of the Peregrine Falcon, *Falco peregrines*, in Canada. *Can. Field-Nat* 104: 273–284.

CADE, T.J. AND D.M. BIRD. 1990. Peregrine Falcons, *Falco peregrinus*, nesting in an urban environment: a review. *Can. Field-Nat.* 104: 209–218.

DAWSON W.L. 1923. *The birds of California*. South Moulton Co., San Diego.

FRANK, S. 1992. General considerations of Peregrine Falcon nesting. *NYCARIES* 92: 209–218.

GILROY, M.L. AND J.H. BARCLAY. 1988. DDE residues and eggshell characteristics of reestablished peregrines in the eastern United States. Pp. 403–412 in T.J. Cade, J.H. Enderson, C.G. Thelander and C.M. White, eds. *Peregrine falcon populations: their management and recovery*. The Peregrine Fund, Boise, Idaho.

GRINNELL, J. AND M.W. WYTHE. 1927. Directory to the bird-life of the San Francisco Bay region. *Pacific Coast Avifauna* 18: 1–160.

JARMAN, W.M., R.J. NORSTROM, M. SIMON, S.A. BURNS, C.A. BACON AND B.R.T. SIMONEIT. 1993. Organochlorines, including chlordane compounds and their metabolites, in Peregrine falcon, Prairie falcon, and Clapper rail eggs from the USA. *Environ. Pollution* 81: 127–136.

MONK, J.G., R.W. RISEBROUGH, D.W. ANDERSON, R.W. FYFE AND L.F. KIFF. 1988. Within-clutch variation of eggshell thickness in three species of falcons. Pp. 377–384, in T.J. Cade, J.H. Enderson, C.G. Thelander and C.M. White, eds. *Peregrine falcon populations: their management and recovery*. The Peregrine Fund, Boise, Idaho.

PEAKALL, D.B. AND L.F. KIFF. 1988. DDE contamination in peregrines and American kestrels and its effect on reproduction. Pp. 337–350 in T.J. Cade, J.H. Enderson, C.G. Thelander and C.M. White, eds. *Peregrine falcon populations: their management and recovery*. The Peregrine Fund, Boise, Idaho.

RATCLIFFE, D. 1980. *The peregrine falcon*. T. & A.D. Poyser, Calton.

TORDOFF, H.B. AND P.T. REDIG. 1988. Dispersal, nest site selection, and age of first breeding in peregrine falcons released in the upper Midwest, 1982–1988. *Loon* 60: 148–151.

APPENDIX 1

Prey items of the three known pairs of peregrine falcons using urban sites in the San Francisco Bay Area, where [1] = SFOBB-East Bay pair; [2] = SFOBB-West Bay pair; and [3] = GGB pair. Identifications based on either collected prey remains or observed hunting episodes.

Prey items	Pair
Common Moorhen	1
Marbled Godwit	1, 3
Whimbrel	1, 3
Willet	1, 3
Greater Yellowlegs	1
Wandering Tattler	3
Phalarope, sp.	3
Dowitcher, sp.	1, 3
Red Knot	3
Dunlin	1, 3
Sanderling	3
Western Sandpiper	1, 3
Least Sandpiper	1, 3
Bonaparte's Gull	3
Western Gull	3
Rock Dove	1, 2, 3
Mourning Dove	1, 2, 3
Cockatiel	3
Budgerigar	2
Amazon parrot, sp.	2
White-throated Swift	3
Northern Flicker	2, 3
Acorn Woodpecker	2
Downy Woodpecker	2
Western Kingbird	3
Violet-green Swallow	1
Varied Thrush	1, 2, 3
American Robin	1, 2, 3
Cedar Waxwing	1
European Starling	1, 2, 3
Western Meadowlark	1, 3
Red-winged Blackbird	1
Northern Oriole	1
House Finch	1, 2

3

Eggshell Thickness and Contaminant Analysis of Reintroduced, Urban Nesting Peregrine Falcons in Wisconsin

Greg A. Septon and Jim B. Marks

Abstract – Between 1989 and 1993, addled and infertile eggs ($N = 9$) from a reintroduced population of peregrine falcons were salvaged from urban nest sites in Wisconsin. The emptied and dried eggshells were measured and weighed to determine the "Ratcliffe thickness index". Since 1991, the egg contents ($N = 7$) have also been analyzed for organochlorine and heavy metal residues. A comparison of eggshell thickness indices was made between pre-DDT (1947) eggshells from Wisconsin ($N = 6$) and our salvaged eggshells. This comparison indicated 10% thinning in the salvaged eggshells ($N = 9$). The findings of our environmental contaminant analyses were compared with similar studies undertaken in North America between 1980 and 1989. For the Wisconsin eggs which were analyzed, geometric mean residue levels of DDE and DDT were found to be the lowest of all compared North American samples. Geometric mean PCB residue levels were lower than levels detected in Canadian *Falco peregrinus pealei* and *F. p. tundrius* subspecies and east coast peregrines but higher than *F. p. anatum* populations in Canada, California and Colorado. The geometric mean level of mercury in Wisconsin eggs was higher than Canadian *F. p. tundrius* levels but lower than that found in Canadian *F. p. anatum*.

Key words: peregrine falcons; reintroduced; eggshell thickness; contaminants; urban; Wisconsin.

From 1930 to 1970, large amounts of chlorinated hydrocarbons and other chemicals were released into the Great Lakes environment. These contaminants included pesticides such as DDT and dieldrin and industrial chemicals such as polychlorinated biphenyls (PCBs) and mercury, all of which are not readily biodegradable. The accumulation of environmental contaminants in some fish-eating birds near the Great Lakes is continuing to affect reproductive success (Sweet 1992). In 1989, we began to assess how these contaminants may be affecting the reintroduced peregrine falcon in urban areas in Wisconsin. In 1993 there were three urban nesting pairs of peregrines in Wisconsin. Two were successful and produced young. These were in Milwaukee and Sheboygan. The third pair in LaCrosse produced one infertile egg.

METHODS

Since 1989, all addled and infertile peregrine eggs ($N = 9$) were salvaged from urban nest sites in Wisconsin. The blown eggshells were allowed to dry and were then measured and weighed to determine their Ratcliffe thickness indices (Ratcliffe 1967). Pre-DDT Wisconsin eggshells ($N = 6$) were also measured and weighed to determine their Ratcliffe thickness indices. The Mann-Whitney U test was used to determine whether differences were significant.

Since 1991, the egg contents have been analyzed for organochlorine and heavy metal residues by Hazleton Environmental Services, Inc. (HES) of Madison, Wisconsin. HES utilized the soxhlet extraction method for gas-liquid chromatography (GLC) employing electron capture detection (Griffitt and Craun 1974, Johnson et al. 1976, Journal of AOAC 1990). Results were reported in mg per kg. The salvaged eggshells in this study were prepared for and accessioned into the collections of the Milwaukee Public Museum (MPM).

RESULTS AND DISCUSSION

To gain an historical perspective of eggshell thickness levels in Wisconsin, we compared pre-DDT Wisconsin peregrine eggshells from the MPM ($N = 5$) and Chicago Academy of Sciences collections ($N = 1$) with those salvaged between 1989 and 1993 ($N = 9$). The data on pre-DDT Wisconsin peregrine eggshells are presented in Table 1. Table 2 presents data on the eggshells salvaged between 1989 and 1993. The pre-DDT eggshells have an average Ratcliffe thickness index of 2.008 while the eggshells salvaged in our study have an average index of 1.806 which, although 10% thinner, did not change significantly ($P > 0.005$)

Table 1. Pre-DDT (1947) eggshell thickness data for Wisconsin peregrine falcons.

Year/location/ egg number	Ratcliffe index
1911 TRY-1	1.997
1933 FYBF-1	2.014
1933 FYBF-2	2.116
1933 FYBF-3	2.041
1933 OTCK-1(A)	1.732
1933 OTCK-1(B)	2.148
Average	2.008
S.D.	0.147
Range	2.148–1.732

TRY = Troy, Sauk County; FYBF = Ferry Bluff, Sauk County;
OTCK = Otter Creek, Sauk County.

Table 2. 1989–93 eggshell thickness data for reintroduced peregrine falcons in Wisconsin.

Year/location/ egg number	Ratcliffe Index
1989 MKE-1	1.898
1989 MKE-2	1.947
1991 MKE-1	1.839
1991 MKE-2	1.663
1991 MKE-3	1.719
1992 SBN-1	1.622
1993 LSE-1	1.770
1993 SBN-1	1.866
1993 SBN-2	1.933
Average	1.806
S.D.	0.118
Range	1.947–1.622

MKE = Milwaukee; SBN = Sheboygan; LSE = LaCrosse.

from the historic indices. This eggshell thinning is still below the 17% level which has been correlated with population declines (Peakall and Kiff 1988). Since we are really looking at two different populations, i.e. the pre-DDT native population and a reintroduced population of wide genetic variation, these indices cannot be compared in a manner one would use for assessing the same subspecies.

Table 3 compares the levels of the most prevalent contaminants found in our study (DDE, DDT, PCBs, Hg) with those found in similar studies across North America (Peakall et al. 1990, Jarman et al. 1993). We note the low mean level of DDT found in the Wisconsin eggs especially when compared with the levels in other regions. In all regions covered in Table 3, the mean levels of DDT are above 5 ppm, except for Wisconsin, where the mean is 0.026 ppm. This could indicate the near complete breakdown of DDT within this region and/or reflect the fact that peregrines nesting at urban sites in Wisconsin are not preying heavily on migrant bird species suspected of carrying high DDT levels. Mean DDE levels in Wisconsin are also lower than levels reported from other regions.

At present, PCB levels in Wisconsin eggs are probably not high enough to warrant immediate concern. Peakall et al. (1990) indicated that the best estimate for critical levels of "total PCBs" at which reproductive effects may occur was egg levels greater than 40 ppm. P. Redig (pers. comm.) made the following comment pertaining to contaminants of peregrine falcons in the reintroduced Midwestern population:

> To date we have no evidence of any reproductive problems, including egg-shell thinning, in any of the pairs we have. Since some of these females have been laying eggs for over six years, cumulative problems would very likely have been expressed by now if they were to occur.

Table 3. Environmental contaminant residue comparisons for North American peregrine falcon eggs in parts per million wet weight.

Country/region Subspecies Collection years	N		DDE	DDT	PCBs		Hg
Canada[a] F. p. pealei 1980–86	4	Mean Min. Max.	5.84 5.26 6.57	5.87 5.27 6.62	8.54* 7.97* 9.80*		n.d. n.d. n.d.
Canada[a] F. p. anatum 1980–87	74	Mean Min. Max.	9.13 1.51 44.10	9.23 1.52 46.53	3.97* 1.58* 30.70*		0.446 0.120 1.140
Canada[a] F. p. tundrius 1980–86	26	Mean Min. Max.	6.84 1.61 18.50	7.27 1.78 19.08	9.82* 1.61* 32.25*		0.092 0.000 0.420
USA, East Coast[b] F. peregrinus 1986–87	9	Mean Min. Max.	n.d. n.d. n.d.	8.8Σ 2.2Σ 25.2Σ	13.8Σ 4.5Σ 25.0Σ		n.d. n.d. n.d.
USA, Colorado[b] F. p. anatum 1986–89	5	Mean Min. Max.	n.d. n.d. n.d.	11Σ 4Σ 24Σ	0.74Σ 0.33Σ 2.6Σ		n.d. n.d. n.d.
USA, California[b] F. p. anatum 1989	9	Mean Min. Max.	n.d. n.d. n.d.	8.8Σ 4.1Σ 29.0Σ	4.1Σ 0.51Σ 9.6Σ		n.d. n.d. n.d.
USA Wisconsin F. peregrinus 1991–93	7	Mean	3.61	0.026	5.71 0.446 0.16	(a) (b) (c)	0.163
		Min.	0.97	<0.01	0.25 <0.05 <0.05	(a) (b) (c)	0.087
		Max.	6.60	1.47	15.3 5.06 5.04	(a) (b) (c)	0.430

* = commercial mixture of PCBs, Σ = Σ DDT & Σ PCBs, a = Aroclor 1260, b = Aroclor 1248, c = Aroclor 1254, n.d. = not determined.
[a]Canadian contaminant data from Peakall et al. (1990).
[b]Data on contaminant levels in USA other than Wisconsin from Jarman et al. (1993).

An example supporting this comment is a female peregrine that nested successfully in Milwaukee for six years, producing 18 offspring.

PCB levels from geographic populations of peregrines across North America are shown in Table 3. Since the methodology varies from study to study and since the analysis of the 209 PCB congeners and mixtures thereof also varies, it is difficult to make meaningful comparisons between these populations and draw conclusions from the available data.

However, when looking at the PCB levels we observed disparity between some of the samples and compared this to the Wisconsin data. In the Canadian studies dealing with *F. p. anatum* and *F. p. pealei*, the minimum PCB levels were less than 7.97 ppm and the maximum levels above 30 ppm for each subspecies. In Wisconsin, the level of disparity was much less. Even though three PCB compounds, Arochlor 1260, 1248 and 1254, were indicated, they showed a minimum level of <0.05 ppm and a maximum level of 15.3 ppm.

The information gathered in this study has allowed us to take a first look at how Wisconsin's reintroduced peregrines have been affected by environmental contaminants. With continued monitoring and analysis, we will be able to document future changes in contaminant levels. Cooperating with other Midwestern states and standardizing analysis procedures, we hope to acquire additional contaminant data enabling us to monitor and better assess the current contaminant levels of peregrines within the region.

ACKNOWLEDGMENTS

We would like to thank Stan Temple for measuring our first and second groups of eggs and for his advice with analysis work which got the study moving. We would also like to express our gratitude to Wally Jarman for answering questions and providing us with current data on contaminant levels of peregrines in the USA. We would like to thank Clayton White for giving us one of his last copies of the April–June 1990 issue of The *Canadian Field Naturalist*, Matt Marks of Hazleton Wisconsin, Inc. for explaining the "many wonders" of PCBs. We would also like to thank the members of the Wisconsin Peregrine Society whose memberships and donations made this analysis work possible, and the Zoological Society of Milwaukee County for funding which has enabled the continuation of this study.

REFERENCES

GRIFFITT, K.R. AND J.C. CRAUN. 1974. *J. Assoc. Off. Anal. Chem.* **57**: 168–172.
JARMAN, W.M., R.J. NORSTROM, M. SIMON, S.A. BURNS, C.A. BACON AND B.R.T. SIMONEIT. 1993. Organochlorines, including chlordane compounds and their metabolites in Peregrine falcon, Prairie falcon and Clapper rail eggs from the USA. *Environ. Pollut.* **81**: 127–136.
JOHNSON, L.D., R.H. WALTZ, J.P. USSARY AND F.E. KAISER. 1976. *J. Assoc. Off. Anal. Chem.* **59**: 174–187.
JOURNAL OF ASSOCIATION OF OFFICIAL ANALYTICAL CHEMISTS. 1990. Mercury in fish. Official methods of analysis, Method 977.15 (modified). 15th ed.
PEAKALL, D.B. AND L.F. KIFF. 1988. DDE contamination in peregrines and American kestrels and its effect on reproduction. Pp. 337–350 in T.J. Cade, J.H. Enderson, C.G. Thelander and C.M. White, eds. *Peregrine falcon populations: their management and recovery*. The Peregrine Fund, Boise, ID.
—, D.G. NOBLE, J.E. ELLIOTT, J.D. SOMERS AND G. ERICKSON. 1990. Environmental contaminants in Canadian peregrine falcons, *Falco peregrinus*: a toxicological assessment. *Can. Field-Nat.* **104**: #2 248.

RATCLIFFE, D.A. 1967. Decrease in eggshell weight in certain birds of prey. *Nature* **215**: 208–210.
—. 1980. *The peregrine falcon*. T.&A.D. Poyser, London.
SWEET, C.W. 1992. International monitoring of the deposition of airborne toxic substances to the Great Lakes. *Proc. 9th World Clean Air Congr.* Montreal, 1992. Paper 11b.07, Air and Waste Management Association, Pittsburgh, Pennsylvania, USA.

4

The Urban Buteo: Red-shouldered Hawks in Southern California

Peter H. Bloom and Michael D. McCrary

Abstract — Red-shouldered hawks nesting in both natural and urban habitats were studied in southern California. Of 170 red-shouldered hawk breeding territories examined, 30 (17.6%) were in urban environments. Although not classified as urban, another 24 territories (14.1%) contained active nests within 100 m of a ranch house, fire station, or water treatment facility. These 24 pairs were frequently exposed to high levels of human activity. Of 77 urban nests, 37.7% were in non-native trees including several species of eucalyptus, California fan palm, and deodara cedar. These observations suggest this species is relatively adaptable to human-altered landscapes. Well-planned parks and reserves will ensure that this adaptable hawk will continue to exist in California even in the midst of large-scale urban development.

Key words: red-shouldered hawk; urban; territory; adaptable.

Several diurnal North American raptors demonstrate high degrees of adaptation to human populations and the pressures they exert on them. Raptors noted for their ability to adapt to human environments include the peregrine falcon (Cade *et al.* 1988), merlin (Oliphant and Haug 1985), American kestrel (Palmer 1988), and osprey (Palmer 1988, Poole 1989). With the exception of the merlin, which to date has been reported using urban environments only in Canada, these four raptors routinely hunt, mate and reproduce in urban areas throughout their range in North America.

In contrast, members of other genera, such as *Cathartes, Elanus, Circus, Accipiter* (PB unpubl. data) and *Aquila* (Scott 1985) are much less tolerant of human activity and rarely, or in some cases never nest in urban environments. Human presence may not entirely exclude these raptors as they sometimes nest in the interface between urban and natural areas, although this relationship is often temporary. Similarly, some nocturnal raptors have not adjusted to urbanization in southwestern California. As a result of urbanization, nesting long-eared owls have declined by at least 55% in that region and are not adapting to urban pressures (Bloom 1994).

Members of the genus *Buteo* exhibit a broad range of adaptability to human-altered environments in California. The western subspecies of the red-shouldered hawk appears to be the most adaptable of the 10 North American breeding *Buteos*, with some pairs using urban habitats almost exclusively (Bloom et al. 1993). The Swainson's hawk in California seems to be the next most flexible *Buteo*. Some pairs spend the majority of their time hunting in agricultural areas and nest inside residential areas (James 1992, England et al. unpubl. data, PB pers. obs.). Red-tailed hawks commonly nest adjacent to occupied ranch houses and forage in agricultural areas. However, only under unique circumstances does the species nest and forage in urban environments (PB unpubl. data). Red-tailed hawk territories in or near urban areas are often transitory and usually disappear as development increases. Ferruginous hawks appear the least tolerant and to our knowledge have never been reported nesting in or adjacent to urban environments.

In this paper we present information on the frequency of urban nesting red-shouldered hawks, habitat, behavior, comparisons with other California-nesting raptors, and evidence that this species has the capacity to adjust to some types of urban development.

STUDY AREA

The 1435 km^2 study area is located in coastal southern California and extends from the San Luis Rey River in San Diego County north to the Santa Ana River in Orange County. The study area is bounded on the west by the Pacific Ocean and extends east about 15–30 km to the crest of the Santa Ana and Santa Margarita Mountains. Principal research areas of pairs nesting in native habitats include Starr Ranch Audubon Sanctuary, Camp Pendleton Marine Corps Base, Rancho Mission Viejo, Irvine Ranch and numerous Orange County Parks (Bloom et al. 1993). In contrast to some of our previous research (McCrary and Bloom 1984a, 1984b, McCrary et al. 1992, Bloom et al. 1993), the focus of this study is on urban and urban-interface nesting pairs and not on pairs nesting in natural areas.

Habitat of most urban-nesting red-shouldered hawks consisted of exotic non-native vegetation, manicured lawns, athletics fields, buildings, roads, parking lots and utility poles. Many however, consisted of a central core area and/or perimeter of natural habitat. Understory grasses were primarily non-native Mediterranean annuals consisting of ripgut brome and foxtail chess. Characteristic non-native trees include numerous species of eucalyptus, with blue gum predominating; California fan palm; and several pine species (*Pinus* spp.). Dominant trees in natural habitats included coast live oak, western sycamore and black willow.

METHODS

We have studied nesting red-shouldered hawks for approximately 25 yrs. During this time most of our focus was on pairs nesting in natural habitats, which has probably resulted in an underestimation of urban red-shouldered hawk nesting territories in our study area. In the case of urban-nesting red-shouldered hawks we noted location, tree species, proximity to nearest human habitations, human activity level, and recorded the fledging success of certain pairs. Two members of two urban pairs were also equipped with radio transmitters (Bloom et al. 1993).

In this study our focus was on the successful adjustment of urban-nesting red-shouldered hawks to human-modified environments. We defined urban-nesting pairs as those pairs with territories that were estimated to be at least 50% urban. Many territories were entirely surrounded by residential, light industrial, or other buildings and were nearly always in view of people or vice versa. Urban territories were also characterized by pairs that selected nest trees in residential areas, college campuses, golf courses, cemeteries, vacant lots, city and county parks, and were frequently observed foraging within these areas.

We made no attempt to identify all natural and urban breeding territories but were probably successful in locating the majority. In addition to our normal yearly survey activities, in 1993 and 1994 we published a request for information on red-shouldered hawk nest sites in the newsletters of two local National Audubon Society chapters. All nest sites were plotted on US Geological Survey (USGS) topographic maps. Nest tree species, habitat, and fledging success were recorded when known.

RESULTS AND DISCUSSION

Urban Territory Types

We identified 170 red-shouldered hawk nesting territories and classified 30 (17.6%) of these as urban. The presence of several urban locations throughout the study area where individuals or pairs were calling suggests more urban nests remain to be found.

While only 30 of 170 territories fit our description of urban nesting pairs, nests in 24 (14.1%) other territories were within 100 m of buildings such as a ranch house, fire station, or water treatment facility. These 24 pairs were frequently exposed to high levels of human activity at close range, and although not classified as urban pairs, they further indicate the adaptability of red-shouldered hawks to human environments.

The urban environment of red-shouldered hawks can be classified into four general types, which may also reflect different processes in the adaptation of this species to human populations. Type 1 consists of situations where development (urban environment) has encroached and totally surrounded a nesting territory complete with its natural habitat. Territories of this type are characterized by large mature trees, water and an adequate quantity of foraging habitat as well as areas of seclusion from people. These hawks regularly interact with people owing to the closeness of hunting perches and their nest to the urban edge. We characterized three territories as Type 1.

Type 2 consists of situations where an urban environment is created in an area not previously occupied by red-shouldered hawks. Red-shouldered hawks do not hunt from the wing and require hunting perches (Bloom et al. 1993). They also require trees for nesting and are not known to nest on cliffs or man-made structures e.g. utility poles. In this situation, suitable nest trees and hunting perches were lacking, but over time exotic trees used for landscaping became available. Hunting perches were provided by other exotic trees and utility structures. Thus, new red-shouldered hawk habitat was created through urban development. These nesting territories may be short-lived, since adjacent requisite hunting areas may eventually be developed. However, we are aware of protected nesting territories that have existed for decades in human-created habitats outside our study area, at the San Diego Wild Animal Park, and San Diego and Los Angeles Zoos. We classified six territories as Type 2.

Type 3 is where development has occurred adjacent to an existing territory, leaving most of the territory intact. In this situation a pair expands its home range by occupying the newly created urban habitat via the use of utility poles and landscape trees used as hunting perches and nest sites. These territories are characterized by mature natural woodlands, small to large exotic trees, water, adequate hunting habitat, and areas of seclusion from people. Twelve territories fit this description.

Type 4 is where hawks reside in heavily used county or city parks including both predominantly natural parks and intensively managed lawn or playing field parks. Nine territories were classified as Type 4. These parks receive low to moderate use five days/week and low to intense use on weekends. Vegetation in the parks is characterized by young to mature exotic and native trees. Some pairs nested in large mature trees found in golf courses and hunted in the surrounding natural habitat. Likewise, some pairs used college campuses in a similar fashion. We classified 140 territories as natural.

Behavior

Red-shouldered hawks nesting in urban environments seem undisturbed by the presence of people; even large crowds (>100) at 7 territories playing athletic sports, conducting equestrian activities, and camping directly underneath their nest trees in county parks did not cause nest abandonment. Most hawks also develop a tolerance for joggers or people climbing in and out of their vehicles and do not abandon hunting perches unless approached within 10–50 m. However, some individuals become very protective of their nests and are as aggressive as some northern goshawks (Bent 1937, PB pers. obs.). Six adults from five urban territories were captured and removed following complaints from people who sustained head injuries. We generally average one phone call per year concerning aggressive red-shouldered hawks in southern California.

The differences in adjustments to human activity is most apparent when attempting to observe radio-tagged and/or color-banded individuals. Those individuals occupying urban environments are readily observed and can be approached quite closely, facilitating the reading of color-bands and even US Fish and Wildlife Service bands. Once located, these individuals can be visually monitored for entire days without disturbing them (McCrary 1981, Bloom 1989). In contrast, those occupying natural environments with low levels of human activity are markedly more difficult to observe. Radio-tagged hawks in natural areas may seldom be seen and will be disturbed if approached.

Nest Trees

In contrast to red-shouldered hawk territories in natural areas where predominantly native trees were used for nest sites, urban pairs frequently nested in non-native species.

A variety of nest trees was selected by urban-nesting red-shouldered hawks, including natives when portions of the natural habitat were left undisturbed. Of 77 nest trees recorded in urban environments the most frequently used native trees were western sycamore (52.9%) and coast live oak (10.4%). Exotic nest trees included several species of eucalyptus (32.5%), fan palm (3.9%) and deodara cedar (1.3%). The introduction of eucalyptus has contributed enormously to the expansion of nesting red-shouldered hawks into otherwise unsuitable breeding habitat in California. Location of nest trees relative to human activity does not seem important as several territories had nests located in trees within the area of highest human use. Importantly however, these trees were among the largest in the territory.

Reproduction

Reproductive success in this study of pairs nesting in urban locations was greater than that previously reported in natural areas in the same region (Wiley 1975). Wiley (1975) reported 1.34 young fledged per nesting attempt and 2.05 young fledged per successful nest from non-urban territories. We found 1.80 young fledged per nesting attempt ($N = 50$) and 2.50 fledged per successful nest ($N = 36$) in urban environments. Nest success was 65.5% on Wiley's study (1975) and 72.0% in this study. The higher fledging success and nest success observed in this study suggests the possibility that predation pressure may be lower, and/or hunting success greater in urban environments. However, the two studies are not directly comparable because nests in our study were usually examined only two to three times per season when chicks were two weeks or older and some mortality may have been missed, particularly in the early nestling stage. Wiley (1975) examined chicks at 1–3 day intervals beginning at hatching, allowing for more accurate assessment of fledging success, but increased potential for higher predation.

Food habits

Although we have not analyzed prey of urban red-shouldered hawks separately from others, the general food habits of this species, i.e. preying upon invertebrates and small vertebrates, contribute to this species' adaptability to urban environments (Bloom et al. 1993). Compared with red-tailed hawks nesting in southern California, red-shouldered hawks tend to consume smaller prey species (authors unpubl. data). Smaller species preyed upon by red-shouldered hawks probably predominate over larger ones in urban environments because populations of larger prey such as Audubon's cottontail and California ground squirrels have been locally reduced or extirpated.

Anecdotal observations of red-shouldered hawk behavior in urban environments also indicate the potential importance of diet in this species' adaptability to human presence. Red-shouldered hawks have been noted feeding on processed food used in zoos (J. Nagata-Lewis pers. obs.) and food discarded around buildings and athletic fields (MM pers. obs.).

Space Use

Red-shouldered hawks have unusually small breeding seasons and annual home ranges (Bloom et al. 1993), and pairs nesting in urban environments are no exception. Home range size (100% minimum convex polygon) of radio-tagged red-shouldered hawks in southern California averaged 1.69 km^2 for 7 males and 1.15 km^2 for 6 females. Home ranges of 2 radio-tagged males classified as urban were 0.45 km^2 and 0.69 km^2. These hawks used tiny home ranges despite the fact that much of the habitat within their home ranges consisted of buildings and asphalt. In comparison, home ranges (100% minimum convex polygon) of 2 male and 1 female radio-tagged red-tailed hawks from the same study area were 3.61, 7.19, and 13.57 km^2, respectively.

Of interest, three radio-tagged red-shouldered hawks used hunting areas that were disconnected from the main body of their home ranges (Bloom et al. 1993). These areas were separated from their main ranges by large tracts of habitat unsuitable for hunting, e.g. devoid of hunting perches. Similarly, we have observed urban-nesting red-shouldered hawks making direct flights to favored disjunct hunting areas while bypassing large expanses of unsuitable commercial or residential areas. The reduced space needs of red-shouldered hawks compared with other *Buteos* may allow this species to occupy small tracts of natural vegetation that may remain after development. This aspect of their adaptability may be further enhanced by their use of isolated hunting areas within the urban environment.

Comparisons with Other Raptors

While 17.6% (30) of 170 red-shouldered hawk territories met our definition of urban territories, only 2.2% (8) of 361 red-tailed hawk territories could be classified as such. Of the eight urban territories six nests were in eucalyptus and two were on utility poles. Importantly, four territories were in existing locations (Type 1) prior to being surrounded by development, and incremental urbanization may ultimately exclude them. These territories presently contain modest amounts of unprotected natural open space in the form of utility easements, vacant lots, or flood channels. The other four territories included pairs with nests in residential areas but adjacent to vast range lands or wilderness parks.

In contrast to red-shouldered hawks which have expanded into urban environments, in our experience red-tailed hawks almost always disappear from these areas. Five territories were abandoned when approximately 50–100% of their home ranges were urbanized. Three other territories were abandoned when hunting areas remained, but nest trees were removed for gravel mining. The latter three territories still have the potential to be reoccupied should trees again reach maturity. Some red-tailed hawk territories may thrive when surrounded by urban environments, but only when large quantities of natural open space are preserved (PB unpubl. data).

Similarly, only 4.8% (3) of 62 Cooper's hawk breeding territories were found in urban environments. However, two of these territories were in existing natural areas prior to park designation and while the parks receive considerable recreational use, both parks have large amounts of intact natural habitat, suggesting not so much adaptation to an urban environment as the retention of a traditional nesting area. The level of urban hunting in the surrounding residential community is unknown but appears limited. The third territory was located 150 m from the urban edge in exotic trees on the interface between coastal sage scrub/riparian habitats and an urban environment. In common with the above two urban Cooper's hawk territories this one had even larger amounts of natural open space in the form of a military reservation next to it. None of 62 Cooper's hawk breeding territories were found inside residential areas, suggesting that basic ecological requirements such as nesting habitat and/or prey are probably limiting.

Of 103 white-tailed kite nesting territories no nests were found inside residential areas, or any closer than 150 m from an urban environment. Likewise, of 12 northern harrier nesting territories no nests were found inside urban environments; the nearest nest being 250 m from the urban edge. As one might predict from the predilection of both species for grasslands (Palmer 1988), urban areas with no grassland or compatible crops such as alfalfa, result in no use by these two species for nesting.

CONCLUSIONS

Red-shouldered hawks are regular components of the urban nesting avifauna in coastal southern California where land managers have preserved adequate open space and habitat. In contrast to other coastal southern California nesting raptors, including the red-tailed hawk, Cooper's hawk, white-tailed kite, and northern harrier, the red-shouldered hawk has a demonstrated flexibility to adapt to human-altered environments. Clearly, red-shouldered hawks can optimize their survival potential by exploiting space and habitat that other raptorial species seem unable to do, and behaviorally, they, like the peregrine falcon, seem almost preadapted to coexistence with people provided certain needs are met. Relative to other Buteonine raptors the western subspecies of the red-shouldered hawk is adapting well to urban environments provided that the appropriate quantity, and habitat types are preserved (Wilbur 1973, Harlow and Bloom 1989, Bloom et al. 1993). However, the eastern subspecies is listed under various protective categories in several states and needs continued monitoring. In terms of its long-term ability to coexist with humanity we believe that *Buteo lineatus elegans* is the *Falco peregrinus anatum* of the buteos.

ACKNOWLEDGMENTS

For suggesting the idea for this manuscript we proudly acknowledge the late Richard R. "Butch" Olendorff. Superb field assistance was provided by Ed Henckel, Judith Henckel, Susan Galaugher, Jeff Kidd, and Donna Krucki. We also thank Richard and Donna O'Neill of Rancho Mission Viejo, Carol Hoffman of The Irvine Ranch, Peter DeSimone of Starr Ranch Audubon Sanctuary, Slader Buck of Camp Pendleton Marine Corps Base, and Tim Miller of Orange County Harbors Beaches and Parks for permission to conduct research on lands under their purview. The Sea and Sage Audubon Chapter and Southcoast Audubon Chapter, and their members, are thanked for publishing and responding to our information requests. Our manuscript benefited from the critiques of Michael M. Morrison, Victor Apanius and Charles T. Collins.

REFERENCES

BENT, A.C. 1937. Life histories of North American birds of prey. *US Nat. Mus. Bull.* 167: 1–409.
BLOOM, P.H. 1989. Red-shouldered hawk home range and habitat use in southern California. M.S. thesis, California State Univ., Long Beach.
—, 1994. The biology and current status of the long-eared owl in coastal southern California. *Bull. South. Calif. Acad. Sci.* 93: 1–12.
—, M.D. MCCRARY AND M.J. GIBSON. 1993. Red-shouldered hawk home-range and habitat use in southern California. *J. Wildl. Manage.* 57: 258–265.
CADE, T.J., J.H. ENDERSON, C.G. THELANDER AND C.M. WHITE, eds. 1988. *Peregrine falcon populations: their management and recovery.* The Peregrine Fund, Inc., Boise, Idaho.
HARLOW, D.L. AND P.H. BLOOM. 1989. Buteos and the golden eagle. Pp. 102–110 in B.G. Pendleton, C.E. Ruibal, D.L. Krahe, K. Steenhoff, M.N. Kochert and M.N. LeFranc, Jr., eds. *Proc. Western Raptor Management Symposium and Workshop.* Nat. Wildl. Fed., Washington D.C.

JAMES, P.C. 1992. Urban-nesting Swainson's hawks in Saskatchewan. *Condor* **94**: 773–774.
McCRARY, M.D. 1981. Space and habitat utilization by red-tailed hawks (*Buteo lineatus elegans*) in southern California. M.S. thesis, California State Univ., Long Beach.
McCRARY, M.D. AND P.H. BLOOM. 1984a. Lethal effects of introduced grasses in red-shouldered hawks. *J. Wildl. Manage.* **48**: 1005.
McCRARY, M.D. AND P.H. BLOOM. 1984b. Observations on female promiscuity in the red-shouldered hawk. *Condor* **86**: 486.
McCRARY, M.D., P.H. BLOOM AND M.D. GIBSON. 1992. Observations on the behavior of surplus adults in a red-shouldered hawk population. *J. Raptor Res.* **26**: 10–12.
OLIPHANT, L.W. AND E. HAUG. 1985. Productivity, population density and rate of increase of an expanding merlin population. *Raptor Res.* **19**: 56–59.
PALMER, R.S. 1988. *Handbook of North American Raptors, Vol. 4, Diurnal Raptors.* Yale University Press, New Haven.
POOLE, A.F. 1989. *Ospreys: A natural and unnatural history.* Cambridge University Press, Cambridge.
SCOTT, T.A. 1985. Human impacts on the golden eagle population of San Diego County from 1928 to 1981. M.S. thesis. San Diego State Univ.
WILBUR, S.R. 1973. The red-shouldered hawk in the western United States. *West. Birds* **4**: 15–22.
WILEY, J.W. 1975. The nesting and reproductive success of red-tailed hawks and red-shouldered hawks in Orange County, California. *Condor* **77**: 133–139.

5
Urban Nesting Biology of Cooper's Hawks in Wisconsin

Robert N. Rosenfield, John Bielefeldt, Joelle L. Affeldt and David J. Beckmann

Abstract — We investigated various aspects of the nesting biology of Cooper's hawks in a central Wisconsin suburb. We found the highest nesting density (272 ha per breeding pair) yet known for this species. Means for clutch size (4.2) and for number of bandable young (4.0) at nine nests were comparable to the highest known for the species. Results from our study do not support the suggestion that Cooper's hawks cannot breed in highly fragmented forests or suburban areas.

Key words: Cooper's hawk; nesting; Wisconsin; suburbs.

Investigators elsewhere have reported instances of Cooper's hawks nesting in urban/surburban settings (Stahlecker and Beach 1979, Dancey 1993), and we have also found numerous urban and suburban nests in Wisconsin. Murphy *et al.* (1988) studied one such nest and suggested that researchers need to document the use of urban environments by breeding Cooper's hawks, a species still assigned Extirpated, Endangered, Threatened, or a species of Special Concern status in 16 eastern states (Mosher 1989, Rosenfield *et al.* 1991b). As part of our 14-yr study of the breeding ecology of the Cooper's hawk in Wisconsin (Rosenfield and Bielefeldt 1993), we investigated aspects of its nesting biology in a central Wisconsin suburb.

To determine nesting density (ha per breeding pair) we searched in 1993 for nests on a 3540-ha area in the city of Stevens Point (Portage County) and its suburbs (population ca. 36 000). For a general description of the study area see Murphy *et al.* (1988). We conducted intensive ground searches of previously occupied woodlots and visited potential nesting areas at dawn in late March through mid-April, when mated pairs were readily detected on a daily basis because they were predictably present and vocal (especially females) at sunrise when building nests (Rosenfield 1990, Rosenfield *et al.* 1991a). We also trained inexperienced observers to recognize calls and other signs, e.g. pluckings and other prey remains, or excrement, of occupancy. Trainees arrived at suggested nesting areas about 30 min before sunrise and sat motionless against a tree for

1 h. Nest finding by trainees was often facilitated by conspicuous nest-building behaviors at dawn, e.g. audible breaking of twigs and frequent flights to nests (Rosenfield et al. 1991a). Clutch size was determined by climbing to nests in mid–late May; young were banded during mid–late June.

We found 13 nests, or 272 ha per breeding pair. This nesting density was, to our knowledge, the highest known for the species. Median inter-nest distance was 0.8 km (range =0.5 to 2.4 km). Twelve of these 13 nests were found at or before incubation; nine (75%) of these 12 nests produced at least one young of bandable age ($\geq$14 days). Means for clutch size (4.2) and for number of bandable young (4.0) at the nine nests were comparable to the highest known for the species, including the pre-DDT era (Rosenfield and Bielefeldt 1993).

Woodlots used for nesting ranged from 1–12 ha. Five nests were in mixed pine (jack, red, and/or white) stands, five were in mixed pine-oak (*Quercus* spp.) stands, two were in oak, and one was in a park-like setting of white pine with a mowed lawn. All stands were second-growth, with trees averaging about 20–25 m in height and 20–40 cm in diameter-at-breast height (dhb). Conifer forests and plantations are frequent nest sites elsewhere in Wisconsin and other midwestern states (Mutter et al. 1984, Rosenfield et al. 1991b, Wiggers and Kritz 1991). Many of these urban sites have not been checked annually, but we know that $\geq$3 sites have been occupied repeatedly by breeding pairs over 13 to 15 yr, while $\geq$7 other sites have been occupied 25 times in 27 search-yr.

Density, productivity and repeated reoccupancy of urban woodlots suggest that these nesting areas are not marginal habitats or population sinks. Dispersal data are also a significant but seldom studied aspect of avian demography (Newton 1991). One urban-hatched male dispersed 2.4 km to a subsequent urban breeding site that it continuously occupied for 5 yr; one marked adult female moved 2.4 km from its rural nesting site in 1990 to breed on our urban study area in 1991.

Some researchers have implied that Cooper's hawks may not breed in highly fragmented forests (Robinson 1991 and references therein, Bosakowski et al. 1993). Results from our study area do not support this suggestion. Bosakowski et al. (1993) implied that nesting Cooper's hawks may not "survive well" in suburban habitat; our results on nesting density, productivity and reoccupancy from our urban site do not agree with this premise. Comparable nesting densities in urban and rural habitats in Wisconsin (Rosenfield and Bielefeldt 1993) suggest that the Cooper's hawk may be one of the most abundant nesting falconiforms in the state. Popular sources and other authors have also called the Cooper's hawk "uncommon" (Robbins 1991) in Wisconsin, and "seldom found" (Peterson 1980) or "gone" (Mackenzie 1986) as a breeding bird in the eastern US. Again, our urban results and other studies, e.g. Bednarz et al. (1990), Conrads (1990), Wiggers and Kritz (1991), do not support their assertions. Popular accounts probably supply much of the public's awareness of bird populations, including a widespread misunderstanding about the numbers of a secretive bird such as the Cooper's hawk. Avian ecologists clearly need to disseminate up-to-date, accurate information on bird populations to a public that does not read the technical journals.

ACKNOWLEDGMENTS

We thank the Wisconsin Department of Natural Resources and the University of Wisconsin at Stevens Point for their financial and logistical support of our studies. T. Aschenbach, E. Jacobs, and D. Wileden helped locate nests. J. J. Negro, R. Bierregaard, and an anonymous reviewer provided helpful comments on this paper.

REFERENCES

BEDNARZ, J.C., D. KLEM, L.J. GOODRICH AND S.E. SENNER. 1990. Migration counts of raptors at Hawk Mountain, Pennsylvania, as indicators of population trends, 1934–1986. *Auk* **107**: 96–109.

BOSAKOWSKI, T., R. SPEISER, D.G. SMITH AND L.J. NILES. 1993. Loss of Cooper's hawk nesting habitat to suburban development: inadequate protection for a state-endangered species. *J. Raptor Res.* **27**: 36–40.

CONRADS, D.J. 1990. Status of the Cooper's hawk in Iowa. Unpubl. Rep., Iowa Dept. Nat. Resour., Des Moines, IA USA.

DANCEY, H. 1993. A pair of subadult Cooper's hawks nest in Indiana. *Indiana Quart.* **71**: 26–34.

MACKENZIE, J.P.S. 1986. *Birds of the world, birds of prey*. Northword Inc., Minocqua, WI USA.

MOSHER, J.A. 1989. Status reports: accipiters. Pp. 47–52 in *Proc. Northeast Raptor Management Symposium and Workshop*. Natl. Wildl. Fed., Washington, D.C.

MURPHY, R.K., M.W. GRATSON AND R.N. ROSENFIELD. 1988. Activity and habitat use by a breeding male Cooper's hawk in a suburban area. *J. Raptor Res.* **22**: 97–100.

MUTTER, D., D. NOLIN AND A. SHARTLE. 1984. Raptor populations on selected park reserves in Montgomery County, Ohio. *Ohio Acad. Sci.* **84**: 29–32.

NEWTON, I. 1991. Concluding remarks. Pp. 637–654 in C.M. Perrins, J-D. Lebreton and G.J.M. Hirons, eds. *Bird population studies, relevance to conservation and management*. Oxford Univ. Press, Oxford, UK.

PETERSON, R.T. 1980. *Eastern birds, a field guide to the birds*. Houghton Mifflin Co., Boston, MA USA.

ROBBINS, S.D., Jr. 1991. *Wisconsin birdlife, population and distribution, past and present*. Univ. Wisconsin Press, Madison, WI USA.

ROBINSON, S.K. 1991. Effects of habitat fragmentation on midwestern raptors. Pp. 195–202 in *Proc. Midwest Raptor Management Symposium and Workshop*. Natl. Wildl. Fed., Washington, D.C.

ROSENFIELD, R.N. 1990. Pre-incubation behavior and paternity assurance in the Cooper's hawk (*Accipiter cooperii* [Bonaparte]). Ph.D. thesis, North Dakota State Univ., Fargo, ND USA.

— AND J. BIELEFELDT. 1993. Cooper's hawk (*Accipiter cooperii*) in A. Poole and F. Gill, eds. *The birds of North America, No. 75*. Philadelphia: The Academy of Natural Sciences, Washington, D.C., American Ornithologists' Union.

—, — AND J. CARY. 1991a. Copulatory and other pre-incubation behaviors of Cooper's hawks. *Wilson Bull.* **103**: 656–660.

—, —, R.K. ANDERSON AND J.M. PAPP. 1991b. Status reports: accipiters. Pp. 42–49 in *Proc. Midwest Raptor Management Symposium and Workshop*. Natl. Wildl. Fed., Washington, D.C.

STAHLECKER, D.W. AND A. BEACH. 1979. Successful nesting by Cooper's Hawks in an urban environment. *Inland Bird Banding News* **51**: 56–57.

WIGGERS, E.P. AND K.J. KRITZ. 1991. Comparison of nesting habitat of coexisting sharp-shinned and Cooper's hawks in Missouri. *Wilson Bull.* **103**: 568–577.

6

Urban Ecology of the Mississippi Kite

James W. Parker

Abstract – The Mississippi kite may be the most abundant urban raptor in North America. In this paper I review aspects of its biology relevant to urban nesting and discuss the possible origin, current condition, and potential future of urban nesting. Studies of this species since 1968 indicate that its choice of food, and foraging and nesting habitats facilitate its urban existence. Other life history characteristics, however, do not seem so conducive to its extraordinary urban success. The species began to draw attention as an urban nester in the mid and late 1970s because of its aggressive behavior toward humans. Since then kites have become common in scores of towns of all sizes in four states. Urban densities are high and roosts often exceed 50 birds. Comparisons of urban versus rural colonies indicate that urban populations are probably more successful in a number of ways and will persist.

Key words: Mississippi kite; urban; nesting; breeding; aggression; humans.

When I began my study of the Great Plains population biology of the Mississippi kite, cases of urban nesting were rare (Parker 1974, 1975). In 1977, after having noted its increasing presence in urban areas, I began examining an urban kite population in Meade, Kansas. One summer later the kite population in Ashlands, Kansas made media headlines when 28 kites were shot in response to their diving at humans. Since then urban nesting has continued to expand geographically, and usually has been described in connection with the phenomenon of nest-defense diving (Parker 1979, 1980, 1988a, Gennaro 1988a). My incomplete collection of newspaper stories on kite diving, 1978–1992, includes 36 articles from 15 papers in 4 Great Plains states, as well as features in the Bangor (ME) Daily News and Boston Globe. The Mississippi kite is now the most conspicuous, and perhaps most numerous North American urban raptor, and two studies (Shaw 1985, Gennaro 1988b) have recently added to my own efforts to monitor and understand urban kites.

   The considerable growth of the Mississippi kite population west of the Mississippi River since the early twentieth century has been described and analyzed (Parker and Ogden 1979, Parker 1988b). It did not represent a population recovery, but rather a rural population expansion from a relatively undisturbed state. Colonization of urban habitat by kites has followed this growth and is a logical, but unexpected extension. However, urban nesting by kites has not been so thoroughly described, nor analyzed. Here, I present a brief, preliminary description of urban kite biology, an equally preliminary effort to explain why and how urban populations have developed, and what the future of urban populations may be.

METHODS

In summer 1977, I executed a full census of the wooded areas of Meade, Kansas, a town of about 2300 people on 2.5 km^2. In later years (1979–80, 1982–87), I monitored variable numbers of nests there without completely censusing the town. I have also selectively monitored kite populations in several other towns, including Ashland (1979), Englewood (1982–85), Fowler (1984–86), Dodge City (1982, 1985), and Newton (1982–85), Kansas, and Altus (1986–87) and Oklahoma City (1985, 1987), Oklahoma. My standard field techniques (based on repeated visits to nests throughout each nesting season) were described by Parker (1974, 1975), as were the large samples of rural populations and nests. These samples, with Glinski and Ohmart (1983), provide recent standard data sets to which the demography of urban populations can be compared. This analysis draws selectively from these data and from studies by Shaw (1985) and Gennaro (1988b), each of which has focused solely on one urban population in Texas and New Mexico, respectively, and did not involve true censuses of complete urban areas.

HISTORY, SCHEDULE AND GEOGRAPHY OF URBANIZATION

Parker and Ogden (1979) offered the first and most comprehensive review of how and why kite numbers in the prairies expanded greatly in the first half of the twentieth century. This was based on first-hand, in-depth knowledge of the species' biology, as well as historical references and should be considered a basic prelude to analysis of urbanization. Bolen and Flores (1989) added interesting historical details about early observations of kites by Europeans, but did not shed further light on the reasons for population expansion. Their review was primarily drawn from the literature, and thus, they misinterpreted or mis-described significant aspects of kite biology, such as characteristics of large kite colonies.

In small numbers kites probably began to nest closer to humans soon after the first trees planted by early settlers reached suitable size (>6 m). Kites nesting in human-planted trees would have come from trees in, or near, riparian and oak prairie woodland or savanna. Nests in these areas are now, and probably were, scattered widely, with only a minor tendency for colonial association (JP pers. obs., M. Ports pers. obs.; see also Sutton 1939, Glinski and Ohmart 1983), especially compared with nesting today in human-established woodlands (Parker 1974, 1975).

By 1960, it is likely that large numbers of colonies existed and were expanding in shelterbelts established by the Civilian Conservation Corps and in farmstead woodlots. Certainly this would have brought large numbers of breeding kites geographically very close to urban areas throughout western Kansas, Oklahoma, and northcentral Texas. Conversations with many long-time citizens of these areas indicate that kites were noticed in small numbers in some towns like Ashland in the late 1960s and early 1970s. Soon thereafter, urban nesting was recognized on a large scale, with urban populations in parts of west Texas and New Mexico probably being established later than those in areas where shelterbelt nesting was common. Up until recently, urbanization had been almost a strictly western phenomenon, but Sweet (1988) reported a small amount of urban nesting in Illinois and Missouri, and suggested it to be increasing. In west Tennessee, as elsewhere in the east, most nesting remains in mature bottomland forest (Kalla and Alsop 1983).

NATURAL HISTORY AND DEMOGRAPHY

The basic biology of the Mississippi kite gives clues to how and why the species began nesting in urban areas. Kites capture large numbers of large insects, their major prey, and in their extreme western range seem to depend on or prefer cicadas (Glinski and Ohmart 1983, Shaw 1985). However, there is no evidence that cicadas are a controlling or dominant dietary element in the central plains or farther east. In some areas kites also take a variety and considerable number of vertebrates (Parker 1974, 1975, 1988b). It may be that the species' food base has expanded as the result of habitat change in parts of its range and that this contributes to a lack of food limitation for both reproduction and population size (Parker and Ogden 1979, Parker 1975, 1988b).

Kites are extremely gregarious. There is no convincing evidence of territoriality around nests. Shaw (1985) observed that nesting adults sometimes chased yearlings from the vicinity of nests, but I have never observed such behavior. Nests are at heights above 3 m in trees of about every available species and condition, and kites often reuse nests. Parker (1974, 1988b) stressed that kites' catholic tastes in nesting habitat indicate that there are no features of potential nesting habitat that would attract kites except for trees more than about 6 m tall, snags for perching, and the presence of other nesting kites. Love et al. (1985) showed only a tendency for kites to use larger shelterbelts in less disturbed prairie with decreased human presence.

This species' life history characteristics somewhat resemble those of a "K-strategist" (Parker 1974; Parker and Ogden 1979; Parker and Ports 1982). It has a small clutch, i.e. 2 eggs, renests infrequently, suffers considerable but unpredictable early mortality and nest failure, and produces only about 0.6 fledglings per nest effort in rural habitat. Young adults seem to have a long life

Table 1. Comparisons for urban versus rural Mississippi kite colonies.

Variables	Urban colonies[a]	Rural colonies[b]
Nest density	0.86 ha/nest ($N = 41$)	1.1 for shelterbelts; far greater for other colonies
Mean nearest-neighbor distance	111 m ($N = 41$)	115 m for shelterbelts; far greater for other colonies
Reuse of nests for third year	28% ($N = 41$)	Maximum of 13%
Age structure (% yearlings)[c]	17–20%	5–8%
Nest attempts hatching eggs	83% ($N = 34$)	57–61%
Nest attempts fledging young	83% ($N = 34$)	49–50%
Fledglings per nest attempt	1.0–1.3 ($N = 34$)	0.62–0.69

[a] Nests primarily in Meade, Kansas in 1977 and 1979.
[b] Summarizing from 280 nests in 35 shelterbelt and similar woodlots, and 95 nests in 6 other colonies of oak prairie, mesquite, and cottonwood/riparian vegetations from Parker (1974). Sample sizes for different variables were different but were usually much larger than for urban samples.
[c] From counts of large numbers of kites in many flocks in Meade, Ashland, and Englewood, Kansas.

expectancy, and yearlings sometimes act as helpers for adult pairs (Parker 1974, 1988b, Parker and Ports 1982, Glinski and Ohmart 1983). Thus, little about the species' demography and life history indicates the capability for rapid population growth.

My 1977 census of the Meade, Kansas kite population showed 41 nesting efforts, and a selective partial census showed many of the nests to be reused in 1979. Nests were slightly clumped within available wooded areas very much like in rural habitat, especially the trees at Meade State Park (Parker 1974). The Meade nests and those from other towns provide the urban data for Table 1. These data will be augmented for more extensive and critical analysis in the future, and are not intended for statistical analysis here. They are meant as a generalized description of trends and tentatively support the following hypotheses:

(1) urban kite populations are at least as dense as the densest rural populations;
(2) urban nests can be more persistent than rural nests; and
(3) urban productivity is considerably greater leading to a change in population age structure favoring yearlings.

All studies of rural populations show very similar production of fledglings per nest effort (Parker 1988b), and the urban populations of Shaw (1985) and Gennaro (1988b) support the hypothesis of elevated urban productivity (1.1 and 1.2 fledglings per nest effort). These latter two studies also show large cohorts of yearlings (24% and 17–20% of total populations, respectively). Parker and Ports (1982) presented weak evidence that yearling helpers raised productivity, but Shaw's (1985) data do not support this.

Finally, of relevance is Parker's (1974) calculation, using the method of Henny *et al.* (1969), of the productivity required per nest for the rural kite population to remain stable. The various estimates, based on different assumptions for the model, were 0.5 or fewer fledglings per nest. This is clearly consistent with the observation of growth of rural populations.

SPECULATION ON THE HOWS AND WHYS OF URBAN NESTING

When settlers first began to influence the North American prairies in the 1800s, Mississippi kite populations almost certainly were below current densities and possibly stable. What caused the change leading to urbanization? Population size was controlled by one or more factors: food availability, predation, aspects of nesting habitat selection not yet understood, and aspects of kite behavior determining prey and habitat selection. Because the Mississippi kite, like its close tropical relative, the Plumbeous kite (but see Palmer's comment for the Mississippi kite in Parker 1988b), has such a small clutch, it is likely that the genus evolved in response to strong natural selection pressure imposed at least partly by limited food during the nesting season. With this in mind, I suggest the following scenario for the development of urban populations, comprised of accepted and speculative elements, primarily to provoke further thought about continuing population trends.

Initial kite response to humans was likely based upon an adjustment to a reduction in the availability of riparian trees and to the maturation of the first tree plantings by settlers. This probably led to the current tendency of kites to tolerate denser nesting within wooded areas that are small, but sufficiently large to accommodate more than one nest (see below and Parker 1974). Like many raptors, kites simultaneously experienced increasing pressure to tolerate human presence and activity, but may not have been sensitized negatively to humans as much as in the east (Parker and Ogden 1979). In other words, kites began to get used to people and fortuitously possessed enough plasticity in habitat selection that they could take advantage of new prairie woodlands. The result was a greatly shifted and expanded nesting distribution tending away from riparian trees. For unknown reasons, this has allowed or caused the kites to nest more densely, and it brought them even closer to humans and urban areas on a wide geographic basis in Kansas and Oklahoma, setting the stage for urban colonization.

By itself, however, this situation does not really explain the great population increase in kites; it simply allowed and encouraged it. To explain the increase in population size, we must turn to changes in patterns of mortality and food availability. It was stated earlier that the kites may have experienced a significant increase in both insect and vertebrate prey. Accepting this and expanding on the details, the following is at least plausible.

As agricultural activity expanded across the prairies, the kites' insect prey base would have been the first prey component to increase. This alone could have raised kite productivity above replacement levels by making survival of two nestlings more frequent, and it could have encouraged expansion of the population into shelterbelts and similar man-made woodlands. The species' tendency to increase its capture of birds and other vertebrates, and to eat road kills, would have begun much later and in that order as man-made woodlands and roads for high-speed auto travel increased in number. Whatever the specifics, productivity of rural kite populations in the late 1960s and 1970s did not seem food limited because starvation of nestlings was rarely observed for any brood size (Parker 1974, 1975, 1988b). Additionally, a field experiment using brood size manipulation (JP unpubl. data) showed that kites in rural areas can raise broods of three nestlings.

As food limitation of productivity has apparently lessened, so too have the impacts and pattern of nest predation changed. Originally, predators of eggs, nestlings and nesting adults would not have encountered kites in dense colonies. The most serious of these predators prior to arrival of settlers, as well as now, would have been nocturnal. The finding of a single pair of kites or an active nest would not often have necessarily pointed the way to another kite nest, but that has now changed. For nocturnal predators, the development of dense rural nesting colonies has made the finding of one nest a predictor of others in the vicinity. Thus, kites now are undoubtedly more conspicuous to their most important predators, including large owls and climbing mammals. Habitat change favoring kites has likely also directly favored and increased the numbers of their predators. For example great horned owls are very common in shelterbelts and other rural prairie woodlands today. However, kites are very aggressive toward their predators, and dense nesting gives them heightened potential for effective group harassment and attack on predators during the day.

The combined result of these predation factors on kite nest success is hard to assess. Local nest failure from predators can be severe and unpredictable in rural colonies, but it is not widespread or intense at all times (Parker 1974) and clearly did not prevent population growth in the years prior to urbanization.

With a regional population benefiting from increased food, unlimited by predators, and nesting near towns, the stage was set for urbanization. Pressed by high densities in rural areas, and maybe by chance alone, small numbers of adult kites from rural colonies relocated to urban trees, and/or young first-time nesters chose urban trees. For these birds, food availability could only have changed for the better. They continued, and perhaps increased, their access to vertebrates like birds as well as road kills, and did not reduce access to their insect prey.

Initial urbanization should have severely reduced nest failure from predation, since kites likely left most of their most important predators in the countryside. This alone may explain the nearly doubled nest productivity of urban versus rural nests. It is also possible that the greater expanse of urban trees reduces weather-related nest loss by reducing wind velocity at nests. Loss of nests in storms has been significant in rural colonies, especially early in the nesting period (Parker 1974). Although their respective contributions would be hard to evaluate, elevated food supplies, reduced predation, and reduced weather impacts together played some role in the higher productivity of urban nests. Reduced predation may have been the most significant factor.

THE FUTURE OF URBAN KITE POPULATIONS

Urban nesting now appears to be self-sustaining. Urban kite populations should be growing rapidly even now in towns where the carrying capacity has not yet been reached. Although rural colonies remain large and productive, a great proportion of fledglings is being produced by urban colonies, and this is probably the reason that yearlings are so numerous in urban populations. In saturated urban areas, productivity should be providing a large pool of emigrants to support the establishment and growth of nearby demes, urban or rural. The fledglings produced by urban adults may well be conditioned to choosing urban nest sites, further increasing the percentage of the Great Plains population using urban trees.

Barring significant changes in food availability, it is unlikely that the future will bring anything but large kite populations in the central Great Plains where rural and urban populations are mutually supporting.

Helping at the nest has been documented only recently. Could it be a result of a glut of yearlings in urban colonies coupled with a behavioral response of these birds to high nesting density?

Actually, urbanization could have taken place in the Central Plains without a prior population shift into nearby rural shelterbelts and similar woodland. This probably occurred in eastern New Mexico, as there are no nearby large rural populations in man-made woodland that could have provided immigrants to Gennaro's (1988b) urban population. The New Mexico population could be a direct result of total regional population expansion to the east. However, it is very unlikely that urbanization would have occurred as fast as it has, or that a New Mexico population would have developed at all, if expansion into rural man-made woodlands had not occurred in the central plains.

Could urbanization become widespread in the east? Probably, but eastern kites do not seem to be as densely colonial, and this may affect the potential for urban nesting. Also, until eastern kites begin to use second growth woodland on a larger scale, they are not as likely to become more urbanized. Perhaps they will shift to using second growth as they urbanize. There is an on-going effort to increase kite numbers in western Tennessee (Parker 1984, Martin and Parker 1991) where nesting is being encouraged by the release of kites in second-growth woodland.

Finally, how is urbanization affected by circumstances in the Mississippi kite's habitat in South America during the austral winter? Because of land use practices, perhaps the prey base there is expanding also. However, little is known of this species' ecology during the austral winter (see Davis 1989). At this time, it is possible to say only that food availability and the behavior of the human population toward kites in South American habitat have not prevented major population increases in urban nesting areas in North America in the immediate past.

REFERENCES

BOLEN, E.G. AND D.L. FLORES. 1989. The Mississippi kite in the environmental history of the Southern Great Plains. *Prairie Nat.* 21, 65–74.
DAVIS, S. 1989. Migration of the Mississippi kite *Ictinia mississippiensis* in Bolivia, with comments on *I. plumbea. Bull. B.O.C.* 109: 149–152.
GENNARO, A.L. 1988a. Extent and control of aggressive behavior toward humans by Mississippi kites. Pp. 249–252 in *Proc. SW Raptor Manage. Symp. and Workshop*. R.L. Glinski, B. Pendleton, M.B. Moss, M.N. LeFranc, Jr., B.A. Millsap and S.W. Hoffman, eds. Natl. Wildl. Fed. Wash; D.C.
—. 1988b. Breeding biology of an urban population of Mississippi kites in New Mexico. Pp. 88–98 in *Proc. SW Raptor Manage. Symp. and Workshop*. R.L. Glinski, B. Pendleton, M.B. Moss, M.N. LeFranc, Jr., B.A. Millsap and S.W. Hoffman, eds. Natl. Wildl. Fed. Wash; D.C.
GLINSKI, R.L. AND R. OHMART. 1983. Breeding biology of the Mississippi kite in Arizona. *Condor* 85: 200–207.
HENNY, C.J., W.S. OVERTON AND H.M. WIGHT. 1969. Determining parameters for populations by using structural models. *J. Wildl. Manage.* 34: 690–703.
KALLA, P.I. AND F.J. ALSOP. 1983. The distribution, habitat preference, and status of the Mississippi kite in Tennessee. *Amer. Birds* 37: 146–149.
LOVE, D., J.A. GRZYBOWSKI AND F.L. KNOPF. 1985. Influence of various land uses on windbreak selection by nesting Mississippi kites. *Wilson Bull.* 97: 561–565.
MARTIN, K. AND J. PARKER. 1991. Mississippi kites reborn in Tennessee. *Tenn. Cons.* 57(3): 5–10.
PARKER, J.W. 1974. The breeding biology of the Mississippi kite in the Great Plains. Ph.D. diss., Univ. Kansas, Lawrence, Kansas.
—. 1975. Populations of the Mississippi kite in the Great Plains. In *Population status of raptors*. Raptor Res. Report, No. 3.
—. 1979. About those kites ... *Kansas Fish and Game* 36(1): 5–6.
—. 1980. Kites of the prairies. *BirdWatcher's Digest* 2(6): 86–95.
—. 1984. Transfer of nesting Mississippi kites from Kansas to Tennessee. Project narrative, correspondence, reports. Cent. for Env. Res. and Ed. Univ. Maine at Farmington.
—. 1988a. The ace dive-bomber of the prairie is a terror on the green. *Smithsonian* 19(4): 54–63.
—. 1988b. Mississippi kite. Pp. 166–86 in Palmer, R.S., *Handbook of North American birds*, Vol. 4. Yale Univ. Press, New Haven.
— AND J.C. OGDEN. 1979. The recent history and status of the Mississippi kite. *Amer. Birds* 33: 119–129.
— AND M. PORTS. 1982. Helping at the nest by yearling Mississippi kites. *Raptor Res.* 16: 14–17.
SHAW, D.M. 1985. The breeding biology of urban-nesting Mississippi kites (*Ictinia mississippiensis*) in west central Texas. MS thesis, Angelo State Univ; San Angelo, Texas.
SUTTON, G.M. 1939. The Mississippi kite in spring. *Condor* 41: 41–53.
SWEET, M.J. 1989. Kites and the northern harrier. Pp. 32–41 in *Proc. Midwest Raptor Manage. Symp. and Workshop*. R.L. Glinski, B. Pendleton, M.B. Moss, M.N. LeFranc, Jr., B.A. Millsap and S.W. HOFFMAN, eds., Natl. Wildl. Fed. Washington: D.C.

7

Costs and Benefits of Urban Nesting in the Lesser Kestrel

J. L. Tella, F. Hiraldo, J. A. Donázar-Sancho and J. J. Negro

Abstract – Causes of mortality, diet and breeding success of an urban population of lesser kestrels in southern Spain were compared with those of a rural population 700 km away. One of the benefits of urban nesting was lower predation. In the rural population, predation was the primary cause of death for adults and nestlings, and involved at least ten different predator species, including one species of snake, four mammals and four birds of prey. Only two predator species were confirmed for urban kestrels. The diet of the two populations was similar. Prey delivery rates however, were lower in urban colonies. This resulted in a significant loss of nestlings from starvation in the urban colonies, where breeding success was significantly lower than in the rural colonies. Reduced food availability, therefore, seems to be one of the drawbacks of urban nesting for lesser kestrels.

Key words: lesser kestrel; predation; diet; urban nesting; Sevilla; Monegros.

Nest-site availability has been identified as an important limiting factor for population numbers and breeding success of birds of prey (Newton 1979, Negro and Hiraldo 1993). In their search for suitable nest places, different species of raptors breed on man-made structures, sometimes within urban areas (Brown and Amadon 1968, Newton 1979). The habit of urban-breeding might impose costs and benefits different from the ones encountered by rural populations of the same species. Nonetheless, there is scarce published information on the consequences of urban breeding for raptors.

One of the benefits associated with urban breeding is likely to be lower predation pressure, except if raptors are locally persecuted by humans. However, if urban-nesting raptors do not find a reliable food supply within the limits of the city (see Oliphant and McTaggart 1977, Sodhi *et al.* 1991) or in nearby degraded habitats, e.g. dumping sites and recreational parks; see Meinertzhagen 1954, Brown and Amadon 1968, Donázar 1992, they could be forced to make longer trips to the foraging grounds than their rural counterparts.

The lesser kestrel is a small Old World falcon that breeds in cavities, often in colonies of up to 100 pairs (Cramp and Simmons 1980). In Western Europe, where the species has suffered a sharp decline since the 1960s (Biber 1990), colonies are usually in old buildings. The strongholds of the lesser kestrel population are in Spain (about 5000 pairs or 70% of the European total), where approximately one-half of the breeding pairs nests in urban areas (Gonzalez and Merino 1991).

Our aims were to compare:

(1) predation pressure;
(2) diet and prey delivery rates; and
(3) breeding success, between an urban population of lesser kestrels in southern Spain (Sevilla) and a rural one located in northeastern Spain (Monegros).

STUDY AREAS

Monegros

This is a geographical region within the Ebro River Basin (Aragon, NE Spain). Cereal cultures have replaced natural shrub-steppes and *Juniperus* woods in the predominantly flat landscape of present-day Monegros. The climate is semi-arid continental Mediterranean, with annual rainfall averaging 350 mm. The elevation is 300–400 m above sea level.

Abandoned farm houses and other buildings are scattered in the Monegros landscape. They harboured the 49 lesser kestrel colonies that we studied in 1993, most of them within an area of approximately 250 km^2. No nests were located in natural cavities.

Sevilla Province

The study was carried out in eight colonies located in several towns in Sevilla province during 1988–93. The study area, of about 10 000 km^2, is a mixed farmland area intensively cultivated with cereals, sunflowers, olive trees and irrigated fields. The climate is typically Mediterranean, with a temperate rainy winter, which allows the wintering of some lesser kestrels (Negro *et al.* 1991). Summers are hot and rainless. Average annual rainfall is about 500 mm. The elevation in the area ranges between 10 and 150 m above sea level. The general structure of the landscape is similar to the one in Monegros, as both are heavily cultivated, but in the Sevilla study area there are far fewer buildings in the countryside. Most of the kestrel colonies were in urban buildings, although there were also a few colonies in isolated farm houses and rocky outcrops.

METHODS

The occurrence of potential predators was determined by observation of the colonies ($N = 8$ in both Sevilla and Monegros) from a distance with spotting scopes, from the time of pair formation in February until the young fledged in July. We observed kestrels for 500 h in Monegros and 2000 h in Sevilla. We considered potential predators as bird-eating animals, e.g. snakes, birds of prey, mammals observed within a radius of <100 m from the colony.

When dead adult or nestling kestrels were found, we tried to determine the cause. In the case of recently dead young, losses were attributed to starvation when they were underweight for their age and if they did not show external infections or parasites (Negro et al. 1993b).

The frequency of prey delivery to the young was recorded in 14 colonies during 199 h in Monegros, and in 2 colonies during 1042 h in Sevilla. Only pairs having ≥3 nestlings were considered in this analysis, to avoid the bias of possibly low feeding frequencies in pairs with small brood sizes. All observations were conducted when the nestlings were 1–5 wk old. Some prey items were identified with the help of a 20–60 × spotting scope, and most were at least assigned to one of the following categories:

(1) invertebrates <3 cm long, e.g. beetles;
(2) invertebrates >3 cm long, e.g. locusts and *Scolopendra* spp.; and
(3) vertebrates.

Colonies have been used as sample units in the analysis of prey delivery because pairs in the same colony share the same hunting grounds and thus observations are not statistically independent.

All nests in the colonies were located and visited to determine brood size and to band the nestlings. In order to minimize disturbance during the incubation period, we visited only a selected number of colonies to determine clutch size. In estimating clutch size, clutches of one egg, probably abandoned before clutch completion and which never hatched, were excluded (Negro et al. 1993b). Eggs that failed to hatch remained in the nest, so eggs disappearing near hatch were assumed to have produced nestlings which died and went undetected. Fledging success was estimated as the number of nestlings surviving until the age of banding, which usually took place a few days prior to fledging. As the colonies were searched for dead nestlings at the end of the breeding season, fledging success was corrected for those nests that had suffered documented losses.

RESULTS

Predation Pressure

In the rural colonies in Monegros, the number of potential predators was higher than in the urban colonies in Sevilla Province (Table 1). Considering those cases in which predation was actually confirmed, the number of species predating lesser kestrels was also higher in rural colonies (nine versus two predator species).

Predation was the main cause of death for adult and young kestrels in rural colonies, while in urban areas human-related activities caused most of the adult

Table 1. Potential[a] (P) and confirmed (C) predators of lesser kestrel eggs (E), nestlings (N), fledglings (F) and adults (A) in rural versus urban colonies in Spain.

Predator species	Rural colonies				Urban colonies			
	E	N	F	A	E	N	F	A
Reptiles								
Ladder snake	C	C	C	C	—	—	—	—
Montpellier snake	P	P	P	P	—	—	—	—
Ocellated lizard	P	P	—	—	—	—	—	—
Mammals								
Black rat	C	C	C	C	—	—	—	—
Norwegian rat	—	—	—	—	P	P	P	P
Garden dormouse	P	P	—	—	—	—	—	—
Stone marten	P	C	P	P	—	—	—	—
Red fox	C	C	C	C	—	—	—	—
Feral cat	—	P	P	C	—	P	P	P
Feral dog	—	P	P	—	—	—	—	—
Birds								
Red kite	—	—	P	—	—	—	—	—
Black kite	—	—	P	P	—	—	—	—
Short-toed eagle	—	—	P	—	—	—	—	—
Common buzzard	—	—	P	—	—	—	—	—
Booted eagle	—	—	P	C	—	—	P	P
Golden eagle	—	—	P	P	—	—	—	—
Egyptian vulture	—	—	P	C	—	—	—	—
Marsh harrier	—	—	P	—	—	—	—	—
Montagu's harrier	—	—	P	—	—	—	—	—
Peregrine falcon	—	—	P	C	—	—	—	—
Eurasian kestrel	—	P	C	C	—	—	P	P
Barn owl	—	P	P	C	—	C	C	C
Little owl	—	—	P	P	—	—	P	P
Eagle owl	—	—	P	P	—	—	—	—
Jackdaw	P	P	—	—	C	P	—	—
Common raven	—	P	P	P	—	—	—	—

[a]Bird-eating predators observed within a radius of 100 m centered on the colony.

deaths. Starvation was the main cause of mortality for urban nestlings (Table 2). In addition, the predation rate for adults while attending the nest was significantly higher in rural colonies (10 predated adults in 131 rural nests versus 3 in 334 urban nests; Fisher Exact Test, $P < 0.01$). The number of predated broods was also higher in rural colonies (15 of 131 rural versus 2 of 334 urban broods; Fisher Exact Test $P < 0.001$).

Diet and Prey Delivery Rates

Invertebrates were the main prey item delivered to nests in both rural and urban colonies (Table 3). Due to their larger size, vertebrates, although delivered

Table 2. Mortality causes for adult and nestling lesser kestrels in urban and rural colonies in Spain.

Cause of death	Adults		Nestlings	
	Rural	Urban	Rural	Urban
Starvation	—	—	1	158
Predation	18	6	15	9
Disease	—	—	1	8
Poaching	—	—	—	10
Accident in nest	1	2	4	2
Hit by car	1	3	—	—
Shot	1	6	—	—
Electrocution	—	2	—	—
Restoration work[a]	—	5	—	—
Unknown	2	15	17	7

[a]Entangled in safety nets installed in scaffolding used for restoration work in the buildings harbouring kestrel colonies.

Table 3. Prey items delivered to nestlings by lesser kestrels in rural and urban colonies in Spain.

Prey items	Rural		Urban	
	N	%	N	%
Invertebrates	203	87.1	113	88.8
<3 cm long	118	50.6	75	59.0
Coleoptera	25		9	
Orthoptera	14		56	
Hymenoptera	—		1	
Caterpillars	3		3	
Unknown	76		6	
>3 cm long	85	36.5	38	29.9
Scolopendra spp.	16		—	
Orthoptera (*Acrididae*)	4		14	
Orthoptera (*Tettigoniidae*)	33		16	
Orthoptera (other)	—		2	
Odonata	—		1	
Caterpillars	19		4	
Unknown	13		1	
Vertebrates	30	12.9	14	11.2
Reptiles	13		1	
Birds	—		2	
Small mammals	17		11	
TOTAL	233		127	

to young at a lower rate than invertebrates, accounted for an important fraction of the biomass ingested by the nestlings in both types of colonies. Prey-size distributions were not statistically different between rural and urban colonies ($\chi^2 = 2.3$, $P = 0.30$, df $= 2$).

Prey delivery rates in the rural colonies (3.98 feedings pair^{-1} h^{-1}, $N = 14$ colonies) were not significantly different from the urban ones (2.04 feedings pair^{-1} h^{-1}, $N = 2$ colonies; $t = 1.64$, $P = 0.12$).

Breeding Parameters

Clutch size was slightly higher, and bordering on statistical significance, in rural colonies (mean = 4.4 eggs, SD = 0.2, $N = 7$ colonies) than in the urban ones (mean = 4.1 eggs, SD = 0.4, $N = 8$ colonies; Mann-Whitney test, $P = 0.052$). Fledging success was significantly higher in rural colonies (3.7 fledgings, SD = 0.4, $N = 6$ rural colonies versus 2.2 fledgings, SD = 0.4, $N = 8$ urban colonies; Mann-Whitney test, $P < 0.001$). The proportion of total nest failure was not significantly different in rural (15.8%, SD = 10.8 $N = 8$ colonies) versus urban colonies (24.4%, SD = 17.3, $N = 8$ colonies; Test of Proportions, $P = 0.33$).

DISCUSSION

Lesser kestrels breeding in urban colonies incurred less predation than their rural counterparts. However, the two study areas are separated by 700 km. Our results might simply reflect differences in diversity and abundance of predators between regions. Both areas are within the Mediterranean ecological region in the Iberian Peninsula and share a similar 'guild' of predatory species at similar densities (unpubl. data). Although we cannot discount the hypothesis mentioned above, differences are more likely due to the fact that most predators avoid urban areas when foraging (Newton 1979, Andrew and Mosher 1982).

Lesser kestrels are more vulnerable to predation when attending the nest. The nest site, which is also commonly used by the pair for roosting during the pre-laying season, is practically the only place where they can be surprised by predators. Urban kestrels were possibly less susceptible because most of their potential predators, including other birds of prey, simply avoid urban areas. Additionally, town colonies are commonly located in tall buildings, such as churches and castles (Negro et al. 1991). Within these buildings kestrels select the highest locations, probably to avoid human disturbance and predation by domestic carnivores (Negro and Hiraldo 1993). Rural buildings in Monegros, however, are smaller and most of the nests were closer to the ground.

The diet and the size of prey were similar in rural and urban colonies. In a previous study however, Negro et al. (1992a) reported a much lower frequency of vertebrate prey items (0.9%, $N = 1113$) in the Sevilla area. The discrepancy probably occurred because in the current study we considered only those pairs having ≥3 nestlings for the comparison with Monegros, while Negro *et al.* (1992a) included those pairs feeding one and two nestlings. In addition, the Negeo *et al.* (1992a) study comprised observations up to the time of fledgling independence and fledglings are mainly fed insects. It seems therefore that the two populations we compared in this study had similar diets. For other birds of prey however, differences in diet have been reported between urban and rural populations, e.g. merlins (Oliphant and McTaggart 1977, James and Smith 1987).

Feeding rates were not significantly lower in the urban colonies. Nonetheless, this lack of a difference may have resulted from the very small number of urban colonies in which we studied this parameter. The fact that breeding success was lower in urban colonies and that nestling mortality was mainly caused by starvation does suggest however, that urban kestrels had more difficulty finding food. In addition, radio-tracked urban kestrels ranged over large areas while foraging during the nestling period (Negro *et al.* 1993a). This resulted in long foraging trips, indicating that food was insufficient close to the colony (Gaston and Nettleship 1981, Wittenberger and Hunt 1985). Although lesser kestrels have not yet been radio-tracked in Monegros, they were usually observed hunting very close to their colonies.

In conclusion, our results suggest that urban kestrels are under different pressures from those faced by rural kestrels during the breeding season. Urban kestrels suffered less predation but on the other hand, they seemed to have problems finding enough food. As a result, nestling mortality owing to starvation was severe. Those differential pressures could possibly bring distinct demographic or behavioral patterns to each population, such as differences in the allocation of parental investment, frequency of kleptoparasitism (Negro *et al.* 1992a), polygyny (Hiraldo *et al.* 1991) and extra-pair copulations (Negro *et al.* 1992b).

ACKNOWLEDGMENTS

We thank M. de la Riva, J.M. Bermúdez and Y. González for helping in Sevilla, and I. Sanchez, M. Villarroel, R. López, M. Pomarol, C. Sánchez, G. Blanco, C. Pedrocchi and Guarderiá Forestal for assistance in Monegros, P.C. James and P. Veiga provided constructive comments on the manuscript. The CSIC-CICYT funded the research (project PB90-1021).

REFERENCES

ANDREW, J.M. AND J.A. MOSHER. 1982. Bald eagle nest site selection and nesting habitat in Maryland. *J. Wildl. Manage.* **46**: 383–390.

BIBER, J.P. 1990. *Action Plan for the Conservation of Western lesser kestrel Populations*. ICBP Study Report No. 41. Cambridge, UK.
BROWN, L. AND D. AMADON. 1968. *Eagles, hawks and falcons of the World*. Country Life Books. Feltham, UK.
CRAMP, S. AND K.E.L. SIMMONS. 1980. *The birds of the Western Palearctic*. Vol. 2. Oxford University Press, Oxford, UK.
DONÁZAR, J.A. 1992. Muladares y basureros en la biologiá y conservación de las aves en España. *Ardeola* 39: 29–40.
GASTON, A.J. AND D.N. NETTLESHIP. 1981. The thick-billed murres of Prince Leopold Island – a study of the breeding biology of a colonial, high arctic seabird. *Can. Wildl. Serv. Monogr.* 6: 1–350.
GONZALEZ, J.L. AND M. MERINO. 1990. El cernícalo primilla (*Falco naumanni*) en la Península Ibérica. ICONA, Serie Tecnica. Madrid, Spain.
HIRALDO, F., J.J. NEGRO AND J.A. DONAZAR. 1991. Aborted polygyny in the lesser kestrel *Falco naumanni*. *Ethology* 89: 253–257.
JAMES, P.C. AND A.R. SMITH. 1987. Food habits of urban-nesting merlins, *Falco columbarius*, in Edmonton and Fort Saskatchewan, Alberta. *Can. Field-Nat.* 101: 592–594.
MEINERTZHAGEN, R. 1954. *Pirates and predators*. Oliver & Boyd. Edinburgh and London, UK.
NEGRO, J.J. AND F. HIRALDO. 1993. Nest-site selection and breeding success in the lesser kestrel *Falco naumanni*. *Bird Study* 40: 115–119.
—, M. de la RIVA AND J. BUSTAMANTE. 1991. Patterns of winter distribution and abundance of lesser kestrels (*Falco naumanni*) in Spain. *J. Raptor Res.* 25: 30–35.
—, J.A. DONAZAR AND F. HIRALDO. 1992a. Kleptoparasitism and cannibalism in a colony of lesser kestrels (*Falco naumanni*). *J. Raptor Res.* 26: 225–228.
—, —, —. 1992b. Copulatory behaviour in a colony of lesser kestrels: sperm competition and mixed reproductive strategies. *Animal Behav.* 43: 921–930.
—, —, —. 1993a. Home range of lesser kestrels (*Falco naumanni*) during the breeding season. Pp. 144–150 in M.K. Nicholls and R. Clarke, eds., *Biology and conservation of small falcons*. Proceedings of the 1991 Hawk and Owl Trust Conference. Hawk and Owl Trust, London, UK.
—, —, —, L.M. HERNANDEZ AND M.A. FERNANDEZ. 1993b. Organochlorine and heavy metal contamination in non-viable eggs and its relation to breeding success in a Spanish population of lesser kestrels. *Environ. Pollu.* 82: 201–205.
NEWTON, I. 1979. *Population ecology of raptors*. T. & A.D. Poyser, Calton, UK.
OLIPHANT, L.W. AND S. McTAGGART. 1977. Prey utilized by urban merlins. *Can. Field-Nat* 91: 190–192.
SODHI, N.S., I.G. WARKENTIN AND L.W. OLIPHANT. 1991. Hunting techniques and hunting rates of urban merlins. *J. Raptor Res.* 25: 127–131.
WITTENBERGER, J.F. AND G.L. HUNT. 1985. The adaptive significance of coloniality in birds. Pp. 2–78 in D.S. Farner, J.R. King and K.C. Parkes, eds. *Avian biology*, Vol. 8. Academic Press, Orlando, USA.

8

Nesting Success of Western Burrowing Owls in Natural and Human-altered Environments

Eugene S. Botelho and Patricia C. Arrowood

Abstract – The reproductive success of 27 pairs of western burrowing owls was studied in human-altered and natural areas on the campus of New Mexico State University. Pairs nesting in human-altered areas had significantly more nestlings and fledged significantly more young than pairs nesting in natural areas. The pairs nesting in natural areas had closer neighbors than did pairs in altered areas. The poorer reproductive success of natural-area pairs could have been due to increased inter-owl disturbance and/or to increased predation on young.

Despite the heightened possibility of disturbance and increased risk of mortality by automobiles, pairs nesting in burrows near frequent human activity may have had access to greater food resources (insects and bats attracted to street and other lights; carcasses of great-tailed grackles, Brewer's blackbirds, white-winged and mourning doves in local campus roosts) than pairs nesting in natural areas.

Key words: western burrowing owl; nesting success; human-altered sites; natural sites.

Western burrowing owls often live in close contact with humans and have been deemed the raptor least affected by human disturbance (Martin 1973). Abbott (1930) reported the occurrence of burrowing owls in culvert drains in 1921 in human-settled parts of San Diego, California. A breeding population of around a hundred western burrowing owls lived at the Oakland Municipal Airport and a surrounding golf course for several years (Thomsen 1971). The Florida burrowing owl, previously restricted to a small area around Cape Coral, is now widespread and secure over most of the Florida peninsula (Wesemann and Rowe 1987, B.A. Millsap pers. comm.). Wesemann and Rowe (1987) found that the number of Florida owls per hectare peaked when land development consumed about 60% of a hectare. Higher concentrations of beetles and lizards, the Florida owl's preferred foods, occurred in human-settled areas.

Burrowing owls utilize man-made structures like culvert drains and pipes, probably as a result of a natural tendency to exploit burrows and cavities, especially those of burrowing mammals (Bent 1938, Coulombe 1971). Burrowing owls will also nest in man-made artificial burrows (Collins and Landry 1977,

Andersen 1979, Landry 1979, Henny and Blus 1981, Olenick 1987, ESB and PCA pers. obs.). Close association with humans, however, has often resulted in owl deaths as a result of collisions with automobiles, human vandalism of burrows and harassment by pet dogs and cats (Chapman 1951, Bue 1955, Konrad and Gilmer 1984, Gleason and Johnson 1985, Wesemann and Rowe 1987).

The ecology of the burrowing owl has been studied in detail in a variety of habitats (Coulombe 1971, Thomsen 1971, Martin 1973), but comparative studies of the nesting success of owls in both human-altered and natural sites are lacking. Since owls will be increasingly forced to utilize human-altered areas in the future due to loss of both their normal habitat and burrowing mammals which provide homes (Lang and Weseloh 1973, Collins and Landry 1977), it is important to determine whether their reproductive success in human-altered areas is significantly impaired. Therefore, this study focuses on the reproductive success of 27 pairs of western burrowing owls nesting in either natural or human-altered areas on the campus of New Mexico State University in Las Cruces, New Mexico.

STUDY AREA AND METHODS

Two human-altered and two natural sites were studied from October 1992 to October 1993. Ten pairs nested in human-altered sites which included the university quadrangle (quad: owls nested among closely spaced buildings interspersed with medium to large expanses of lawn) and football stadium (stadium: owls nested at all four corners of the stadium and at the base of cement walls on the interior). Owls nesting in these areas often utilized burrows under cement walkways or abutments in the vicinity of street or other lights or along fences between recreational areas (Table 1). Soil in these areas consists of rich loamy topsoil which provides durable burrows resistant to collapse. Vegetation includes cultivated grass on well-manicured lawns with a few trees and shrubs. Lawns are artificially watered daily throughout the nesting season.

Seventeen pairs nested in natural sites located in remote parts of the campus where natural desert vegetation was present, along with natural arroyos. Here human access was rare or prohibited, but did sometimes occur. One natural site was an abandoned landfill whose initial contents had been covered with soil. The landfill had not been used (nor further filled) for at least 10 yr prior to this study. Since it had never been completely filled, the landfill is essentially a broad arroyo with low grasses and bushes on the flat bottom and steep north and south facing dirt sides overlooked by desert vegetation. Rock squirrels, resident in the landfill, appeared to excavate many of the burrows used by the owls. The other natural site was an earthen dam and an adjacent large natural arroyo. The dam and arroyo are a catchment basin for floodwater. Both were covered by desert grasses, forbs and bushes. Most burrows in the landfill were located 3–10 m high in the sides of the banks (Table 1). The two burrows on the dam were 1–2 m

Table 1. The type of area in which individual burrows of burrowing owls were situated, whether the burrow was videotaped (*), whether the burrow was an artificial one (+), and the number of nestlings and fledglings produced (nestlings, fledglings) or (abandoned).

Human-altered	Natural
1 – on lawn under fence*+ (3,2)	A – on earthen dam (3,3)
2 – under cement bleachers (3,3)	B – on earthen dam (abandoned)
3 – under cement bleachers (3,3)	C – flat on ground+ (abandoned)
4 – under walkway* (4,2)	D – flat on ground*+ (3,0)
5 – on lawn under fence+ (abandoned)	E – flat on ground (3,3)
6 – under cement foundation (3,2)	F – base of arroyo wall (2,2)
7 – under sidewalk (2,2)	G – base of arroyo wall (abandoned)
8 – under cement foundation* (5,5)	H – 3 m up in arroyo wall (abandoned)
9 – under walkway (5,5)	I – 6 m up in arroyo wall (2,2)
10 – under cement staircase (4,2)	J – base of arroyo wall (abandoned)
	K – base of arroyo wall (1,1)
	L – 6 m up in arroyo wall (1,1)
	M – base of arroyo wall (abandoned)
	N – 10 m up in arroyo wall (abandoned)
	O – 10 m up in arroyo wall (3,1)
	P – 6 m up in arroyo wall (abandoned)
	Q – base of arroyo wall (2,2)

from the top. Two burrows in the arroyo were on level or sloping sites (Table 1). The soil type in the natural sites was a gravel and sand mixture which could be extremely hard and durable when dry, but highly susceptible to erosion during intense rainstorms when rainwater percolated rapidly through the gravel and sand. A few burrows ($N = 4$) in natural sites did collapse after intense rain storms. Vegetation on natural sites consisted of typical arid Chihuahuan Desert flora, including creosote bush and mesquite. Several species of short bushes and short grasses were also common.

All burrows were individually marked with wooden stakes. More than 120 adult and juvenile owls observed during this study were banded with a USFWS aluminum band and a combination of two colored plastic leg bands for individual identification. Owls were captured using a 61 cm cube-shaped cage covered with plastic screen. A one-way door of clear plexiglass at the end of a short tube was placed over the burrow entrance, thus trapping the owls in the cage as they left the burrow (Banuelos 1993). This method of capture proved to be harmless to the owls and allowed all the nestlings at a particular burrow to be captured together once they began moving out of the burrow tunnel.

The number of nestlings at each burrow was determined at the time of capture and during subsequent observation periods. Observations lasting 5–10 min were conducted at 23 burrows using a 20–60× spotting scope at least three times per week. Hour-long observation sessions using a video camera set on a tripod at the burrow entrance were conducted seven times per week at four burrows (Table 1). The identity of the parents was confirmed during each observation period. A nestling was classified as fledging upon its first successful sustained flight. The exact date of fledging was recorded for the four burrows which were videotaped and estimated at the others.

The distance between each nesting burrow and its closest neighbor was determined using a string (10 m in length) tied to two stakes that made a straight line between burrows. We used a meter stick to record the actual distances. All distance measures were rounded to the nearest meter.

Since an assumption of normality of values could not be made in this study, the nonparametric Mann-Whitney U test was used to compare the sites in terms of the number of nestlings and fledglings produced and the distance between nesting pairs. Spearman rank correlation coefficients (r_s) examine the relationship between nearest neighbor distance and the number of nestlings and fledglings produced. All probabilities are two-tailed and significance is designated at the 0.05 probability level.

RESULTS

The number of nestlings produced in human-altered areas ($\bar{X} = 3.20 \pm 1.47$, $N = 10$ nests) was significantly higher than at natural sites ($\bar{X} = 1.05 \pm 1.23$, $N = 17$ nests, $U = 24.5$, $n_1 = 10$, $n_2 = 17$, $P < 0.05$). Similarly, the number of fledglings produced in human-altered areas ($\bar{X} = 2.50 \pm 1.43$, $N = 10$ nests) was significantly higher than at natural sites ($\bar{X} = 0.68 \pm 0.98$, $N = 17$ nests, $U = 31.0$, $n_1 = 10$, $n_2 = 17$, $P < 0.05$).

Eight of 17 nests on natural sites were abandoned by both adults. Since abandonment occurred when we knew other clutches were hatching, we presumed that those abandonments occurred prior to or around the time of hatching of eggs. All pairs in human-altered areas (with the exception of one in which the male died prior to completion of the clutch) produced some young. The number of nestlings produced in human-altered areas remained significantly higher than in natural areas even when the nine abandoned burrows were excluded from the analysis ($\overline{X} = 2.22 \pm 0.79$ nestlings/successful burrow on natural sites, $U = 11.5$, $n_1 = 9$, $n_2 = 9$, $P < 0.05$). Similarly, the number of fledglings produced in human-altered areas had a tendency to be significantly higher than in natural areas when the nine abandoned burrows were excluded from the analysis ($\overline{X} = 1.56 \pm 0.83$ fledglings/successful burrow on natural sites, $U = 18.5$, $n_1 = 9$, $n_2 = 9$, $0.05 < P < 0.10$).

The proximity of pairs in the natural areas to each other could have contributed to their overall lower reproductive rate and higher rate of abandonment. We measured the distance from one nesting burrow to its closest neighbor in both areas. For the two natural sites, the average distance between nesting burrows was 52.83 ± 19.37 m at the landfill and 86.6 ± 62.00 m at the dam. Among human-altered sites, the average distance between nesting burrows at the stadium was 93.75 ± 36.42 m, and 833.75 ± 332.77 m on the quad. Impenetrable concrete, roads and buildings separated burrows in the stadium and quad, respectively. Overall, the distance between burrows in human-altered areas ($\overline{X} = 417.89 \pm 460.28$ m) was significantly higher than at natural sites ($\overline{X} = 62.76 \pm 41.64$ m) ($U = 27.5$, $n_1 = 10$, $n_2 = 17$, $P < 0.05$). The distance between nests in human-altered sites remained significantly higher than in natural sites even with the omission of the nine abandoned burrows ($\overline{X} = 71 \pm 49.18$ m between successful burrows on natural sites, $U = 14.0$, $n_1 = 9$, $n_2 = 9$, $P < 0.05$). A significant positive correlation exists between the distance separating all nesting burrows, i.e. those on all 4 sites, and the number of nestlings ($r_s = 0.430$, $N = 27$, $P < 0.05$) and fledglings ($r_s = 0.385$, $N = 27$, $P < 0.05$) produced.

DISCUSSION

The number of nestlings and fledglings produced in the one year of this study was significantly lower among burrows on natural sites than among burrows in human-altered areas. Nest abandonment occurred more often in natural areas, but it was not in itself responsible for the differences in young produced in the two areas. Since humans were, for the most part, restricted from natural sites, human disturbance was probably a minor contributing factor in inducing nest abandonment among pairs in natural areas. Burrow collapse, as a result of heavy rain, in natural areas became a factor late in the nesting season when most young had fledged. Those burrows which contained nestlings at the time of collapse were re-opened, presumably from the outside by adults (ESB and PCA pers. obs.). Burrow collapse was not suspected to have resulted in the death of any owls during the study period.

Since burrows in natural areas were situated closer together than those in human-altered areas, inter-owl disturbance, rather than human–owl disturbance, may have contributed to the failure or abandonment of some nests. All burrows on each of the human-altered sites (with the exception of one pair in an artificial burrow in which the male died prior to completion of the clutch) produced at least some fledglings. Green and Anthony (1989) concluded that nest abandonment was the primary cause of nesting failure in a population of burrowing owls in Oregon, with 32.5% of nests abandoned; one of two nesting burrows was abandoned if nesting burrows were less than 110 m apart. We also found greater nest abandonment in more densely occupied areas.

The success of burrows in human-altered sites is interesting since many studies have shown that owls are often negatively affected by man (Chapman 1951, Bue 1955, Konrad and Gilmer 1984, Gleason and Johnson 1985, Wesemann and Rowe 1987). One reason for the success of pairs on human-altered sites could be artificial lighting. Lights attracted large numbers of both insect and mammalian prey (bats) throughout the night. Owls were regularly observed perched under lights hawking and diving after insects, even on roadways. The stadium was irregularly used by humans at night, but even when unused some lights remained on all night. Also, the part of the campus between the stadium and the quad contained large roosting concentrations of birds and many nesting pairs, including white-winged doves, mourning doves, Brewer's blackbirds and great-tailed grackles. The remains of both nestlings and adults of each of the above species were commonly found at burrow entrances. Also, the remains of several Brazilian free-tailed bats were found inside the nest cavity of a pair that utilized an artificial burrow on a human-altered site (see Table 1 for a list of pairs which used artificial burrows). Owls were never seen capturing bats, but examination of the bats within the nest cavity revealed them to be freshly killed.

The poor performance observed in nests in natural areas could have been partly due to predation. Predators appeared to be uncommon on the study sites with the exception of the landfill, in which a pair of barn owls nested in a burrow just 2 m from an active burrowing owl nest. Barn owls were not observed to prey on burrowing owls during the study, but they were mobbed by the latter when they flew near a burrow. Feral cats, dogs, coyotes and several species of snakes were also on the sites, but their contribution to burrowing owl predation is unknown. Nestlings in the landfill occasionally fell from the steep ledges outside the burrow entrances. One such nestling was recovered and replaced in the burrow, but no dead nestlings were recovered below burrows that were situated high in arroyo banks.

CONCLUSIONS

The results of this study are similar to those of Green and Anthony (1989) in that short inter-burrow distances were associated with greater burrow abandonment and smaller brood sizes. We still found a difference in reproductive success between natural and human-altered sites, however, when the abandoned nests were excluded from the analyses. The high incidence of nest failure and subsequent desertion by adults may have been due to direct (agonistic) and/or indirect (reduced food supply) competition from neighboring pairs. The higher number of nestlings produced on human-altered sites suggests that owls may be able to cope with human disturbance more effectively than with competition from close neighbors, even though burrowing owls are often discussed as successfully nesting in loose colonies. We do not know the relative contribution of predation to chick loss in the two areas. This study also suggests that the hunting behavior of owls nesting in high-density natural conditions should be studied in order to determine the extent of competitive interactions and the role of these interactions in determining reproductive success. To develop successful methods of managing this species, competition and territoriality in contrasting sites must be studied.

ACKNOWLEDGMENTS

We thank Owen Lockwood, Ben Woods and Pat Montoya for allowing access to burrow locations on the NMSU campus. Betsy Botelho helped with field work. Dennis Clason offered suggestions on statistical analyses. Dan Howard, Brian Millsap and Jeffrey Lincer provided constructive reviews of the manuscript. This research is supported in part by the NMSU Department of Biology, New Mexico Department of Higher Education and a Frank M. Chapman Award from the American Museum of Natural History.

REFERENCES

ABBOTT, C.G. 1930. Urban Burrowing owls. *Auk* **47**: 564–565.
ANDERSEN, J.W. 1979. The Burrowing owl in Sacramento. *Bulletin of the Sacramento Zoological Society* **16**: 9–12.
BANUELOS, G.H.T. 1993. An alternative trapping method for Burrowing owls. *J. Rapt. Res.* **27**: 85–86.
BENT, A.C. 1938. Life histories of North American birds of prey. Part 2. *U.S. Nat. Mus., Bull.* 170.
BUE, G.T. 1955. Recent observations of Burrowing owls in Lyon and Yellow Medicine Counties, Minnesota. *Flicker* **27**: 40–41.
CHAPMAN, H.F. 1951. Western Burrowing owls. *South Dakota Bird Notes* **3**: 60.
COLLINS, C.T. AND R.E. LANDRY. 1977. Artificial nest burrows for Burrowing owls. *N. Am. Bird Bander* **2**: 151–154.
COULOMBE, H.N. 1971. Behavior and population ecology of the Burrowing owl, *Speotyto cunicularia*, in the Imperial Valley of California. *Condor* **73**: 162–176.
GLEASON, R.S. AND D.R. JOHNSON. 1985. Factors influencing nesting success of Burrowing owls in southeastern Idaho. *Great Bas. Nat.* **45**: 81–84.
GREEN, G.A. AND R.G. ANTHONY. 1989. Nesting success and habitat relationships of Burrowing owls in the Columbia Basin, Oregon. *Condor* **91**: 347–354.
HENNY, C.J. AND L.J. BLUS. 1981. Artificial burrows provide new insight into Burrowing owl nesting biology. *Rapt. Res.* **5**: 82–83.

KONRAD, P.M. AND D.S. GILMER. 1984. Observations on the nesting ecology of Burrowing owls in central North Dakota. *Prairie Nat.* **16**: 129–130.

LANDRY, R.E. 1979. Growth and development of the Burrowing owl *Athene cunicularia*. M.A. thesis, California State Univ., Long Beach, CA.

LANG, V. AND D.V. WESELOH. 1973. Burrowing owl inquiry. *Calgary Field Nat.* **4**: 197–198.

MARTIN, D.J. 1973. Selected aspects of Burrowing owl ecology and behavior. *Condor* **75**: 446–456.

OLENICK, B. 1987. Reproductive success of Burrowing owls using artificial burrows in Idaho. *Eyas* **10**: 38.

THOMSEN, L. 1971. Behavior and ecology of Burrowing owls in the Oakland Municipal Airport. *Condor* **73**: 177–192.

WESEMANN, T. AND M. ROWE. 1987. Factors influencing the distribution of the Burrowing owl (*Athene cunicularia*) in Cape Coral, Florida. *Proc. Nat. Symp. Urban Wildl.* 129–137.

9

Eastern Screech Owls in Suburbia: A Model of Raptor Urbanization

Frederick R. Gehlbach

Abstract – Concurrent studies of eastern screech owls in a rural environment and nearby young and older suburbs suggest that such ecologically "plastic" raptors respond to the urbanization process with increases in traits such as egg and chick survival. The reproductive features interact to make city populations more productive and denser than rural populations, yet urban populations are also more stable, unlike natural counterparts. Enhanced urban resources and climatic stability cause the changes making urbanization a unique process that promotes a combination of early and late successional attributes in raptor populations.

Key words: eastern screech owl; cities; life history; raptors; urbanization.

A recent symposium on the population ecology of urban birds concluded that many species are more productive and have denser populations in cities than in the countryside (Tomialojc and Gehlbach 1988). Generally this is due to milder weather, more food, and fewer predators and competitors in the city. Although data on urban raptors have been scarce, information from disparate sources suggests that these populations are enhanced by one or more of the same environmental features, e.g. Wendland 1980, Wesemann and Rowe 1987, Warkentin and James 1988, Galeotti 1990, Peske 1990.

Here I briefly review urbanization in the eastern screech owl, based on Gehlbach (1994b), and propose a model for alterations in the reproductive life history and population ecology of any urban raptor. Although based primarily on screech owls, this model includes my experiences with other raptors in the same rural and suburban environments, especially the barred owl and broad-winged hawk. I am concerned with shifts in adaptive repertoire caused by climatic and resource changes during the age and growth of a city, not preadaptations that facilitate city life (Gehlbach 1988).

BACKGROUND

I investigated three breeding populations of eastern screech owls within 8 km of each other in McLennan County, Texas, 1976–91. They inhabited a rural

Table 1. Average annual environmental features in a 1976–87, rural-urban continuum housing eastern screech owls in Central Texas (percent coefficient of variation in parentheses except for the younger suburb, added to the study in 1979, and first three features measured only in 1979 and 1987).

	Rural	Suburbia 10-yr old	Suburbia 30-yr old
Human density – n km^{-2}	<1	402	508
Green space – %	86	63	26
Tree density – n ha^{-1}	1082	428	327
Avian prey – kg ha^{-1}	0.03 (102)	not measured	0.1 (87)
Insect prey – g trap-night^{-1}	7 (72)	7	8 (53)
Avian predators – kg ha^{-1}	0.04 (6)	not measured	0.03 (4)
Temperature – °C	16 (18)	17	18 (12)
Precipitation – total cm	30 (116)	31	35 (85)

Table 2. Average annual life history and population ecological features of eastern screech owls in a 1976–87 rural–urban continuum in Central Texas (percent coefficient of variation in parentheses except for the younger suburb, added to the study in 1979).

	Rural	Suburbia 10-yr old	Suburbia 30-yr old
Life History			
Female nesting mass – g	176 (3)	177	175 (2)
First egg date – March	27 (29)	21	22 (25)
Clutch size – n	4 (14)	4	4 (12)
Egg survival – %	45 (87)	51	70 (28)
Brood size – n	2 (73)	2	3 (17)
Chick survival – %	72 (82)	86	86 (12)
Fledging mass – g	109 (7)	115	123 (6)
Nestling days – n	26 (3)	27	28 (6)
Population			
Recruited fledglings – %	0	5	6 (168)
Nesting yearlings – %	64 (65)	58	47 (32)
Fledglings/eggs – %	28 (87)	42	49 (27)
Fledglings/pair – n	1 (89)	2	2 (33)
Successful nests – %	31 (78)	64	62 (20)
Nesting density – n ha^{-1}	0.02 (52)	0.07	0.11 (17)

area (270-ha study plot on the Middle Bosque River), a small 10-yr old suburb surrounded by countryside (45-ha plot at Harris Creek), and a larger 30-yr old suburb (270-ha plot in Woodway) contiguous with the city of Waco. All successful and unsuccessful nests were evaluated and equivalent information from nest boxes and natural-cavity nests was obtained (Gehlbach 1994a). Tables 1 and 2 contain only selected, illustrative data; more information, including field and analytical methods, is in Gehlbach (1994b).

In developing the model I assume that the three spatially different populations illustrate a temporal continuum representative of any one place in the range of screech owls or other species preadapted to live in cities. Also I assume that culturally altered habitats remain suitable for such raptors throughout the urbanization process, and the birds are protected from undue human interference. The model illustrates progressively altered life history features and their consequences for population dynamics. Thus, rural to urban trends rather than absolute distinctions are of primary importance (though most means in Tables 1 and 2 differ significantly among sites, Gehlbach 1994b).

RESULTS

Environment (Table 1)

In the larger older suburb, human density increased at the expense of green space and arboreal habitat required by eastern screech owls. By contrast to the rural site, however, the older suburb had more avian prey more continuously available. Also, there were fewer potential avian competitors (*Bubo, Buteo, Caprimulgus, Corvus, Strix* spp.). Urban insect prey were more numerous and less fluctuating than rural equivalents, and the consuming biomass of potential competitors was less in the city. Annual temperatures and precipitation were higher in the older suburb compared with the other two sites, together with increased climatic stability in the older suburb.

Life History and Population Ecology (Table 2)

Female screech-owls had nearly the same nesting mass at all three locations but tended to be older in suburbia because of improved adult survival with urbanization. The owls produced equal-size clutches on all plots, though nearly a week earlier in both suburbs. More eggs survived to hatch and brood sizes were larger in the older suburb compared with the younger one and the countryside. Size at fledging increased during urbanization, possibly because of an increased nestling period. Chick survival was higher in suburbia, so relatively more suburban chicks fledged per egg. Of equal importance is the fact that all but the nesting period among the eight life history traits were more stable annually in the older suburb compared with the rural site.

Fledgling output per breeding pair increased in the city as did the frequency of successful nests, and only in suburbia was productivity sufficient for population maintenance (average 1.8 fledglings per breeding pair, Gehlbach 1994b). Also, both productivity and nesting success were consistently higher in the older suburb. In fact, increased stability characterized all six population consequences of the life history changes. Recruitment as breeders was recorded only for those offspring produced and recruited in suburbia except for a single rural owlet also recruited into suburbia. Finally, nesting eastern screech owls in the older suburb were denser, and their population was more stable than in the countryside.

THE MODEL (Table 3)

Habitat supporting owl and hawk populations declines with city growth, but the deficit is more than offset by increased and more stable prey populations, fewer natural predators and competitors, and climatic moderation. For instance, older suburban green space was only 30% of the rural value in my screech owl study, but avian prey populations were nearly 300% larger (the 26% green space reported in Table 1 is representative of urban sites; Landsberg 1981). Artificial nest sites or unusually dense natural sites, e.g. corvid nests (Warkentin and James 1988), plus permanent surface water in bird baths, fish ponds, etc. also foster urban raptors.

Variable and hence modifiable reproductive and population attributes increase during urbanization, and some like first egg date and chick survival may stabilize with any degree of urbanization. Other features like egg survival and fledging mass may continue to increase. Mean and variance shifts in most life-history traits are relatively small (average 19% for means, 50% for variances), but together the individual features interact to produce greater population changes (average 226% and 68%, respectively). Thus, the older suburban population of eastern screech owls was about five times denser with three times more annual stability than the rural population.

A few attributes like adult body size do not respond to urbanization in the time-span represented by this study. Other constant traits like clutch size may not represent urbanization very well, because they are more influenced by adult age and experience than environmental change (Gehlbach 1994b). Yet only two of the eight reproductive features in Table 2 were unchanged by urbanization, so the majority are 'plastic'. This flexibility must typify urbanizable raptors, e.g.

Table 3. A general model of the urbanization of owls, hawks, and other native species.

FAVORABLE ENVIRONMENTAL FACTORS PRODUCE
(A) Climatic Moderation, (B) More Permanent Surface Water and Nest Sites, (C) Fewer Predators and Competitors, (D) Increased Prey Population Size and Stability.

INCREASES IN REPRODUCTIVE TRAITS AND HENCE THERE IS
(A) Greater Adult Survival and thus Breeding Experience, (B) Earlier Nesting, (C) Greater Egg Survival, (D) Greater Mass and Survival of Chicks.

POSITIVE POPULATION RESPONSES
(A) More Successful Nests, (B) Greater Fledgling Production, and (C) More Fledglings Recruited as Breeders, resulting in (D) Increased Population Density and Stability.

broad-winged hawk, by contrast to others unable to live in cities because of more rigid lifestyles, e.g. red-shouldered hawk.

In the paradigm of natural succession, species of young (pioneer) communities are highly productive with large, unstable populations in the fluctuating environment (r-selection). Eventually, they are replaced by or evolve into less productive species with comparatively small, stable populations in an older, more constant climax environment (K-selection, e.g. Odum 1969). Does the urbanization of an owl or hawk mirror this sequence, which is appropriate for any species that can occupy a wide range of natural and cultural landscapes?

Compared with natural succession, urbanization is regressive in some respects. The birds become more productive, not less, and denser, not sparser, in cities. Of course cities feature many pioneer environmental attributes conducive to the change in lifestyle, including large populations of relatively few prey species, so they partly resemble pioneer natural communities instead of the climax natural landscapes they may replace. When confronted with urbanization, ecological plasticity permits certain raptors to assume those population attributes exhibited under pioneer natural conditions.

The feature that characterizes urbanization as a unique process for native owls, hawks, and associated biota is the increase in climatic and resource stability with the consequent greater stability of urban populations. Such stability, manifested in reproductive traits as well as population features, is not exclusive of high productivity and density as it is in natural communities. Thus, the city environment is unique in fostering population attributes associated with opposite ends of the natural successional continuum.

While the model described herein (Table 3) may imply that urbanization benefits raptors and other wildlife, there are provisos. For instance, vehicle collisions and biocide poisonings were the first and second most frequent causes of screech owl mortality in the older suburb and seemed to be greater threats there than in the rural setting (Gehlbach 1994b). Moreover, some related raptors do not or rarely nest in cities, hence decline with urbanization, while others urbanize readily. Of forest nesting congeners in my study area, for example, the red-shouldered hawk nested only in rural habitat by contrast to the largely suburban nesting broad-winged hawk.

Previous studies of raptors in cities substantiate only parts of the urbanization model, because they were not designed to test it (references in introduction). Simultaneous investigations of city and adjacent rural populations are needed for such studied urban species as the burrowing owl, tawny owl, Eurasian sparrowhawk and merlin. However, these may be more r-selected than other city nesters, and the different lifestyles of open and cavity nesters must also be included among raptors requiring comparative study to test the model.

A common list of population attributes to be assessed might reveal interesting geographic-cultural patterns. Also, standard definitions of urban, suburban, and rural are necessary because they often mean different things in Europe and North America (Tomialojc and Gehlbach 1988). Human population density and amount of natural or semi-natural (green) space are basic to the definitions. Means make useful comparisons, but knowledge of environmental and life history flux is also necessary to assess the species' plasticity and stability hypotheses. These are just a few suggestions for a field of inquiry that needs many more comparative, long-term studies.

ACKNOWLEDGMENTS

Paolo Galeotti, Lubomir Peske and Ludwig Tomialojc discussed aspects of the urbanization model with me. Cheryl McCollough, Juan Jose Negro, Ian Warkentin and Stanley Wiemeyer provided editorial suggestions on the manuscript.

REFERENCES

GALEOTTI, P. 1990. Territorial behavior and habitat selection in an urban population of the tawny owl *Strix aluco* L. *Boll. Zool.* 57: 59–66.
GEHLBACH, F.R. 1988. Population and environmental features that promote adaptation to urban ecosystems: the case of eastern screech-owls (*Otus asio*) in Texas. *Acta XIX Congr. Internatl. Ornithol.* 2: 1809–1813.
—. 1994a. Nest-box versus natural-cavity nests of the eastern screech-owl: an exploratory study. *J. Raptor Res.* 28: 154–157.
—. 1994b. *The eastern screech owl: life history, ecology, and behavior in the suburbs and countryside.* Texas A. and M. Univ. Press. College Station, Texas, USA.
LANDSBERG, H.E. 1981. *The urban climate.* Academic Press, New York, USA.
ODUM, E.P. 1969. The strategy of ecosystem development. *Science* 164: 262–270.
PESKE, L. 1990. The population of sparrowhawks living in Prague: the changes of nesting bionomy in an environment with high human influence. *Papers 2nd South Bohemia Ornithol. Conf.* 1989: 293–300 (Czech with English Summary).
TOMIALOJC, L. AND F.R. GEHLBACH, eds. 1988. Avian population responses to man-made environments. *Acta XIX Congr. Internatl. Ornithol.* 2: 1777–1830.
WARKENTIN, I.G. AND P.C. JAMES. 1988. Nest-site selection by urban merlins. *Condor* 190: 734–738.
WENDLAND, V. 1980. Der waldkauz (*Strix aluco*) im bebauten Stedgebiet von Berlin (West). *Beitr. Vogelkd. Jena* 26: 157–171.
WESEMANN, T. AND M. ROWE. 1987. Factors influencing the distribution and abundance of burrowing owls in Cape Coral, Florida. Pp. 129–137 in L.W. Adams and D.L. Leedy, eds. *Integrating man and nature in the metropolitan environment.* Nat. Inst. of Urban Wildlife, Columbia, Maryland, USA.

Raptors and Artificial Nest Sites

10
Red-tailed Hawks Nesting on Human-made and Natural Structures in Southeast Wisconsin

William E. Stout, Raymond K. Anderson and Joseph M. Papp

Abstract — Raptors commonly nest on powerline towers in the western United States (US). This phenomenon usually occurs on open plains, prairie or savannah, and is attributed to the absence of suitable natural nest sites. In the eastern US, red-tailed hawks nest predominantly in deciduous trees and less frequently in evergreens; in southeast Wisconsin, they nest almost exclusively in deciduous trees. Red-tailed hawks will nest on human-made structures in the eastern US. However, there is only one published report of two successful nests on power poles in Polk County, Florida. We documented 15 successful red-tailed hawk nests in 4 yr on five human-made structures in southeast Wisconsin, and compared them with nests on natural substrates. Four structures were high voltage transmission towers (three different structure types), and one was a billboard. The use of human-made structures for nest substrates may be a local occurrence, and may be related to growing urban populations of red-tailed hawks in locations such as Milwaukee, Wisconsin.

Key words: red-tailed hawk; powerline; billboard; urban; suburban; rural.

Raptors commonly nest on human-made nest substrates such as powerline poles and high voltage transmission towers, in the western United States (US; Olendorff and Stoddart 1974, Gilmer and Wiehe 1977, Olendorff *et al.* 1981, Bechard *et al.* 1990). This usually occurs on open plains, prairie or savannah, and is attributed to the absence of suitable natural nest sites (Olendorff *et al.* 1981, 1989).

In the eastern US and Canada, the osprey is the only raptor that commonly nests on powerline towers and poles, at least in local areas, including the upper Midwest (Stocek 1972, S. Postupalsky pers. comm.). Osprey and, less frequently, bald eagles will nest on platforms (Postupalsky and Stackpole 1974, Postupalsky 1978, Gieck 1991, S. Postupalsky pers. comm.). American kestrels readily nest in boxes placed on powerline towers or poles (Hamerstrom *et al.* 1973, Stahlecker and Griese 1979, Varland and Loughin 1993), and have nested in the partially damaged crossarm of a pole (Illinois Power Company 1972). Peregrine falcons have been reintroduced extensively in metropolitan areas, nesting on buildings, bridges, towers and on other artificial structures (Temple 1988). In the eastern US, red-tailed hawks nest predominantly in deciduous trees and less frequently in evergreens (Bent 1937, Titus and Mosher 1981, Bednarz and Dinsmore 1982, Palmer 1988, Speiser and Bosakowski 1988). In southeast Wisconsin, they nest almost exclusively in deciduous trees (Orians and Kuhlman 1956, Gates 1972, Petersen 1979). Red-tailed hawks will nest on human-made structures in the eastern US (Brett 1987, Speiser and Bosakowski 1988, Preston and Beane 1993). However, Toland's (1990) account of two nesting attempts on power poles in Polk County, Florida appears to be the only published report to document reproductive success and describe the sites of red-tailed hawk nests on powerline towers in the eastern US. They were built on temporary towers carrying 230-kV lines, and were about 12 m high; each successfully fledged one young.

Here, we describe 15 successful red-tailed hawk nests on five human-made structures in southeast Wisconsin, and compare them with 84 sites with natural nest substrates.

STUDY AREA

The study area consists of the metropolitan Milwaukee area in southeast Wisconsin. It includes Milwaukee County (43° north, 88° west) and parts of Waukesha, Washington and Ozaukee Counties. Milwaukee and Ozaukee Counties are bordered by Lake Michigan to the east. Land within the study area includes urban, suburban and rural habitat.

METHODS

Nests were located by a vehicle survey between 1 February and 30 April from 1987 through 1993 (Craighead and Craighead 1969). Nests were visited periodically to determine reproductive success (Postupalsky 1974). Nest site (microhabitat), habitat, and area (macrohabitat) data were recorded according to Titus and Mosher (1981) and Mosher et al. (1986, 1987). Nest site data were collected when nestlings were 2–5 wk old. Closure at the nest was ranked as: open, mostly open, mostly closed, or closed (0, 1, 2, and 3, respectively). Data on habitat were collected within a 0.04-ha circular plot (11.3 m radius centered on the nest tree) after fledging through September for new nesting pairs found each year. Canopy, understory, shrub and ground cover and slope of the plot were recorded. Shrub structure was determined by shrub density, shrub index and density board (Mosher et al. 1986).

We cover-typed land-use within a 1.5-km radius (706.9 ha) of the nests for macrohabitat analysis and classified nest sites as urban, suburban and rural based on these data. The amount of natural, agricultural, residential and industrial land within the macrohabitat area was determined from 1990 aerial photos (1 cm = 48 m) with a compensating polar planimeter. Natural areas included fallow fields, grassland, shrubland, woodlots and open water. The mean area of open water was <1% (maximum = 6%) and primarily consisted of natural pothole ponds. Open water was therefore included in the natural category. Agricultural areas included row crops, e.g. corn, cover crops, e.g. alfalfa, pastures, and tree nurseries. Residential areas included human dwellings, and other buildings and land associated with them, e.g. farm homesteads. Subdivisions comprised most of the residential area. Industrial areas included non-residential buildings, pavement, graded land, e.g. gravel pits, mowed land, e.g. cemeteries, airports and land surrounding industrial buildings, and non-mowed land closely associated with human activity, e.g. freeway intersections. A nest site was classified as urban if ≥70% of the macrohabitat was used for industrial or residential purposes (developed), rural if ≤30%, and suburban if between 30% and 70% was developed. The Baxter-Wolfe interspersion index (Baxter and Wolfe 1972) was determined by counting the number of changes in habitat type along the north–south and east–west median lines within the macrohabitat (Mosher et al. 1987). The distance to the nearest active red-tailed hawk nest also was recorded (Clark and Evans 1954).

Non-parametric statistical analysis (Kruskal-Wallis test, Chi-square approximation: Sokal and Rohlf 1981) was used to compare artificial substrate nest sites (billboard and transmission towers) to natural substrate nest sites (trees). Tests were considered significant when $P \leq 0.05$. We utilized the Statistical Package for the Social Sciences (Nie et al. 1975) for statistical analyses.

RESULTS

We documented 15 successful red-tailed hawk nests on human-made structures in five separate territories in southeast Wisconsin (Fig. 1). Two nesting attempts occurred in 1990, three in 1991, five in 1992, and five in 1993. Each territory was active in successive years after the nest was discovered. Four nest substrates were high voltage transmission towers (three different structure types), and one was a billboard. The powerlines on two towers carried 138 kV, one carried 345 kV, and one carried both 230 and 345 kV. The red-tailed hawk pair using the billboard nested in three different locations on the same billboard in three successive yr: 1991, 1992 and 1993 (mean nest height: 18.8 m; range 17.4–19.5 m, N = 3). All nesting attempts on artificial structures were successful (fledged at least one young).

In 1989 and 1990, we found 84 red-tailed hawk nests in different territories on natural nest substrates (trees). Sixty-five of these territories were successful (23, 38, and four sites fledged 1, 2, and 3 young, respectively). For 83 of these sites, 23 were in the main crotch of the tree, 47 were in a secondary crotch (angle from vertical of the supporting limb <45°), four were braced against the main trunk, and nine were in the crotch of a horizontal branch (angle from vertical of

Figure 1. Red-tailed hawk nests on high voltage powerline towers (a–d) and a billboard (e–h). Red-tailed hawks nested in three different locations on the same billboard in three successive yr: 1991 (f-1, g), 1992 (e, f-2), and 1993 (f-3). Three nestlings (5–6 wk old) with an adult perched on the lower, outside catwalk (g). The urban billboard-based nest site (h).

the supporting limb >45°). Ten nests were found in white oaks, 13 in burr oaks, eight in red oaks, 19 in green ashes, nine in white ashes, four in willow spp., three in each of sugar maples, basswoods and slippery elm, two in each of silver maples and American elms, and one each in American beech, shagbark hickory, catalpa, cottonwood, red birch, black cherry, and white pine.

Table 1. Nest site characteristics ($\bar{x} \pm$ SE, N, [range]) of red-tailed hawks nesting on artificial (billboard and transmission towers) and natural substrates (trees). Chi-square values and probabilities are given (Kruskal-Wallis test, Chi-square approximation).

Variable	Artificial nest Substrates	N	Natural nest Substrates	N	χ^2	P
Nest substrate height (m)	34.3 ± 5.2 (20.7–45.7)	5	22.1 ± 0.4 (14.1–32.6)	82	6.018	0.014*
Nest height on substrate[a] (m)	23.4 ± 2.0 (17.4–30.8)	7	17.4 ± 0.4 (9.1–24.0)	83	9.215	0.002*
Percent nest height[a]	81.6 ± 6.3 (51.0–94.2)	7	79.3 ± 0.9 (57.2–95.1)	82	1.758	0.184
Nest closure[a]	0.43 ± 0.20 (0–1)	7	1.39 ± 0.08 (0–3)	82	10.519	0.001*
Nest length (m)	0.79 ± 0.03 (0.65–0.92)	8	0.63 ± 0.01 (0.42–0.86)	83	13.454	<0.001*
Nest width (m)	0.41 ± 0.03 (0.03–0.50)	8	0.48 ± 0.01 (0.30–0.68)	83	3.303	0.069
Nest height (m)	0.32 ± 0.02 (0.26–0.37)	8	0.38 ± 0.01 (0.16–0.78)	82	2.852	0.091
Cup length (m)	0.28 ± 0.02 (0.22–0.36)	5	0.27 ± 0.00 (0.21–0.40)	76	0.582	0.445
Cup width (m)	0.21 ± 0.02 (0.18–0.28)	5	0.23 ± 0.00 (0.16–0.29)	76	2.051	0.152
Productivity	1.73 ± 0.15 (1–3)	15	1.32 ± 0.10 (0–3)	84	2.599	0.106

[a] Includes values for the three billboard nest locations.
* Indicates statistically different means.

Nests were higher above the ground, closure at the nest was more open by rank, and nest length was significantly longer for nests on artificial substrates compared with nests on natural substrates (Table 1). Nesting success (nests fledging at least one young) for artificial substrate nests ($N = 15$) and natural substrate nests ($N = 84$) was 100% and 77%, respectively. However, productivity for nests on human-made structures was not significantly different than natural sites (Table 1).

No overstory trees were present within the habitat area of nests on artificial substrates. As a result, five overstory variables for artificial substrate sites (canopy height, number of overstory tree species, number of overstory trees, basal area of tress, and percent canopy cover; all values = 0; $N = 5$) were different from those for natural substrate sites (mean ± SE, range; 19.9 ± 0.4 m, 12.6–26.5 m; 2.4 ± 0.1, 1–7; 8.3 ± 0.6, 1–23; 1.02 ± 0.05 m^2, 0.23–2.04 m^2; 58.7 ± 2.8%, 5–100%; respectively, $N = 75$). Since the towers and billboard were not in or adjacent to a woodlot, the distance to nearest forest opening for these sites (0 m) also was different from that for natural substrate sites (17.0 ± 3.0 m, 0–106 m, $N = 58$) for this study. Habitat for nests with natural substrates include the nest tree and usually a woodlot. In 1989 and 1990, 88 nest sites for different red-tailed hawk pairs were found on natural substrates. Forty-

Table 2. Habitat characteristics for ($\bar{x} \pm$ SE, N, [range]) red-tailed hawk nests on artificial (billboard and transmission towers) and natural substrates (trees). Chi-square values and probabilities are given (Kruskal-Wallis test, Chi-square approximation).

Variable[a]	Artificial nest Substrates	N	Natural nest Substrates	N	χ^2	P
Slope of plot (%)	6.9 ± 2.7 (0–15.6)	5	3.1 ± 0.5 (0–26.7)	74	3.100	0.078
No. of understory/sapling stems ≥1 cm diameter	18.4 ± 13.8 (0–71)	5	57.6 ± 4.4 (0–183)	75	5.507	0.018*
Understory/sapling cover (%)	24.5 ± 15.9 (0–78)	5	67.8 ± 2.7 (5–100)	75	6.741	0.009*
Ground cover (%)	40.0 ± 17.9 (0–100)	5	55.8 ± 3.5 (3–100)	75	1.112	0.291
No. of shrub species	1.4 ± 0.9 (0–4)	5	5.8 ± 0.4 (0–16)	75	8.229	0.004*
Shrub density	246.6 ± 149.7 (0–690)	5	150.3 ± 15.3 (0–606)	75	0.135	0.712
Shrub index	433.2 ± 261.6 (0–1180)	5	416.4 ± 36.2 (0–1486)	75	0.321	0.570
Density board – 0–0.3 m	42.6 ± 9.0 (15–60)	5	46.2 ± 1.9 (0–60)	75	0.017	0.896
Density board – 0.3–1.0 m	90.2 ± 28.5 (16–140)	5	85.9 ± 5.0 (0–140)	75	0.443	0.505
Density board – 1.0–2.0 m	77.8 ± 45.6 (0–194)	5	100.7 ± 7.3 (0–200)	75	0.267	0.604
Density board – 2.0–3.0 m	64.0 ± 38.8 (0–165)	5	96.1 ± 7.0 (0–199)	75	1.029	0.310

[a]Variables are described in Titus and Mosher (1981), and Mosher et al. (1986).
*Indicates statistically different means.

three nests were in woodlot interiors (the tree crown did not reach the edge of the woodlot), 12 on interior woodlot openings, 20 on woodlot edges, eight in forest savannahs, four in hedgerows, and one in a lone tree. Two understory/sapling variables for artificial substrate sites were significantly different from those for natural substrate sites (Table 2). Shrub diversity for artificial substrate sites was significantly less than for natural substrate sites. However, six other variables for shrub structure and ground cover were not significantly different (Table 2).

Land use adjacent to nests on human-made structures varied from 98% developed (industrial and/or residential land use) for urban sites to 20% developed for rural sites (Table 3). For sites with natural nest substrates, nest sites ranged from 92% to 6% developed. Three artificial substrate sites were urban, one suburban, and one rural. Eleven natural nest sites were urban, 21 suburban, and 17 rural. The amount of natural, industrial and residential land within the macrohabitat of artificial substrate sites was not significantly different from those with natural substrates. However, agricultural land use surrounding nests on human-made structures was significantly less than for nests on natural substrates (Table 3). Also, the distance to the nearest active red-tailed

Table 3. Macrohabitat characteristics for ($\bar{x} \pm$ SE, N, [range]) red-tailed hawk nests on artificial (billboard and transmission towers) and natural substrates (trees). Chi-square values and probabilities are given (Kruskal-Wallis test, Chi-square approximation).

Variable	Artificial nest Substrates	N	Natural nest Substrates	N	χ^2	P
Distance to nearest water (m)	300.8 ± 177.2 (30–984)	5	255.6 ± 46.8 (0–1008)	38	0.116	0.733
Distance to nearest building (m)	154.6 ± 48.7 (48–331)	5	194.4 ± 16.4 (30–571)	49	0.625	0.428
Distance to nearest road (m)	150.8 ± 68.2 (24–403)	5	232.2 ± 24.1 (34–878)	49	1.571	0.210
Distance to nearest active RTH nest (m)	2790.8 ± 556.5 (1865–4968)	5	1930.9 ± 163.1 (403–5700)	45	4.284	0.038*
Baxter-Wolfe interspersion index	22.2 ± 2.3 (17–28)	5	25.6 ± 0.9 (8–40)	48	1.488	0.223
Natural area (ha)	195.95 ± 82.82 (16.3–457.4)	5	203.28 ± 12.89 (55.1–431.2)	49	0.026	0.869
Agricultural area (ha)	47.50 ± 26.63 (0.0–116.6)	5	170.00 ± 21.04 (0.0–534.4)	49	5.150	0.023*
Industrial area (ha)	262.83 ± 66.09 (48.1–401.5)	5	153.84 ± 17.52 (0.0–499.1)	49	2.316	0.128
Residential area (ha)	200.48 ± 42.12 (92.6–289.1)	5	179.81 ± 16.89 (21.9–537.2)	49	0.556	0.455

*Indicates statistically different means.

hawk nest for artificial substrate nest sites was significantly farther than for natural substrate sites (Table 3).

In metropolitan areas, cemeteries and freeway intersections probably provided suitable hunting habitat for red-tailed hawks, and comprise 13%, 39% and 40% of land use within the macrohabitat for the three urban nests on artificial structures, respectively.

DISCUSSION

Over the past 20 yr concern has grown over the effects of electro-magnetic fields (EMFs) from high voltage transmission lines, electrochemical oxidants from EMFs, and noise on humans and wildlife (Ellis et al. 1978). Our results show no apparent harmful effects on red-tailed hawk productivity or physiology due to nesting on high voltage transmission towers. Nesting on towers and other human-made structures may increase nest success by minimizing mammalian predation, and possibly by reducing predation by great horned owls. Mammals probably have limited access to these nests because of the steel construction, and great horned owls may not approach the nest because of the exposure and height of the nest on the structures (nest height mean ± SE, and range for nests on powerline towers only; 26.9 ± 2.2 m, 22.8–30.8 m, N = 4), and the urban industrial location (billboard nest, Fig. 1).

Nest sites on artificial substrates may be significantly farther from other active red-tailed hawk nests than natural sites because of their urban location. Three artificial substrate nest sites were urban, the suburban site was 43% developed, and the rural site (20% developed) was an agrarian peninsula almost completely surrounded by urban land. In a comparison of urban, suburban and rural nest sites, excluding the nests on artificial substrates, the nearest neighbor distance for urban sites (mean ± SE, 2763 ± 352 m, range 1490–4920 m, $N = 11$) was significantly farther than suburban (1753 ± 124 m, 799–2904 m, $N = 19$; one-way ANOVA: $F = 9.461$, $P = 0.004$) and rural sites (1249 ± 163 m, 403–2138 m, $N = 14$; $F = 17.123$, $P < 0.001$). Suburban sites also were significantly farther apart than rural sites ($F = 6.247$; $P = 0.018$).

Nesting on human-made structures may be a response to decreased availability of natural nest sites and other changes in land-use patterns such as increased urbanization and monotypic agricultural practices. Nesting red-tailed hawk populations in the Midwest have increased over the last 40 yr (Petersen 1979, Castrale 1991, Bosakowski et al. 1992, S. Postupalsky pers. comm.). In some parts of the US, red-tailed hawks apparently are becoming more tolerant of human activity (Bechard et al. 1990). These factors and/or an increase in availability of human-made structures also could promote nesting on artificial substrates. As with osprey, red-tailed hawks nesting on artificial substrates may be a local phenomenon (Olendorff et al. 1981).

Successful red-tailed hawk nests in urban areas were reported by Valentine (1978) and Hull (1980). Juvenile red-tails may be forced to hunt in urban and other marginal areas because adults occupy prime habitats (Brinker and Erdman 1986). At least 15 red-tailed hawk pairs nested in urban habitat in the metropolitan Milwaukee area of southeast Wisconsin annually during this study. For the past 15 yr, urban developers have been incorporating greenbelt corridors into their plans to allow for wildlife in cities (Leedy et al. 1978). River and highway corridors, airports, golf courses, city parks, cemeteries, and freeway intersections provide suitable habitat for many wildlife species, including red-tailed hawks.

ACKNOWLEDGMENTS

We thank J. C. Bednarz, S. Postupalsky, D. E. Varland and an anonymous reviewer for comments that improved this manuscript, and J. A. Reinartz for statistical advice.

REFERENCES

BAXTER, W. L. AND C. W. WOLFE. 1972. The interspersion index as a technique for evaluation of bobwhite quail habitat. Pp. 158–165 in J.A. Morrison and J.C. Lewis,

eds. *Proc. First Natl. Bobwhite Quail Symp.* Okla. State Univ. Res. Found., Stillwater, Oklahoma USA.
BECHARD, M.J., R.L. KNIGHT, D.G. SMITH AND R.E. FITZNER. 1990. Nest sites and habitats of sympatric hawks (*Buteo* spp.) in Washington. *J. Field Ornithol.* 61: 159–170.
BEDNARZ, J.C. AND J.J. DINSMORE. 1982. Nest sites and habitat of Red-shouldered and Red-tailed Hawks in Iowa. *Wilson Bull.* 94: 31–45.
BENT, A.C. 1937. Life histories of North American birds of prey. US National Museum Bull. No. 167, Washington, D.C. USA.
BOSAKOWSKI, T., D.G. SMITH AND R. SPEISER. 1992. Status, nesting density, and macrohabitat selection of Red-shouldered hawks in northern New Jersey. *Wilson Bull.* 104: 434–446.
BRETT, J. 1987. Northeast Region Report. *Eyas* 10: 18–22.
BRINKER, D.F. AND T.C. ERDMAN. 1985. Characteristics of autumn Red-tailed Hawk migration through Wisconsin. Pp. 107–136 in Michael Harwood, ed. *Proc. Hawk Migration Conference IV*. Hawk Migration Association of North America.
CASTRALE, J.S. 1991. Eastern woodland buteos. Pp. 50–59 in *Proc. Midwest Raptor Management Symposium and Workshop*. Natl. Wildl. Fed., Washington, D.C. USA.
CLARK, P.J. AND F.C. EVANS. 1954. Distance to nearest neighbor as a measure of spatial relationships in populations. *Ecology* 35: 445–453.
CRAIGHEAD, J.J. AND F.C. CRAIGHEAD. 1969. *Hawks, owls and wildlife*. Dover Publications, Inc., New York, New York USA.
ELLIS, D.H., J.G. GOODWIN AND J.R. HUNT. 1978. Wildlife and electric power transmission. Pp. 81–104 in J.L. Fletcher and R.G. Busnel, eds. *Effects of noise on wildlife*. Academic Press, Inc., New York, New York USA.
GATES, J.M. 1972. Red-tailed hawk populations and ecology in east-central Wisconsin. *Wilson Bull.* 84: 421–433.
GIECK, C.M. 1991. Artificial nesting structures for Bald eagles, ospreys and American kestrels. Pp. 215–221 in *Proc. Midwest Raptor Management Symposium and Workshop*. Natl. Wildl. Fed., Washington, D.C. USA.
GILMER, D.S. AND J.M. WIEHE. 1977. Nesting by Ferruginous hawks and other raptors on high voltage powerline towers. *Prairie Nat.* 9: 1–10.
HAMERSTROM, F., F.N. HAMERSTROM AND J. HART. 1973. Nest boxes: an effective management tool for Kestrels. *J. Wildl. Manage.* 37: 400–403.
HULL, C.N. 1980. Additional successful nesting of a Red-tailed hawk in an urban subdivision. *Jack Pine Warbler* 58: 30.
ILLINOIS POWER COMPANY. 1972. IP helps maintain sparrow hawks. *Hi-Lines* November, 1972. p. 14.
LEEDY, D.L., R.M. MAESTRO AND T.M. FRANKLIN. 1978. Planning for wildlife in cities and suburbs. US Dept. Interior, US Fish Wildl. Serv., FWS/OBS-77/66, 64 pp.
MOSHER, J.A., K. TITUS AND M.R. FULLER. 1986. Developing a practical model to predict nesting habitat of woodland hawks. Pp. 31–35 in J. Verner, M.L. Morrison and C.J. Ralph, eds. *Wildlife 2000: modeling habitat relationships of terrestrial vertebrates*. University of Wisconsin Press, Madison, Wisconsin USA.
—, —, AND —. 1987. Habitat sampling, measurement and evaluation. Pp. 81–97 in B.A. Giron Pendleton, B.A. Millsap, K.W. Cline and D.M.Bird, eds. *Raptor management techniques manual*. Nat. Wildl. Fed., Washington, D.C. USA.
NIE, N.H., C.H. HULL, J.G. JENKINS, K. STEINBRENNER AND D.H. BENT, eds. 1975. *Statistical package for the social sciences*. McGraw-Hill, Inc., New York, New York USA.
OLENDORFF, R.R. AND J.W. STODDART. 1974. Potential for management of raptor populations in western grasslands. Pp. 47–88 in F.N. Hamerstrom, B.E. Harrell and R.R. Olendorff, eds. *Management of raptors*. Raptor Res. Rep. No. 2.
—, A.D. MILLER AND R.N. LEHMAN. 1981. *Suggested practices for raptor protection on powerlines – the state of the art in 1981*. Raptor Res. Rep. No. 4.

—, D.D. BIBLES, M.R. DEAN, J.R. HAUGH AND M.N. KOCHERT. 1989. *Raptor habitat management under the US Bureau of Land Management multiple-use mandate.* Raptor Res. Rep. No. 8.

ORIANS, G. and F. KUHLMAN. 1956. Red-tailed hawk and Horned owl populations in Wisconsin. *Condor* **58**: 371–385.

PALMER, R.S., ed. 1988. *Handbook of North American birds.* Vol. 5. Yale Univ. Press, New Haven, Connecticut USA.

PETERSEN, L. 1979. *Ecology of Great horned owls and Red-tailed hawks in southeastern Wisconsin.* Tech. Bull. No. 111, Wisconsin Dept. Nat. Res., Madison, Wisconsin USA.

POSTUPALSKY, S. 1974. Raptor reproductive success: some problems with methods, criteria, and terminology. Pp. 21–31 in F.N. Hamerstrom, B.E. Harrell and R.R. Olendorff, eds. *Management of raptors.* Raptor Res. Rep. 2.

—. 1978. Artificial nest platforms for ospreys and bald eagles. Pp. 35–45 in S.A. Temple, ed. *Endangered birds: management techniques for preserving threatened species.* Univ. Wisconsin Press, Madison, Wisconsin USA.

—, AND S. M. STACKPOLE. 1974. Artificial nest platforms for ospreys in Michigan. Pp. 105–117 in F.N. Hamerstrom, B.E. Harrell and R.R. Olendorff, eds. *Management of raptors.* Raptor Res. Rep 2.

PRESTON, C.R. AND R.D. BEANE. 1993. Red-tailed hawk (*Buteo jamaicensis*). In A. Poole and F. Gill, eds. *The birds of North America.*, No. 52. Philadelphia: The Academy of Natural Sciences; Washington, D.C. USA.

SOKAL, R.R. AND F.J. ROHLF. 1981. *Biometry.* W.H. Freeman and Co., New York, New York USA.

SPEISER, R. AND T. BOSAKOWSKI. 1988. Nest site preferences of Red-tailed hawks in the highlands of southeastern New York and northern New Jersey. *J. Field Ornithol.* **59**: 361–368.

STAHLECKER, D.W. AND H.J. GRIESE. 1979. Raptor use of nest boxes and platforms on transmission towers. *Wildl. Soc. Bull.* **7**: 59–62.

STOCEK, R.F. 1972. Occurrence of Osprey on electric power lines in New Brunswick. *New Brunswick Naturalist* **3**: 19–27.

TEMPLE, S.A. 1988. Future goals and needs for the management and conservation of the Peregrine falcon. Pp. 843–848 in T.J. Cade, J.H. Enderson, C.G. Thelander and C.M. White, eds. *Peregrine falcon populations: their management and recovery.* The Peregrine Fund, Inc., Boise, Idaho USA.

TITUS, K. AND J.A. MOSHER. 1981. Nest-site habitat selected by woodland hawks in the central Appalachians. *Auk* **98**: 270–281.

TOLAND, B.R. 1990. Use of power poles for nesting by Red-tailed hawks in south-Central Florida. *Florida Field Nat.* **18**: 52–55.

VALENTINE, A.E. 1978. The successful nesting of a Red-tailed hawk in an urban subdivision. *Jack Pine Warbler* **56**: 209–210.

VARLAND, D.E. AND T.H. LOUGHIN. 1993. Reproductive success of American kestrels nesting along an interstate highway in central Iowa. *Wilson Bull.* **105**: 465–474.

11

Documentation of Raptor Nests on Electric Utility Facilities Through a Mail Survey

Roberta Blue

Abstract – A two-part questionnaire was distributed to the electric utility industry to document the utilization of electric utility facilities for nesting by raptors. The survey was designed to solicit information on (1) the number and species of raptors nesting on power line structures in the United States and (2) participation by electric utility companies in various raptor enhancement programs. The overall response rate was 62% (88 of 141 companies returned questionnaires). Twelve species of raptors were reported nesting on power line support structures with ospreys and red-tailed hawks most commonly reported. Twenty-four respondents reported other use of their facilities for nest sites, including both natural sites and power plant facilities. Fifty-eight respondents described working cooperatively on raptor enhancement projects. Installation of artificial nest platforms was the most commonly reported enhancement project. Participation in hacking projects was also described, primarily for peregrine falcon releases. Respondents indicated that a great deal of cooperative effort has occurred between the industry and various government agencies and private groups.

Key words: survey; electric utility; raptors; enhancement; nesting platforms.

Use of power line structures by nesting raptors has been well documented in the United States as well as in other parts of the world. Comprehensive reviews of the association between raptors and power lines and/or linear rights-of-way, including both adverse and beneficial impacts, have been provided by Olendorff *et al.* (1981) and Williams and Colson (1989). Studies that have specifically documented the use of power lines support structures for nesting in particular regions include Stahlecker and Griese (1979), Gilmer and Stewart (1983), Roppe *et al.* (1989), Knight and Kawashima (1993) and Steenhof *et al.* (1993). However, I located no documentation of a survey of the electric utility industry to provide an overview of the use of power line structures across the United States, specifically for nesting. Additionally, as a result of the association between raptors and power lines, many authors have suggested or prescribed various habitat enhancement procedures (Nelson and Nelson 1976, Olendorff *et al.* 1981, Postovit and Postovit 1987). I conducted a survey of the electric utility industry to attempt to provide an overview of raptor use of power line structures as nest sites as well as to describe the involvement of various electric utility companies in raptor enhancement projects.

METHODS

I designed a two-part questionnaire to solicit raptor nest data from electric utility companies in the United States. I mailed questionnaires to individuals at 141 electric utility companies throughout the continental United States. Whenever possible, I sent the questionnaire to an individual in the company's environmental services department.

In the first part of the questionnaire (Part A), I asked respondents to record the number of raptor nests observed on power line structures by species during 1992–93, to indicate whether the number of nests recorded was obtained through a direct count or an estimate, and to provide any additional notes or comments that might be useful. In the second part of the questionnaire (Part B), I asked respondents to provide information on raptor enhancement programs conducted by their companies, to provide observations on incidental raptor use of utility facilities (other than power line support structures) for nesting, and to list any agencies (county, state or federal) or conservation groups they had cooperated with on their raptor enhancement projects.

A second copy of the questionnaire, along with a self-addressed envelope was mailed to those companies which had not returned responses four weeks after the first survey was mailed. Additional information about raptor enhancement programs was obtained from articles in *Biocurrents*, a newsletter published by the Biologists' Task Force (of the Edison Electric Institute), and in some cases by follow-up phone calls.

RESULTS AND DISCUSSION

Respondents from 88 electric utilities (both investor-owned and public) returned questionnaires, resulting in an overall response rate of 62%. These 88 electric utilities accounted for approximately 45% of the total circuit km of overhead electric line 220 kV or greater in the United States (excluding Alaska). Of the questionnaires returned, 68 respondents (77%) either reported raptor nests on power line structures and/or described raptor enhancement programs or incidental nesting of raptors on company lands or other facilities.

Raptor Nests Observed on Power Line Structures

Fifty-one companies reported at least one raptor nest on a power line structure. Overall, a total of 12 raptor species were reported nesting on transmission or distribution structures (Table 1). In most cases, no information was given by the respondent on the type of structure used.

Transmission and distribution systems are routinely inspected for mechanical or electrical reliability. However, routine biological surveillance to identify raptor nests would result in a substantial increase in time and cost required to conduct these surveys and is generally not done. For this reason, over 40% of the respondents labeled their numerical data as estimates. The numbers reported (Table 1) only index raptor use of these structures and probably tend to greatly underestimate actual use by some raptor species.

Osprey and red-tailed hawk nests were reported most frequently (Table 1) and in the highest numbers. The greatest concentrations of osprey nests on distribution or transmission structures occurred in Florida with two companies reporting as many as 500 nests each. These observations included nests on platforms mounted directly to the power line support structure.

The raptor species reported in this survey, except for the zone-tailed hawk, have been previously documented using power line support structures for nesting (Stahlecker and Griese 1979, Olendorff et al. 1981, Gilmer and Stewart 1983, Bohm 1988, Roppe et al. 1989, Steenhof et al. 1993). Although no direct reference to eastern screech owl nests in power line poles was found in the literature, pines with woodpecker cavities are a preferred nesting site for this species in the Eastern United States (Bent 1938). The eastern screech owl observations that I report here (Table 1) were nests in woodpecker cavities in power line poles. Several observations were also recorded for American kestrel use of cavities in poles. The zone-tailed hawk was reported nesting on a 115 kV wood H-frame structure in southwestern New Mexico near the Arizona border. Two Harris' hawk nests and a prairie falcon nest were also observed in New Mexico. The Harris' hawk nests were reported from southeastern New Mexico from a 345 kV line and a 230 kV line support structure. The prairie falcon nest was located in the crossarm of a 230 kV wood H-frame structure in the San Juan basin of northwestern New Mexico. Prairie falcons were observed nesting on a high-voltage transmission line in northcentral Nevada by Roppe et al. (1989). In their study they identified two prairie falcon nests located in abandoned raven nests on a lattice steel transmission structure. The respondent from New Mexico indicated frequent use of power line structures for nesting by raptors in his company's service area due to the lack of other nesting sites.

Bald eagle use of power line structures or artificial nest platforms is rare (Olendorff et al. 1980, Bohm 1988, Gieck 1991, Marion et al. 1992). Details of

---

Notes to Table 1

[a]Median number of nests was calculated only when 3 or more companies reported numerical data for this species.
[b]The sample size for calculating the median number was different from the number of companies reporting nests because in several cases no attempt was made to estimate a number.

Table 1. Raptor nest observations on power line structures based on annual counts or estimates from the 1992–93 period reported in a survey of electric utility companies in the United States.

Species	No. of companies reporting a nest observation	Median number of nests reported[a]	Sample size for median number[b]	Additional notes
Osprey	30	4	29	Range 1–500+, greatest concentrations reported from Florida
Red-tailed hawk	25	4	23	Range 1–200+, greatest concentrations reported from California, New Mexico, and Minnesota
Great-horned owl	9	10	8	Range 1–25+
Golden eagle	6	7.5	5	
Bald eagle	2			Details on these nests reported by Marion et al. (1992) and Bohm (1988)
American kestrel	5	—	—	Some kestrel observations from nest boxes on poles, in most cases number was not estimated
Swainson's hawk	3	—	—	
Ferruginous hawk	3	—	—	
Unidentified hawk	3	—	—	
Eastern screech owl	2	—	—	Reported as using abandoned woodpecker cavities in wooden poles
Harris' hawk	1	—	—	Reported from New Mexico
Prairie falcon	1	—	—	Reported from New Mexico
Zone-tailed hawk	1	—	—	Reported from New Mexico

the two observations of eagle nests on power line structures reported in this survey have been previously described by Bohm (1988) and Marion et al. (1992).

Raptor Enhancement Programs

Fifty-eight respondents reported information under Part B of the survey on raptor enhancement programs or use of structures other than power line supports by nesting raptors. Electric utility companies become involved in raptor enhancement efforts for several reasons. Cooperative efforts may be initiated in response to requests for assistance from state or federal wildlife agencies or private conservation groups, as mitigation for detrimental impacts, or as the result of special interest by utility biologists or other company personnel. Many electric utility companies have policies which promote environmental stewardship of the natural resources associated with their generation facilities and associated lands and waters.

Hacking Projects

Nineteen utility companies reported participation in hacking programs to restore a raptor species to their service area. Four of these companies reported participation in hacking projects for more than one species and two companies reported participation in projects involving the same species at two release sites. Peregrine falcons were most often released; however, participation in hacking projects for the bald eagle, osprey, golden eagle and California condor were also reported. Most peregrine falcon projects described have been recent (1988–present), whereas the described bald eagle and osprey projects occurred primarily in the 1970s or early 1980s. Because hacking projects involve a "considerable investment of money, time, and personnel" (Barclay 1987), these projects generally involved several cooperators and the level of participation by each utility varied. Participation ranged from indirect assistance through provision of funding to more direct assistance (ranging from construction and placement of hack boxes to provision of labor and materials, consulting fees, and salaries for paid staff members). Additionally, the electric utility often provided a secure site for the release. Project participants were able to come and go as necessary, but control of access to the general public protected the birds from unnecessary disturbances. Six of the peregrine falcon projects described in the survey occurred in urban areas with five hacking sites atop multi-story office buildings (three of these were electric utility corporate headquarter sites) and one from the second story window of a substation facility. Three releases were at power plant sites, four were in remote areas (not enough information was provided to determine whether these remote sites were utility-owned lands), and three respondents did not provide details on the release site. At one of the power plant locations, hacking boxes were located on one of the power plant decks (floors) at an elevation of 70 m above ground. A camera was mounted near one hack box so that power plant employees could observe the birds on a monitor located in the cafeteria. This undoubtedly helped to foster much interest and enthusiasm for the peregrine falcon among employees at this facility. Hacking projects described involved release of up to 15 young falcons from a single site with variable long-term survival reported. In some cases projects were recent with not enough time elapsed to determine whether nesting pairs will become established in the area.

Nine electric utility companies reported installation of peregrine falcon nest boxes on power plant chimneys to attract nesting falcons. In several cases the nest boxes were installed as a follow-up to a hacking project conducted in that same area. At least two of these companies reported successful nests in one or more of these nest boxes. Use of power plant chimneys has been described as "a developing relationship" since the structures provide the necessary height and may provide protection from predation problems that have occurred at natural sites (Orr and Anderson 1993).

Platform Installation

Installation of nest platforms was the most frequently reported raptor enhancement project with 40 respondents reporting involvement in this activity. Respondents noted that nest platforms were installed for two primary reasons. The first was to ensure a safer situation for both the nesting bird and the power line. Nests in situations where there was high probability of interruption to service on the line or of injury to the bird were often relocated to nest platforms. In other instances, nest platforms were installed to provide nesting substrates where natural nest sites were limited or to attract nesting raptors to areas such as power plant impoundments or wildlife management areas. In most cases (38 of 40 respondents) the platforms were erected for osprey. Four utilities reported erecting nest platforms for hawks, three for bald eagles, and three for golden eagles. One respondent reported that nest platforms are "stock items" installed regularly by line maintenance crews. Permits to allow electric utility companies to relocate nests under the Migratory Bird Treaty Act are obtained from state and federal wildlife agencies.

Other Raptor Enhancement Programs

Ten respondents reported projects to install kestrel nest boxes around their facilities and four companies provided funding and/or allowed private groups or individuals to install nest boxes along their rights-of-way. Although not all respondents provided details on box locations, four reported programs where boxes were mounted directly on power line poles. In two of these cases, respondents reported that agreements had been reached with state and/or federal regulatory agencies allowing them to remove the nest boxes in the event of the need to conduct emergency maintenance on the line.

Other raptor enhancement efforts described by one or more respondents included:

(1) creation of a "raptor runners" group to transport injured raptors to treatment facilities and construction of structures for wildlife rehabilitation center;
(2) assistance with installation of barn owl houses;
(3) management of bald eagle habitat on company lands and reservoirs;
(4) assistance with relocation efforts to move young osprey to establish a nesting population;
(5) provision of research funds for various raptor studies;
(6) participation in regional bald eagle midwinter surveys;
(7) participation in program to monitor osprey nesting success and use of platforms and to collect samples of blood and feathers from successful nests to analyze for organochlorines, PCB isomers, and heavy metals;
(8) provision of educational materials and/or exhibits including construction of observation areas to observe eagle or other raptors;
(9) installation of perch poles;
(10) donation of poles and other materials for nest platforms;
(11) assistance to local governments and airports with installation of nest platforms and moving of nests from undesirable locations;
(12) creation of raptor protection program to educate employees on avoiding impacts to raptors;
(13) creation of burrowing owl nest burrows;
(14) retention of bridge support structures to provide nest area for raptors;
(15) reconfiguration of power lines to reduce hazards to nests; and
(16) creation of rigorous raptor electrocution prevention programs.

Nesting Associated with Utility Sites Other Than Power Line Structures

Twenty-four respondents reported incidental observations of raptor nests on lands or man-made structures (other than power line support structures) associated with electric generating facilities. Five species of raptors were observed using various man-made structures associated with power plants (Table 2). With the exception of the barn owl these species had also been observed nesting on power line support structures.

Six respondents reported bald eagle nests on utility-owned lands associated with impoundments. Several respondents mentioned observations of osprey, red-tailed hawk, barred owl, screech owl, and great-horned owl nests on company-owned lands. These observations were anecdotal and actual use of lands and waters associated with electric generating facilities is certainly much greater than indicated by this survey.

Table 2. Raptor nest observations on other structures reported in a survey of electric utility companies in the United States.

Species	Structure	No. of observations	Additional notes
Osprey	Meteorological monitoring tower	2	
	Piers and cranes	2	
	Spillgate structure	1	
	Microwave tower	1	
	Power plant roof area	1	
	Oil boom platforms	1	At power plant site
Great-horned owl	Substation	2	
	Power plant site	1	Exact location of nest not given
	Coal conveyor	1	At power plant site above active coal silo
Barn owl	Power plant	1	Exact location of nest not given
	Dam	1	Exact location of nest not given
American kestrel	Metal electrical pull box	1	On power plant deck 45 m above ground
	Boiler tower	1	
Red-tailed hawk	Substation	2	

Cooperating Agencies

A high level of cooperation among the electric utility industry and various agencies and private conservation groups was indicated by the number of participants listed by respondents. Various raptor enhancement projects involved cooperation with 35 state natural resource agencies representing 33 states, 6 federal agencies (US Fish & Wildlife Service – 18 respondents, US Forest Service – 7 respondents, Bureau of Land Management – 5 respondents, National Park Service – 3 respondents, NASA – 1 respondent, and Canadian Wildlife Service – 1 respondent), 9 local government or community agencies, 33 private/nonprofit conservation groups, and 6 universities. The primary focus of many of the conservation groups listed was raptor conservation.

ACKNOWLEDGMENTS

I would especially like to thank Dick Thorsell for his guidance and support in designing the questionnaire and conducting this survey. Bobby Ward also provided guidance and support throughout the project. The Biologists Task Force of the Edison Electric Institute and Mr. Joel Mazelis provided endorsement, a mailing list for the questionnaires, and a review of the questionnaire. Rick Yates, Brenda Brickhouse and David Schiller provided helpful comments on the manuscript and Gary Breece provided valuable background information. I am also very grateful to all the respondents who took the time to complete and return the questionnaire and to share information on their companies' activities relating to raptors.

REFERENCES

BARCLAY, J.H. 1987. Augmenting wild populations. Pp. 239–247 in B.A. Giron Pendleton, B.A. Millsap, K.W. Cline and D.M. Bird, eds. *Raptor management techniques manual*. Natl. Wildl. Fed., Washington, D.C. USA.

BENT, A.C. 1938. *Life histories of North American birds of prey*. Vol. II. U.S. Natl. Mus. Bull. 170, Washington, D.C. USA.

BOHM, R.T. 1988. Three bald eagle nests on a Minnesota transmission line. *J. Raptor Res.* 22: 34.

GIECK, C.M. 1991. Artificial nesting structures for bald eagles, ospreys and American kestrels. Pp. 215–221 in *Proc. Midwest raptor management symposium and workshop*. Natl. Wildl. Fed., Washington, D.C.

GILMER, D.S. AND R.E. STEWART. 1983. Ferruginous hawk populations and habitat use in North Dakota. *J. Wildl. Manage.* 47: 146–157.

KNIGHT, R.L. AND J.Y. KAWASHIMA. 1993. Responses of raven and red-tailed hawk populations to linear right-of-ways. *J. Wildl. Manage.* 57: 266–271.

MARION, W.R., P.A. QUINCY, C.G. CUTLIP, Jr. AND J.R. WILCOX. 1992. Bald eagles use artificial nest platform in Florida. *J. Raptor Res.* 26: 266.

NELSON, M.W. AND P. NELSON. 1976. Power lines and birds of prey. *Idaho Wildl. Rev.* 28: 3–7.

OLENDORFF, R.R., R.S. MOTRONI AND M.W. CALL. 1980. *Raptor management – the state of the art in 1980*. US Dept. Interior, Bureau Land Manage., Tech. Note 345.

—, A.D. MILLER AND R.N. LEHMAN. 1981. *Suggested practices for raptor protection on power lines – the state of the art in 1981*. Raptor Res. Rep. No. 4.

ORR, D.J. AND R.J. ANDERSON. 1993. Use of artificial nest boxes by peregrine falcons at electric power generating facilities. Pp. 24-1–24-10 in *Proceedings: Avian Interactions with Utility Structures, International Workshop*. Electric Power Research Institute, Palo Alto, CA.

POSTOVIT, H.R. AND B.C. POSTOVIT. 1987. Impacts and mitigation techniques. Pp. 183–213 in B.A. Giron Pendleton, B. A. Millsap, K.W. Kline and D.M. Bird, eds. *Raptor management techniques manual*. Natl. Wildl. Fed., Washington, D.C., USA.

ROPPE, J.A., S.M. SIEGEL AND S.E. WILDER. 1989. Prairie falcon nesting on transmission towers. *Condor* 91: 711–712.

STAHLECKER, D.W. AND H.J. GRIESE. 1979. Raptor use of nest boxes and platforms on transmission towers. *Wildl. Soc. Bull.* 7: 59–62.

STEENHOF, K., M.N. KOCHERT AND J.A. ROPPE. 1993. Nesting by raptors and common ravens on electrical transmission line towers. *J. Wildl. Manage.* 57: 271–281.

WILLIAMS, R.D. AND E.W. COLSON. 1989. Raptor associations with linear rights-of-way. Pp. 173–192 in B.G. Pendleton, ed. *Proc. western raptor management symposium and workshop*. Natl. Wildl. Fed. Sci. Tech. Ser. 12, Washington, D.C., USA.

12

Osprey Population Increase along the Willamette River, Oregon, and the Role of Utility Structures, 1976–93

Charles J. Henny and James L. Kaiser

Abstract – The population of ospreys nesting along the Willamette and lower Santiam Rivers in western Oregon increased from an estimated 13 pairs in 1976 to 78 pairs in 1993. The number nesting on trees (live and dead) was similar in 1976 (13 pairs) and in 1993 (12 pairs). Ospreys were first observed nesting on utility structures (poles and towers) in 1977, and that nesting segment increased at a rapid rate (from 1 pair in 1977 to 66 pairs in 1993). A logistic growth curve was fitted to the data and, assuming that the logistic growth curve was correct, the osprey population nesting on utility structures was estimated to stabilize at 86 pairs in 2004; however, the population data fitted the exponential growth curve nearly as well. The latter model does not permit the estimation of an upper population limit. Ospreys in 1993 were producing young at about twice the rate necessary to maintain a stable population. Improved water conditions and fish numbers in the Willamette River, a new enlightened attitude toward birds of prey which resulted in less shooting, and the osprey's release from DDT-related reproductive problems after the 1972 DDT ban probably contributed to the population increase. The first osprey that nested on a utility pole in the study area may have been produced on a man-made nesting platform established in 1973 at Crane Prairie Reservoir (first US Osprey Management Area) about 160 km to the southeast. A shortage of tree nest sites along the Willamette River may have limited the osprey nesting population in earlier years, but the seemingly learned response to nest on utility structures has resulted in nest sites being almost unlimited now.

Key words: osprey; Oregon; utility structures; population dynamics; nesting success.

Early records of osprey abundance in Oregon were few, and the species was only designated as either "rare" or "common", with no attempt at enumeration, e.g. Woodcock 1902, Jewett and Gabrielson 1929). After reviewing the historical records, Gabrielson and Jewett (1940:199) reported the osprey as:

"formerly common along the Columbia and Willamette Rivers, in the Klamath Basin, and about the larger Cascade lakes, must now be considered one of the rarer Oregon hawks. It is still present in the Klamath Basin but in sadly diminished numbers. A few are found along the coast, and scattered pairs occur along the larger streams, such as the Rogue, the Umpqua, the Deschutes, the John Day, and the Columbia Rivers."

The reference to "formerly common" in the Klamath Basin is supported by a Vernon Bailey unpublished report (in Henny 1988) of 500 osprey nests (estimated 250–300 nesting pairs) in extreme southern Oregon at the northeast corner of Tule Lake in 1899. Historical numbers nesting along the Willamette River remain unknown. Henny et al. (1978) estimated 308 ± 23 pairs nesting in Oregon in 1976, 94.7% of which were in live or dead trees. Most of the "other" nest sites (13) were platforms constructed in 1973 for ospreys in the central Cascade Mountains at Crane Prairie Reservoir, the first Osprey Management Area created in the United States in 1969 (Roberts 1969). In contrast to Oregon, 69% of ospreys in Chesapeake Bay and coastal New Jersey, Delaware, Maryland, and Virginia (Henny and Noltemeier 1974, Henny et al. 1977) nested on man-made structures including utility poles and towers by the mid-1970s and records of such use in the east date back to 1881.

We discuss the population increase of ospreys nesting along the Willamette River of western Oregon from 1976–93 (only a segment of the nesting osprey population west of the Cascade Mountains in Oregon). We also discuss the role of utility poles and towers in the population increase and we suggest a possible reason why the transition to utility structures occurred during this time period.

STUDY AREA AND METHODS

The main-stem Willamette River is a ninth-order river and the tenth largest river in the conterminous United States in terms of total discharge (Sedell and Froggatt 1984). It is the largest river in the United States with restored water quality (Huff and Klingeman 1976). Historically, high loadings of organic wastes produced critically low dissolved oxygen concentrations, floating and benthic sludge, and *Sphaerotilus natans* beds that reduced salmon migration, recreational use, and aesthetic value. Water quality improved dramatically, salmon runs returned, and recreational uses increased after low-flow augmentation from upstream reservoirs and basinwide secondary sewage treatment began in the 1950s (Hughes and Gammon 1987). Huff and Klingeman (1976) and Hines et al. (1977) documented improvements in water quality and Dimick and Merryfields (1945) and Hughes and Gammon (1987) documented fish assemblages in 1944 and 1983, respectively.

The Willamette River flows into the Columbia River at Portland, Oregon and is fed by a number of smaller rivers that originate primarily in the Cascade Mountains to the east. For this study, we included 286 km of the main stem of the Willamette from Eugene-Springfield (at the southern end of the Willamette Valley) to Portland plus the lower 18 km of the McKenzie River (Fig. 1). In addition, we surveyed the 19 km main stem Santiam River and the lower 20 km of the North Fork and lower 10 km of the South Fork. Additional ospreys are known to nest in the Willamette Valley outside the study area and in other portions of western Oregon (see Henny et al. 1978, Witt 1990). The river banks in the study area primarily support black cottonwood, with a few Douglas fir,

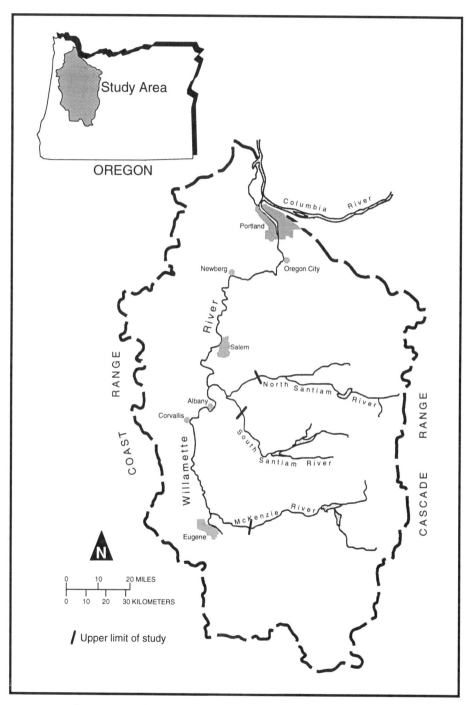

Figure 1. The osprey study area in the Willamette Valley of western Oregon. Only rivers with all or a portion surveyed are shown.

true fir (*Abies* spp.) and bigleaf maple. The cottonwoods, although large in many places, are generally inadequate to support osprey nests. Snags and broken top live trees are scarce. Most of the river in this study area flows through fertile privately owned farmland.

In 1976, osprey nests were located from a Cessna 206 flown about 60–100 m above the ground along the Willamette and the Santiam Rivers. An adjustment factor was used to estimate the total nesting population based on air:ground visibility rates (see Henny *et al.* 1978). On 21 April 1993, an aerial survey was made with a Cessna 182 at the same altitude to locate occupied nests. The area of coverage was similar in 1976 and 1993, i.e. up to 2 km from the river. Once it became apparent that many of the nests were on utility poles and towers, we contacted the utility companies to check their records for nesting ospreys. An essentially complete ground count of the study area was also made. Nest sites were visited (not climbed) at least six to eight times during the nesting season, and the number of young produced at each nest was determined from the ground with the aid of a spotting scope. Active and occupied nests followed the definitions of Postupalsky (1977). Young were conspicuous when prey was delivered, therefore the count of young present near fledging time in each nest was made after a prey delivery.

To provide additional information on the growth rate and the timing of the transition to utility structure nest sites, we asked each landowner how long the nest site(s) on their property had been occupied. If alternate nests, i.e. several power poles nearby, were used in different years presumably by the same pair, the territory was considered always occupied. Although osprey territories occupied in 1993 were not necessarily occupied annually from the initial date of occupancy, the population increase over time was estimated by assuming annual territory fidelity. Therefore, we provide an estimate of the number of nesting osprey pairs using utility structures for each year from 1977–93. The number of nests in trees was essentially the same in 1976 and 1993, and we assume it remained unchanged during the interim. When landowners indicated an inexact number of years that a territory was occupied, we approximated the initial occupation dates, e.g. for three nests first occupied 8–10 yr ago, we assigned one nest at 8 yr, one 9 yr, and one 10 yr. As a quality check, we compared individual nest records with information in the Oregon Department of Fish and Wildlife data base and other sources, including nearby farmers. These records may not be entirely accurate, but we believe occupancies were correct within a few years. No source indicated that utility structures were occupied in 1976 or earlier.

RESULTS

Nesting along Willamette River, 1976

An estimated 13 pairs nested in our defined Willamette Valley study area in 1976 (Table 1). Two pairs on the upper Santiam River were outside the study

Table 1. Distribution of nesting ospreys in the Willamette Valley study area in 1976 (Henny et al. 1978) and 1993 (this study).

Location	1976		1993	
	Nests	Nests/km	Nests	Nests/km
Springfield to Corvallis[a]	5	0.05	24	0.26
Corvallis to Salem (excluding Santiam R.)	4	0.05	20	0.26
Santiam River[b]	0	–	20	0.41
Salem to Newberg	0	–	12	0.22
Newberg to Portland	2	0.04	2	0.04
Total	11 (13)[c]	0.03–0.04	78	0.24

[a] Includes lower 18 km of McKenzie River.
[b] Within boundary established in Study area and methods.
[c] Observed nests from air (estimated nests).

area and excluded. All nests were in live or dead trees. The following year (1977) a nest on a power pole near the Ankeny National Wildlife Refuge adjacent to the Willamette River was observed and its notoriety warranted a sentence in Henny et al. (1978), although their survey was completed a year earlier.

Nesting along Willamette River, 1993

There were 78 pairs of nesting ospreys in the study area in 1993, a six-fold increase since 1976 (Table 1). The increase occurred throughout most of the study area, and was more pronounced in the farmland along the lower Santiam River. A large population increase also occurred on the farmland along the Willamette River. The non-farmland downstream from Newberg to Portland (mostly suburban and urban) showed no increase in ospreys. Perhaps equally important as the population increase was the change in types of nesting sites occupied (Table 2). Most nests (85%) were on utility poles or towers. The utility companies responded to osprey nesting attempts over the years by modifying over half of the nest sites (32 of 58) on power poles both to reduce adverse effects on power delivery and to accommodate the nesting needs of ospreys. Usually, platforms were built above the crossarms and powerlines or new poles with nesting platforms were placed nearby.

All nests in the study area were located within 2 km of the rivers, but 83.3% were within 1 km (Table 3).

Osprey population increase, 1976–93

The number of nests in trees (live or dead) in the study area were similar in 1976 (11 observed, 13 estimated pairs), and 1993 (12). However, those nesting

Table 2. Nesting sites used by ospreys in the Willamette Valley study area, 1993.

Nest structure	Number of nests
Transmission towers	8
Power Poles (total 58)	
Non-energized pole with platform[a,b]	13
Energized pole with platform[a,c]	19
Energized not modified	26
Natural Nests (total 12)	
live trees	8
dead trees	4
Total number of nests	78

[a]Poles modified by utility company.
[b]Adjacent wooden pole with platform placed nearby to attract nesting osprey away from energized powerline.
[c]Extension built on top of pole to minimize interference with energized powerline.

Table 3. Distance from osprey nest to closest river in Willamette Valley study area in 1993.

Distance to river (km)	Nests	
	Number	Percent
≤0.5	45	57.7
0.6–1.0	20	25.6
1.1–1.5	10	12.8
1.6–2.0	3	3.8
Total	78	99.9

on utility poles or towers increased from 0 in 1976 (Henny 1978) to 66 in 1993, and the population increase fits the standard logistic growth equation very well ($r^2 = 0.995$) (Fig. 2). The upper asymptote of the curve (86 osprey pairs nesting on utility structures) was estimated to occur in 2004; however, the data set fit the exponential growth curve nearly as well ($r^2 = 0.968$). The exponential growth curve does not predict a plateau. The growth parameters of the exponential, $r = 0.270 \pm 0.013$ (se), and logistic, $r = 0.374 \pm 0.026$ (se), curves imply estimated annual rates of increase of 27% and $37\%/(1 + \exp[37\%(t - 76)]/154)$, respectively.

The observed production rate in 1993 (1.64 young/active nest) was about twice the rate estimated required to maintain a stable population (0.80 young/active nest) in the northeastern United States (Spitzer 1980, Spitzer et al. 1983). The collection of one freshly laid egg from each of 10 nests during the contaminant phase of this study (observed young/active nest, 1.50 at 10 nests

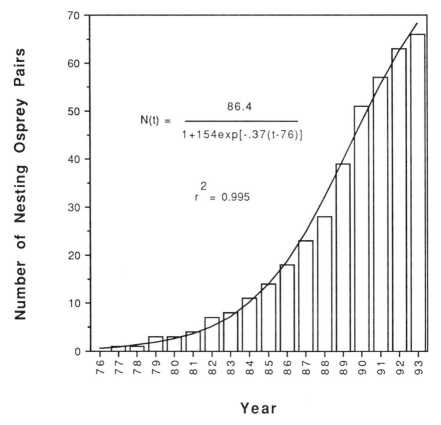

Figure 2. The logistic growth curve and equation for the osprey population nesting on utility structures in the Willamette Valley study area from 1976 to 1993. Information in 1976 and 1993 from actual surveys, and intervening years from landowner observations recorded in 1993.

with egg collected) has a slight negative impact on production. However, the observed production rate in those 63 active nests without an egg collected (1.67 young/active nest) was nearly identical to that for all 73 active nests (1.64 young/active nest).

DISCUSSION AND CONCLUSIONS

The size of the historic osprey population nesting along the Willamette River is unknown, although nesting was reported as early as 1854–55 (Newberry 1857). At that time the valley contained small farms, and an abundance of trees in many regions. Logging and clearing land for farms undoubtedly resulted in a general loss of potential osprey nest sites (broken top trees, snags and live trees)

Table 4. Osprey reproductive parameters in the Willamette Valley Study Area in 1993.

Reproductive parameters	Location		Nest site type		Total
	Willamette R.[a]	Santiam R.	Trees	Utility structure	
Occupied nests	58	20	12 (15.4)[b]	66 (84.6)[b]	78
Active nests	54	19	12 (16.4)	61 (83.6)	73
% Active	93.1	95.0	100	92.4	93.6
Successful nests	39	17	9 (16.1)	47 (83.9)	56
% Nest success, occupied	67.2	85.0	75.0	71.2	71.8
% Nest success, active	72.2	89.5	75.0	77.0	76.7
No. advanced young	82	38	20	100	120
Productivity, occupied	1.41	1.90	1.67	1.52	1.54
Productivity, active	1.52	2.00	1.67	1.64	1.64
Productivity, successful	2.10	2.24	2.22	2.13	2.14

[a] Includes 18 km segment of lower McKenzie River.
[b] % of nest total.

adjacent to the river over the last 100 yr (Sedell and Froggatt 1984). The use of DDT on agricultural crops in the valley also may have affected nesting ospreys during the period after World War II, e.g. see Henny and Anthony (1989). DDT was banned nationwide in 1972, and osprey production at many locations began improving by the mid to late 1970s (Henny 1977). The impact of DDT on osprey populations in western Oregon during the period 1950–75 is unknown because no eggs were obtained for residue analysis, and no historical population data were available. By the early 1980s, DDE in a small series of eggs collected in the Pacific Northwest (none directly from this study area, but some from western Oregon) generally decreased compared with that reported pre-1980 in adjacent states, although some eggs still contained DDE at concentrations sufficiently high to reduce productivity (Henny and Anthony 1989). Therefore, we believe that the population increase associated with the use of utility structures was at least partially due to a gradual release from the effects of DDT and its metabolites.

The high productivity in 1993 was uniformly found at tree nests as well as the utility structures; therefore the utility structures themselves did not seem to be a factor in the high productivity. This finding contrasts with many other studies (see Poole 1989), where higher productivity was reported at more stable man-made nesting structures likes power poles, towers and nesting platforms. The use of man-made structures theoretically should result in higher rates of population increase. However, since only 12 tree nests were in the study area, the 1993 findings concerning relative success at utility structure nests and tree nests seem inadequate to reach firm conclusions.

Why would nesting osprey begin using the utility structures in 1977 and not decades earlier? The first osprey nesting platform program in Oregon began at Crane Prairie Reservoir in 1973, when 39 platforms were established in the

Cascade Mountains (Henny *et al.* 1978) about 160 km from the first power pole nest (1977) discovered in the Willamette Valley. Could ospreys that used man-made platforms at Crane Prairie Reservoir (first used in 1974) have produced a female (the sex most likely to disperse long distances) that pioneered pole nesting in the Willamette Valley in 1977? Ospreys are known to first nest as 3 year olds (Henny and Wight 1969), and some female first-time breeders are known for their long dispersal distances from natal areas, especially in the western United States (Johnson and Melquist 1991). After nesting the first time, they generally remain faithful to the nesting territory in subsequent years.

Hughes and Gammon (1987) reported a considerable change in the fish assemblages of the Willamette River since 1944. In their study, 16 of 18 sites formerly sampled by Dimick and Merryfield (1945) had more species, more intolerant species, i.e. to organic pollution, warm water, and sediment (see Hughes and Gammon 1987), and fewer tolerant species. However, the most important prey species of the osprey in 1993 was the largescale sucker (authors unpubl. data), which is a native omnivorous species classified as tolerant. Largescale suckers were collected at all 14 of our collection sites in 1993 (from river km 80 to km 269), at all Hughes and Gammon (1987) collection sites, and nearly all Dimick and Merryfield (1945) collection sites except below km 82. The overall change in the fish assemblage since 1944 may be of minor importance to the recent osprey increase since the tolerant largescale sucker (the key prey species) was always present. Although consistently present in the river, the largescale sucker now may be more abundant. Since the first installation of a revetment in the Willamette River in 1888, there has been a tremendous loss of secondary side channels, backwaters, and oxbows; but the largescale sucker is among five fish species that appear to benefit from revetments (Hjort *et al.* 1984, Li *et al.* 1984). Largescale suckers graze on diatoms and stone revetments provide good substrate for periphyton (Li *et al.* 1987).

Because of intensive farming, natural nesting sites (trees and snags) may have limited the number of pairs nesting along the river in the 1950s through the early 1970s prior to ospreys adopting power poles and towers. The utility poles used since 1977 were primarily those poles in fields related to farmers' electric irrigation pumps. Many times, the terminal pole at the pump was chosen for the nesting site. Irrigation systems on farmland became common in the study area during the mid-to-late 1950s (about 20 years before use by ospreys). Once it became apparent to the first osprey that utility poles and towers provided adequate alternate nest sites, others seemed to learn quickly.

Although we have no specific quantitative information about osprey shooting in the Willamette Valley, shooting by fisherman, hunters and farm boys may have been a factor in keeping the population low in earlier years. The ospreys' presence at conspicuous nests makes shooting them easy. Some shooting still occurs, but ospreys banded in the United States and bands reported to the Bird Banding Laboratory as "bird shot" represented a high percentage of those reported in earlier years (Henny and Wight 1969). Attitudes towards birds of prey have changed dramatically over the last several decades, and now, most farmers along the Willamette River refer to nesting ospreys as their birds.

Ospreys nesting at their farm appear to be a status symbol for most farmers. One farmer even has an original oil painting of his osprey at its nest.

In summary, the observed osprey population increase probably resulted from a combination of factors including:

(1) Willamette River cleanup and associated fish response;
(2) the banning of DDT in 1972 and associated improvement in osprey production;
(3) a change in human attitude toward birds of prey, and
(4) the apparent learned response of ospreys to nest on utility structures (decades after similar structures were first used in eastern North America) which resulted in nest site availability improving from perhaps being a limiting factor, to being essentially unlimited.

Contaminants in osprey eggs (10 were collected), more detailed information on fish species captured by ospreys, and contaminant burdens in fish will be the subject of a future report.

ACKNOWLEDGMENTS

We appreciate the assistance of R. Goggans, C. Heath and J. Pesek of the Oregon Department of Fish and Wildlife, and thank J. Gunter for providing background data on ospreys nesting in the study area. Logistical support and additional information were provided by personnel from Consumers Power, Philomath; Emerald Peoples Utility District, Eugene; Eugene Water and Electric, Eugene; Pacific Power and Light, Albany and Independence; Portland General Electric, Salem; and Salem Electric, Salem. The landowners were most helpful; we appreciate access to their land and the historical information about occupied osprey nests. J. Hatfield of Patuxent Wildlife Research Center provided statistical advice and assistance. The manuscript was improved by the thoughtful comments of J. Zingo, N. Vyas, and D. Ellis.

REFERENCES

DIMICK, R.D. AND F. MERRYFIELD. 1945. The fishes of the Willamette River system in relation to pollution. *Engineering Expt. Sta. Bull. Series* **20**: 7–55 (Oregon State College, Corvallis, OR).

GABRIELSON, I.N. AND S.G. JEWETT. 1940. *The birds of Oregon*. Oregon State College Press, Corvallis.

HENNY, C.J. 1977. Research, management and status of the Osprey in North America. Pp. 199–222 in R.D. Chancellor, ed. *World Conference on Birds of Prey*. ICBP, Vienna, Austria.

—. 1988. Large Osprey colony discovered in Oregon in 1899. *Murrelet* **69**: 33–36.

— AND R.G. ANTHONY. 1989. Bald Eagle and Osprey. Pp. 66–82 in B.G. Pendleton, ed. Western Raptor Management Symposium and Workshop. *Scientific and Technical Series* No. 12, Natl Wildl. Fed., Washington, D.C. 317 pp.

— AND A.P. NOLTEMEIER. 1974. Osprey nesting populations in the coastal Carolinas. *Am. Birds* 29: 1073–1079.

— AND H.M. WIGHT. 1969. An endangered Osprey population: estimates of mortality and production. *Auk* 86: 188–198.

—, M.M. SMITH AND V.D. STOTTS. 1974. The 1973 distribution and abundance of breeding Ospreys in the Chesapeake Bay. *Chesapeake Sci.* 15: 125–133.

—, J.A. COLLINS AND W.J. DEIBERT. 1978. Osprey distribution, abundance, and status in western North America: II. The Oregon Population. *Murrelet* 59: 14–25.

—, M.A. BYRD, J.A. JACOBS, P.D. MCLAIN, M.R. TODD AND B.F. HALLA. 1977. Mid-Atlantic coast Osprey population: present numbers, productivity, pollutant contamination, and status. *J. Wildl. Manage.* 41: 254–265.

HINES. W.G., S.W. MCKENZIE, D.A. RICKERT AND F.A. RINELLA. 1977. Dissolved-oxygen regime of the Willamette River, Oregon, under conditions of basin-wide secondary treatment. *U.S. Geol. Surv. Circ.* 715-1.

HJORT, R.C., P.L. HULETT, L.D. LABOLLE AND H.W. LI. 1984. *Fish and invertebrates of revetments and other habitats in the Willamette River, Oregon.* Technical Report E-84-9. Oregon State University for US Army Engineer Waterways Experiment Station, Vicksburg, MS.

HUFF, E.S. AND P.C. KLINGEMAN. 1976. Restoring the Willamette River: Costs and impacts of water quality control. *J. Water Pollut. Control Fed.* 48: 2410–2415.

HUGHES, R.M. AND J.R. GAMMON. 1987. Longitudinal changes in fish assemblages and water quality in the Willamette River, Oregon. *Trans. Am. Fisheries Soc.* 116: 196–209.

JEWETT, S.G. AND I.N. GABRIELSON. 1929. Birds of the Portland area, Oregon. *Pacific Coast Avifauna* 19, 54 pp.

JOHNSON, D.R. AND W.E. MELQUIST. 1991. Wintering distribution and dispersal of northern Idaho and eastern Washington Ospreys. *J. Field Ornithol.* 62: 517–520.

LI, H.W., C.B. SCHRECK AND R.A. TUBB. 1984. *Comparison of habitats near spur dikes, continuous revetments, and natural banks for larval, juvenile, and adult fishes of the Willamette River.* WRRI-95, Water Resources Research Institute, Oregon State Univ., Corvallis.

—, C.B. SCHRECK, C.E. BOND AND E. REXTAD. 1987. Factors influencing changes in fish assemblages of Pacific Northwest streams. Pp. 193–202 in W.J. Matthews and D.C. Heins, eds. *Community and evolutionary ecology of North American stream fishes.* Univ. Oklahoma Press, Norman.

NEWBERRY, J.S. 1857. Reports of explorations and surveys to ascertain the most practicable and economical route for a railroad from the Mississippi River to the Pacific Ocean, 1854–55. Vol. VI. Part IV. Zoological Report No. 2, Report upon the Zoology of the route. Chapt. II. Report upon the birds, pp. 73–110, Secretary of War, Washington, D.C.

POOLE, A.F. 1989. *Ospreys: A natural and unnatural history.* Cambridge Univ. Press, Cambridge, England.

POSTUPALSKY, S. 1977. A critical review of problems in calculating Osprey reproductive success. Pp. 1–11 in J.C. Ogden, ed. Transactions of the North American Osprey Research Conference. *Trans. and Proc. Series* No. 2, Natl. Park Serv., Washington, D.C.

ROBERTS, H.B. 1969. *Management plan for the Crane Prairie Reservoir Osprey management area.* US Forest Serv., Deschutes National Forest, Bend, OR.

SEDELL, J.R. AND J.L. FROGGATT. 1984. Importance of streamside forests to large rivers: the isolation of the Willamette River, Oregon, USA, from its floodplain by snagging and streamside forest removal. *Verh.-Internat. Limnol.* 22: 1828–1834.

SPITZER, P.R. 1980. Dynamics of a discrete coastal breeding population of Ospreys in the northeastern United States during the period of decline and recovery, 1969–1978. Ph.D thesis, Cornell Univ., Ithaca, NY.

—, A.F. POOLE AND M. SCHEIBEL. 1983. Initial recovery of breeding Ospreys in the region between New York City and Boston. Pp. 231–241 in D.M. Bird, Chief Ed. *Biology and management of bald eagles and ospreys*. Harpell Press, Ste. Anne de Bellevue, Quebec.

WITT, J.W. 1990. Productivity and management of Osprey along the Umpqua River, Oregon. *Northwestern Nat.* 71: 14–19.

WOODCOCK, A.R. 1902. Annotated list of the birds of Oregon. *Oregon Agric. Expt. Sta. Bull.* 68. Corvallis. OR.

13

The Use of Artificial Nest Sites by an Increasing Population of Ospreys in the Canadian Great Lakes Basin

Peter J. Ewins

Abstract – The Great Lakes drainage basin is inhabited by over 36 million people, and osprey breed in many parts of the basin. Considerable changes in habitats utilized by osprey have occurred, particularly during the twentieth century, associated with urban, industrial and recreational development. Prior to 1945, osprey in Canadian parts of the basin bred only in trees. Since then an increasing proportion have bred on a wide range of artificial structures, including hydro poles, transmission line towers, navigation and communication towers, buildings, and customized artificial platforms (single poles, tripods and quadropods). In the period 1988–93, nests on artificial structures occurred significantly more often along the Great Lakes shorelines (48%) than farther inland (29%). On Lake Huron, 82% of artificial platforms were occupied within one year of installation, suggesting a shortage of suitable natural nest sites on the main Great Lakes. Nests on artificial structures fell down only slightly less often (9% per annum) than did those in trees (12% p.a.), and reproductive output was only slightly higher (1.14 versus 1.06 younger per nest occupied in mid-May, respectively). Osprey population increases, following restrictions on the use of organochlorine pesticides, appear to have been assisted by nesting on artificial structures, as well as by a general high degree of tolerance to human activities near nests.

Key words: ospreys; artificial structures; nests; Great Lakes; increases.

The osprey is a widespread breeding species between subarctic and subtropical latitudes, in North America, Europe, Asia and Australasia (Cramp and Simmons 1980, Poole 1989). Canada and Alaska probably support about 32–50% of the world breeding population, mostly at relatively low densities throughout the extensive lake and river systems in areas remote from human habitation (Wetmore and Gillespie 1976, Scott and Houston 1983, Weir 1987, Poole 1989, Hughes 1990). The main requirements for nest-site selection appear to be proximity to shallow-water foraging areas, inaccessibility to mammalian predators, support structures capable of bearing the weight of the large stick nest, and unimpeded aerial access to the nest on account of this bird's relatively long wings and limited manoeuvrability (Poole 1989).

The Great Lakes drainage basin is one of the most industrialized and densely populated regions of North America, with over 36 million people living there today (Government of Canada 1991). Historically, ospreys appear to have bred throughout most of the basin north of the lower Great Lakes, usually in large trees close to water (Bell 1861, McIlwraith 1886, Fleming 1901), although there is little documentation prior to forest clearance by European settlers. Since the early nineteenth century, widespread timber extraction and subsequent conversion of land and wetlands to agricultural uses (Lawrie and Rahrer 1973, Cameron 1978, Holla and Knowles 1988, Sly 1991), have undoubtedly reduced the number of suitable nesting sites for osprey. Additionally, dramatic habitat changes associated with shoreline development for industry, habitation or recreation have probably contributed to general decreases of Great Lakes osprey populations over the past 150 years or so, particularly around the lower Great Lakes where most humans have settled (Sly 1991).

The widespread use and discharge of organochlorine pesticides and other toxic contaminants between the 1940s and early 1970s, led to widespread reproductive failure and population declines of osprey and other fish-eating birds in the Great Lakes (Postupalsky 1969, 1971, Gilman et al. 1977, Weseloh et al. 1983, Ewins et al. in press). Following the introduction of severe restrictions on the use of these persistent contaminants (in particular DDT) in the early 1970s, osprey populations have increased in the Great Lakes basin (Postupalsky 1977a, 1988, Gieck et al. 1992, Ewins 1992, Ewins et al. 1995), and many other parts of North America (Henny 1983, Spitzer et al. 1978). In parts of the North American range where ospreys have been studied in recent years, they appear to have taken readily to nesting on man-made structures such as hydro poles, navigation markers, and custom-built nesting platforms, often in close proximity to humans (Postupalsky and Stackpole 1974, Carrier and Melquist 1976, Reese 1977, Henny 1986, Poole 1989, Postupalsky 1978, 1989).

This paper provides the first published review of ospreys utilizing artificial nest structures in Ontario. It examines changes in the type of nest structure used, and considers what implications these have had for reproductive output and population changes. These changes are discussed in relation to other human alterations to the Great Lakes basin environment, and the extent and manner in which osprey have adapted to them.

STUDY AREA AND METHODS

Detailed studies in the Canadian Great Lakes basin were carried out from 1991 to 1993 in five main study areas (Fig. 1). Nests within 5 km of the main Great Lakes shoreline were considered as "Great Lakes" nests, i.e. St. Marys River, Georgian Bay, and the St. Lawrence River study areas, all being oligotrophic or mesotrophic waters with many islands and embayments. Reference nests farther inland were at Ogoki Reservoir (an oligotrophic, deep-water reservoir with shallow margins), and in the Kawartha Lakes (a series of shallow,

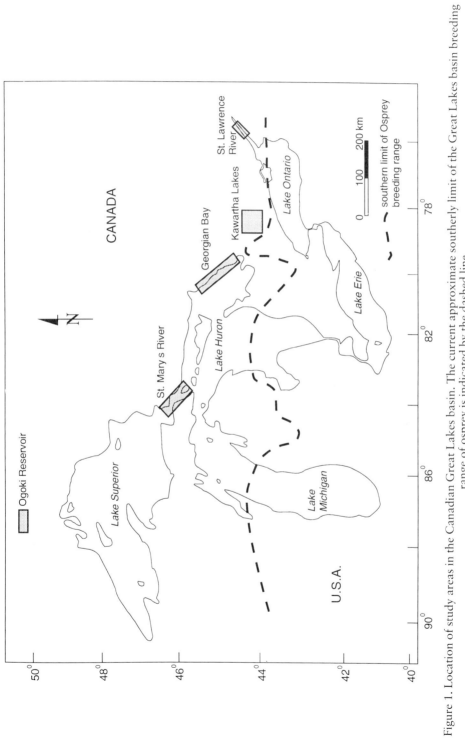

Figure 1. Location of study areas in the Canadian Great Lakes basin. The current approximate southerly limit of the Great Lakes basin breeding range of osprey is indicated by the dashed line.

eutrophic lakes with flooded margins). Shoreline residential development and recreational boating were intense at times in all four of the southern study areas. Only Ogoki Reservoir was remote from human habitation and roads. Additional recent data were collected from various parts of Ontario, but mostly areas between Georgian Bay and the St. Lawrence River. For some analyses, data from the Michigan parts of the St. Marys River were included with the Canadian data.

Information on types of nest structures used by ospreys since 1874 was obtained from a variety of sources: the Ontario nest records scheme (based in the Royal Ontario Museum, Toronto), District offices of the Ontario Ministry of Natural Resources; the Georgian Bay Osprey Society; published and unpublished reports and regional natural history accounts; public response to media requests for information; and this detailed study. Reproductive statistics were not calculated for pre-1991 data, due to the biases present in incidental observations based on casual records at nests.

Numbers of occupied nests were determined in mid-May using aerial surveys, and boat-based nest checks. Although these surveys were not completed on a rigid grid basis, they involved coverage of different habitat types, and were believed to locate a high proportion of occupied nests in each area. Reproductive output was assessed in mid-July for those occupied nests identified during the May surveys, again using aerial survey (rotor-winged and fixed-wing) and direct inspection of nest contents, according to the protocol established by Postupalsky (1977b). Old nests from previous years, which had not been built up in the current year, were not considered.

Ospreys often occupy the same nest site in successive years (Poole 1989, Postupalsky 1989, PE pers. obs.). Each year of occupancy of a given nest was scored as a separate occupation. For example, six successive years of occupation at the same hydro pole was scored as six occupied hydro pole nests, i.e. nest-years. Artificial nest structures were taken to be any occupied nest which was built on a structure partially or completely constructed or modified by humans. Thus, a wooden platform secured to the top of a tree or low stump was regarded as an artificial site. Inactive hydro poles on which a wooden platform had been installed were scored as platform nests, to distinguish them from nests balanced only on cross-arms at the top of active (or inactive) hydro poles.

Rates of population change over n years, using an initial count (IC) and a recent count (RC), were calculated using the following formula, adapted from Ricklefs (1980)

$$[\ln (RC) - \ln (IC)]/n.$$

Statistical methods follow Sokal and Rohlf (1981), significance being accepted at the $P < 0.05$ level. The term "significant" is used in its statistical sense only.

RESULTS

Between 1874 and 1944, every osprey nest recorded in the Canadian Great Lakes basin was in a tree. Since 1945 the proportion of recorded nests which

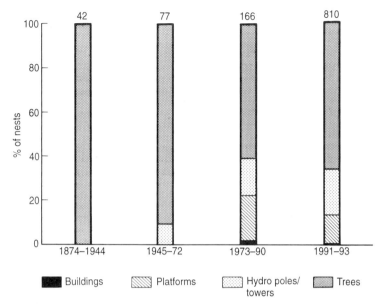

Figure 2. Changes in types of structure supporting occupied osprey nests in Ontario, 1874–1993. The number of nests in each period are given at the top of each histogram.

were in trees has declined steadily, as increasing numbers of nests were found on hydro poles, various types of towers, buildings and on customized artificial nesting platforms (Fig. 2). An unquantified, but probably relatively small proportion, of this decline is attributable to the greater visibility and accessibility of nests on artificial structures. Systematic aerial surveys were not conducted prior to the 1990s, so nests in dead or live trees are under-represented between 1945 and 1990 in Fig. 2. Most of the nests on hydro poles/towers category were placed on the wooden double cross-arms at the top of hydro poles. This configuration has been used on Ontario hydro poles since at least 1915 (Ontario Hydro, unpubl. photographs), so osprey apparently did not begin nesting on such structures until many years after they were first installed. Custom-built artificial nesting platforms for osprey have been erected in various parts of Ontario since at least the early 1970s. Initially this was done mostly by hydroelectric companies in response to power failures caused by osprey or sticks from their nests, which shorted the suspended hydro cables. Square, wooden platforms were either mounted at the top of the existing live hydro pole, or on an alternative pole installed within 10–20 m of the line. Alternative measures were also taken, such as regular destruction of the entire nest and contents, or re-siting of the hydro lines farther down the pole. In general these steps have reduced the frequency of power-cuts, and osprey have usually accepted the alternative nesting arrangements, and been able to reproduce more successfully.

Since the mid-1980s, various individuals and organizations have installed artificial nesting platforms specifically for osprey in parts of southern Ontario, assisted financially by the Ontario Ministry of Natural Resources Community

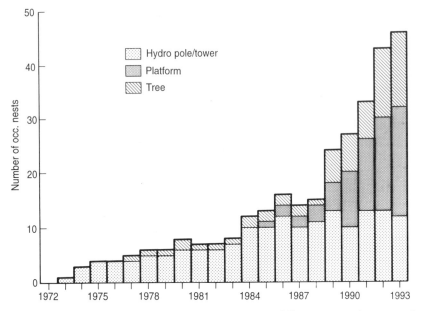

Figure 3. Numbers of occupied osprey nests occurring on different types of structure in Georgian Bay, 1972–93.

Wildlife Improvement Program. In the Kawartha Lakes, most platforms are now on low quadropods or tripods embedded in mud toward the edge of shallow lakes, whereas on Georgian Bay, platforms are on small, rocky islands, either on disused hydro poles, or on 10–15 m poles secured to granitic bedrock with rock-mounts and guys.

On Georgian Bay, osprey were extirpated as a breeding species during the years of heaviest use of organochlorine pesticides, but since the mid-1970s numbers breeding within 5 km of the main lake shoreline appear to have increased dramatically (Fig. 3), at 13% per annum on average (Ewins *et al.* 1995). However, the apparent increase in numbers of natural nest sites used in the early 1990s largely reflects the introduction of aerial surveys, which were able to detect nests in dead trees in otherwise inaccessible swamps (Fig. 3). When only nests on hydro poles and artificial platforms were considered, the average annual rate of population increase since the 1970s has been 11.6%. During the initial stages of this recolonization, hydro poles were the main nest structure used. Since 1989 the number of available customized nesting platforms has increased four-fold, and on Georgian Bay these platforms accounted for 37–43% of occupied nests during the early 1990s (Fig. 3). In the Canadian Great Lakes basin overall, 16% of occupied nests in 1991–93 were on artificial platforms, and 65% were in trees (Fig. 2).

In recent years (1988–93), osprey in the Canadian Great Lakes basin have built their nests on a wide range of different structures. Overall, natural supporting structures accounted for 64% of the occupied nests, and artificial

Table 1. Types of structure supporting osprey nests in the Canadian Great Lakes basin, 1988–93. No. of nestings (and % of total) are given for each category.

	Great Lakes[a]	Inland	Total
No. of nestings	304	597	901
Natural			
Tree[b]	156 (51.3)	422 (70.7)	578 (64.2)
Ground	1 (0.3)	0	1 (0.1)
Total	157 (51.6)	422 (70.7)	579 (64.3)
Artificial			
Platform	58 (19.1)	68 (11.4)	126 (14.0)
Hydro pole	37 (12.2)	81 (13.6)	118 (13.1)
Transmission tower	10 (3.3)	4 (0.7)	14 (1.6)
Other tower[c]	42 (13.8)	10 (1.7)	52 (5.8)
Building[d]	0	12 (2.0)	12 (1.3)
Total	147 (48.4)	175 (29.3)	322 (35.7)

[a] nests within 5 km of Great Lakes, includes Michigan parts of St. Marys River
[b] live or dead tree, or stump
[c] navigation, microwave or telecommunication towers
[d] windmill, feedmill, duck blinds

structures for 36% (Table 1). However, a significantly greater proportion of osprey nests within 5 km of the Great Lakes shoreline occurred on artificial structures, compared with occupied nests farther inland (48% vs. 29%, respectively, $G_1 = 31.3$, $P < 0.001$). Natural sites used were at the top of dead or live trees, mostly white pine or other coniferous species, or aspens, or on tree stumps in flooded lake margins or reservoirs.

In Georgian Bay, one (unsuccessful) nest was placed on the ground on a small rocky island in 1993, and a number of nests have been built (but not occupied) in other unusual sites, such as on TV aerials and on chimneys. These observations suggest a shortage of preferred supporting nest structures there.

Artificial structures used in the Canadian Great Lakes basin in recent years include: customized platforms (on poles, and a few in trees or on stumps); hydro poles (live or disused); wooden (but not metal) high voltage transmission towers/pylons; towers (usually 30–40 m high) supporting navigation lights, microwave, or telecommunication installations; duck blinds; an abandoned metal windmill; and an elevator on a feedmill. In Lake Huron, both Canadian and US Coastguard agencies now adopt a sympathetic approach to osprey that attempt to nest among active navigation light equipment; this often involves installation of metal platforms above solar panels and other electrical equipment. Among artificial sites, platforms and hydro poles were the most frequently used, but along Great Lakes shorelines various towers were also important (Table 1). The proportion of occupied nests that occurred on artificial structures varied significantly among areas on the Great Lakes, broadly reflecting relative availability (Georgian Bay 71%, St. Marys River 44%, other areas 17%;

Table 2. The incidence of osprey nest collapses in three parts of the Great Lakes basin, 1988–93, at artificial and natural nest sites. Data refer to the number of occupied nests which collapsed between successive breeding seasons, expressed as a percentage of the total number of successive nest-years within each area.

Area	Artificial sites		Natural sites	
	No. of occ. nests	No. (%) collapsing	No. occ. nests	No. (%) collapsing
St. Marys River	45	2 (4.4)	69	4 (5.8)
Georgian Bay	183	15 (8.2)	68	11 (16.2)
Kawartha Lakes	93	12 (12.9)	69	10 (14.5)
Total	321	29 (9.0)	206	25 (12.1)

$G_2 = 49.8$, $P < 0.001$). For inland nests, there was similar significant variation among areas, largely reflecting availability of platforms (Kawartha Lakes 55%, other areas 9%; $G_1 = 154.3$, $P < 0.0001$).

In many areas of Ontario where osprey breed, artificial platforms of suitable dimensions, and in appropriate locations, have been occupied very rapidly. In the St. Lawrence River, five platforms were installed on small rocky islets in autumn 1992, and three (60%) were occupied (all were successful) in the 1993 breeding season. On Georgian Bay, of 32 artificial platforms installed since 1985, 22 (69%) are now occupied by osprey. Of the 10 which have not been occupied, seven were installed only in 1992, and the other three are within 100 m of a cottage or an occupied osprey nest. Among these 22 occupied platforms, 18 (82%) were occupied the first breeding season after installation, two (9%) after 2 yr, one after 3 yr (5%) and one after 4 yr. A platform erected in early May 1992 supported a nest within three days, and birds were incubating by the end of the month!

Osprey nests on artificial structures collapsed, i.e. fell down, or were blown down, or disappeared, between years, slightly less often (9%) than did those on natural structures (12%), but this difference was not significant overall ($G_1 = 1.3$). Nests in the St. Marys River appeared to be more stable than in Georgian Bay or the Kawartha Lakes. Georgian Bay nests in trees collapsed or disappeared between years twice as often (16%) as did those on artificial structures (8%), but this difference was not quite significant ($G_1 = 3.1$, $P > 0.05$; Table 2). However, there was considerable variation in the incidence of nest loss between years among types of artificial structure on Georgian Bay, at least between 1988 and 1993. Nests were lost between years at only 2.3% of platforms and at 1.9% of disused hydro pole sites, compared with 23.3% loss at live hydro poles, and 28.6% at towers (mostly microwave or telecommunications). This difference between operational and inactive artificial sites was highly significant ($G_1 = 20.3$, $P < 0.0005$). Many of these losses were due to nest removal by operators of the equipment.

Overall, reproductive output from nests on artificial structures averaged 7.5% higher than at natural sites, but the only marked difference was in the St.

Table 3. Mean reproductive output at occupied osprey nests on natural and artificial structures in five study areas in the Great Lakes basin, 1988–93. Locations of study areas are shown in Figure 1.

Area	Year	Nest structure	No. occ. nests	No. young fledged	No. young occ. nest^{-1}
Ogoki Reservoir	1992	Natural	15	12	0.80
St. Marys River	1991–93	Natural	78	75	0.96
		Artificial	44	57	1.30
Georgian Bay	1991–93	Natural	37	35	0.95
		Artificial	91	83	0.91
Kawartha Lakes	1991–93	Natural	85	106	1.25
		Artificial	110	133	1.21
St. Lawrence River	1993	Natural	4	10	2.50
Total	1991–93	Natural	219	238	1.06
		Artificial	245	273	1.14

Marys River study area, where artificial sites fledged 35.4% more young than did natural ones (Table 3).

DISCUSSION

Osprey in the Great Lakes basin have probably been affected by a variety of human-induced changes to the environment. Some of the most important changes have been: habitat modification; provision of artificial nesting structures; release of anthropogenic chemicals; changes to fish communities; increases in human disturbance; increases in raccoon populations; declines of bald eagles; and cultural shifts in attitudes toward raptors and general environmental issues. Ospreys appear to have adapted reasonably well to some of these changes, based on evidence from recent studies, but in other cases the effect has been a reduction in breeding range and population size. In other words, osprey have been unable to adapt (at least at the population level) to some of these anthropogenic changes. Due to the scarcity of detailed accounts prior to the 1970s, it has been difficult to quantify accurately the effect on osprey of the major habitat changes which have occurred in the Great Lakes basin over the past 150 years.

Habitat modification has affected both nesting sites and foraging areas, via conversion of wetlands and shallow lake margins to agricultural or recreational uses (Lawrie and Rahrer 1973, Sly 1991). Similarly, removal of preferred large nesting trees (either via selective logging, or clear-cutting) has presumably caused breeding osprey to relocate to alternative sites if they were available. Although there have been a few cases in recent years of birds breeding successfully at nearby alternative natural nest sites, the rapid occupation of artificial structures (in Georgian Bay, the St. Lawrence River, and the Kawartha

Lakes) suggests that high quality nest sites close to, or over, water are in short supply. Similar rapid occupation of artificial platforms has occurred elsewhere (Reese 1977, Postupalsky 1978, Poole 1989). In the absence of suitable alternative nest structures, breeders have presumably vacated the area completely. The shortage of suitable nest structures appears to be most acute along the main Great Lakes shoreline (particularly in the lower Great Lakes), in areas with few beaver swamps, and where raccoon populations have increased in recent years (usually close to human settlements). Raccoons are important predators of osprey eggs in some parts of North America (Poole 1989, PE pers. obs.).

The post-DDT era population increases of ospreys in the Great Lakes basin have clearly been assisted by the adaptability of this species to nesting on a wide range of artificial structures. Comparable rates of population increase to that seen in Georgian Bay have occurred in other areas with programmes to install nesting platforms: 10% per annum in the Kawartha Lakes (Ewins 1992, PE pers. obs.); 5% per annum in central Michigan (Postupalsky 1977a); and 7% per annum in Wisconsin (Gieck et al. 1992). During the past 20 years, more than half the occupied nests in Wisconsin, and on wildlife floodings in central Michigan, were on artificial platforms (Postupalsky 1978, 1989, Gieck et al. 1992). Along many parts of the US Atlantic coast, the majority of osprey nests during the past 20 yr have been on artificial structures, mainly on duck blinds, channel markers, and custom-built platforms (Henny et al. 1974, Reese 1977, Henny 1986, Poole 1989). In the Canadian Great Lakes basin overall, from 1991 to 1993, 35% of occupied osprey nests were on artificial structures, a relatively low proportion compared with some parts of the US (Poole 1989, Gieck et al. 1992, Henny and Kaiser Chapter 12). Poole (1989) noted that the most dramatic population increases since the mid-1970s had occurred in areas where the majority of nests were on artificial structures. In line with Poole's summaries from the US, the number of nests on Georgian Bay increased 300% from 1976 to 1986, in years when 75–100% of known occupied nests occurred on artificial structures.

Hydro poles appear to be more commonly utilized by ospreys in Ontario than in other areas studied (Henny et al. 1974, Henny 1977, 1986, Stocek and Pearce 1978, Austin-Smith and Rhodenizer 1983). Despite considerable shifts in attitudes and technology in favour of ospreys, nests on energized hydro poles and towers still had significantly higher failure rates than those on other types of artificial structure, mainly due to nest destruction by humans. In the past most nests on energized hydro poles were destroyed (Stocek and Pearce 1978, Olendorff et al. 1981, Stocek 1983, PE unpubl. data).

On the Canadian side of the Great Lakes basin, nests on artificial structures were slightly less likely than tree nests to be lost between years, but have differed little in terms of average reproductive output, at least in recent years. Custom-built platforms usually provide better support for the nest than do trees, and reproductive output has often been higher, but not always so (Henny et al. 1974, Reese 1977, Postupalsky 1978, Eckstein et al. 1979, Westall 1983, Henny 1986, Poole 1989). Factors such as siting (in relation to potential predators like raccoons), human disturbance, and age structure of the local population raccoons), human disturbance, and age structure of the local population probably have an important influence on overall productivity (Poole 1981, 1989, Henny 1986, PE pers. obs.). It is also important to consider the type of artificial structure, since custom-built platforms in Ontario had much lower rates of nest loss, and overall productivity, than live hydro poles or towers. Poole (1989) noted that storms often blew down nests. In various North American studies 5–70% of tree nests were lost each year, whereas nests on artificial structures rarely succumbed to wind. The lack of any significant difference in the likelihood of nest collapse or blow-down between artificial and natural sites in the Canadian Great Lakes basin, and the relatively low incidence of nest loss in more northerly parts of the breeding range, may reflect the smaller size of nesting trees (and hence reduced nest stability) farther south, in the US and Mexico (A. Poole pers. comm.).

Osprey were heavily persecuted earlier this century in parts of the Great Lakes basin (Saunders and Dale 1933, Todd 1940), and birds were probably wary of humans near nests. In most parts of southern Ontario, osprey appear now to be remarkably tolerant of human disturbance near nests, and most people feel proud to have a pair nesting nearby (PE pers. obs.). On Georgian Bay, interested cottagers formed the Georgian Bay Osprey Society in the late 1980s, and there is now a waiting list for custom-built artificial platforms there. Ospreys in remote areas however, are more sensitive to human intrusion, particularly during the early stages of the breeding season (Poole 1981, PE pers. obs.). French and Koplin (1977) found that some pairs abandoned nests when logging operations began nearby, but in general recreational disturbance appeared to be tolerated. Many young osprey in the Canadian Great Lakes now experience cottagers, fishermen, water-skiers, pleasure boats, and light aircraft before they fledge. If they return to breed in these areas, they may be more accustomed to such activities close to nests than birds raised in locations remote from people. Continual disturbance from the start of nesting is the key factor here: more vulnerable nests are those subject to sudden disturbance later in the breeding season (A. Poole pers. comm.).

Bald eagles steal fish from osprey, and chase them from nesting territories, so the subordinate osprey usually nests well apart (McIlwraith 1886, Ogden 1975, Gerrard *et al.* 1976, Poole 1989, PE pers. obs., E.M. Addison pers. comm.). In many parts of the Canadian Great Lakes basin, bald eagles were extirpated earlier in this century by encroachment of human habitation, land-use changes, as well as persecution and organochlorine pesticides (McKeane and Weseloh 1993). Bald eagles appear to be generally more sensitive than ospreys to human disturbance at the nest, particularly during the early stages of the breeding season, although in a few areas today they nest in surprisingly close proximity to humans (Whitfield *et al.* 1974, Andrew and Mosher 1982, Fraser 1983, Bortolotti *et al.* 1985). The osprey's greater adaptability to human activities near nests, as well as to nesting on a variety of artificial structures, has enabled it to occupy areas which formerly held breeding bald eagles.

Over the past 100 yr, there have been dramatic changes in the size and composition of Great Lakes fish stocks (Christie 1974, Baldwin *et al.* 1979,

Hartman 1988). Human introductions of exotic species (accidental and deliberate), as well as over-exploitation, have had major effects on indigenous fish resources, and a number of native species have been extirpated locally, or altogether. Osprey seem to have adapted well to these relatively rapid changes in fish communities, and now consume exotic species such as carp, goldfish, white perch, alewife, and some salmonids (Ewins 1992, unpubl. data). The adaptability of foraging osprey in the Great Lakes basin is further emphasized by the regular use of about 5% of Ontario's fish-farms, especially during migration periods (Kevan and Weseloh 1992), and the habit of taking dead fish, discarded or left on the ice surface, by fishermen (Ewins and Cousineau 1994).

In summary, osprey in the Canadian Great Lakes basin have been faced with some major environmental changes over the past 150 yr, caused ultimately by humans. They have adapted well to nesting on artificial structures, often in remarkably close proximity to human activities, and this has facilitated the recent population increases in many areas. The diet has probably shifted in an opportunistic manner as the structure of fish communities has changed. However, osprey have often been unable to adapt to loss of foraging areas and preferred natural nest sites, especially when suitable alternatives were scarce. Anthropogenic contaminants and persecution formerly exerted major impacts on Great Lakes osprey, to which the birds were unable to adapt. Relatively recent changes in human attitudes towards the environment have resulted in substantially reduced levels of persecution and chemical contamination.

ACKNOWLEDGMENTS

Funding for much of the recent work was provided by Environment Canada's Great Lakes Action Plan. Logistical support was supplied by the Technical Operations Division of the National Water Research Institute. I am grateful to the following for assistance with various aspects of this study: Ed Addison, Mark Bacro, Mike Barker, Ron Black, the Georgian Bay Osprey Society, Ontario Hydro, Ontario Ministry of Natural Resources, Martin McNicholl, Mike Miller, Sergej Postupalsky, Barry Snider, Ron Tozer, Chip Weseloh, and the Canadian and US Coastguards. A previous draft manuscript was kindly improved by Ed Addison, Alan Poole, Jeff Robinson, Donna Stewart, Chip Weseloh and an anonymous reviewer.

REFERENCES

ANDREW, J.M. AND J.A MOSHER. 1982. Bald Eagle nest-site selection and nesting habitat in Maryland. *J. Wildl. Manage.* **46**: 382–390.

AUSTIN-SMITH, P.J. AND G. RHODENIZER. 1983. Ospreys, *Pandion haliaetus*, relocate nests from power poles to substitute sites. *Can. Field-Nat.* **97**: 315–319.

BALDWIN, N.S., R.W. SAALFELD, M.A. ROSS AND H.J. BUETTNER. 1979. Commercial fish production in the Great Lakes, 1867–1977. *Great Lakes Fisheries Comm. Tech. Rep.* **3**: 1–187.

BELL, R. 1861. Birds collected and observed around Lake Superior and Huron. *Can. Nat. and Geol.* **6**, 270–275.

BORTOLOTTI, G.R., J.M. GERRARD AND D.W.A. WHITFIELD. 1985. Minimizing investigator-induced disturbance to nesting Bald Eagles. Pp. 85–103 in J.M. Gerrard and T.N. Ingram, eds. *The Bald Eagle in Canada. Proceedings of Bald Eagle Days, 1983.* Whitehorse Plains Publ., Headingley Manitoba, and The Eagle Foundation, Apple River, Illinois.

CAMERON, D.A., ed. 1978. Proceedings of White and Red Pine symposium, Chalk River Ontario, 1977. Canadian Forestry Service, Sault Ste. Marie, Ontario.

CARRIER, W.D. AND W.E. MELQUIST. 1976. The use of rotor-winged aircraft in conducting nesting surveys of Ospreys in northern Idaho. *J. Raptor Res.* **10**: 77–83.

CHRISTIE, W.J. 1974. Changes in the fish species composition of the Great Lakes. *J. Fish. Res. Bird. Can.* **31**: 827–854.

CRAMP, S. AND K.E.L. SIMMONS, eds. 1980. *The birds of the Western Palearctic,* Vol. 2. Oxford Univ. Press, Oxford.

ECKSTEIN, R.G., P.V. VANDERSCHAEGEN AND F.L. JOHNSON. 1979. Osprey nesting platforms in north-central Wisconsin. *Passenger Pigeon* **41**: 145–148.

EWINS, P.J. 1992. Ospreys and contaminants on the Canadian Great Lakes, 1991: an interim report to CWS and OMNR. Can. Wildl. Serv. and Ontario Ministry Natl. Res.

— AND E. COUSINEAU. 1994. Ospreys (*Pandion haliaetus*) scavenging fish on ice. *J. Raptor Res.* **28**: 120.

—, S. POSTUPALSKY, T. WEISE AND E.M. ADDISON. 1995. Changes in the status, distribution and biology of Ospreys (*Pandion haliaetus*) breeding on Lake Huron. In M. Munawar, T. Edsall and J. Leach, eds. *The Lake Huron ecosystem: Ecology, fisheries and management.* Ecovision World Monograph Series, S.P. Academic Publishing, Amsterdam.

FLEMING, J.H. 1901. A list of the birds of the districts of Parry Sound and Muskoka, Ontario. *Auk* **18**: 33–45.

FRASER, J.D. 1983. The impact of human activities on Bald Eagle populations – a review. Pp. 68–84 in J.M. Gerrard and T.N. Ingram, eds. *The Bald Eagle in Canada. Proceedings of Bald Eagle Days, 1983.* Whitehorse Plains Publ., Headingley Manitoba, and The Eagle Foundation, Apple River, Illinois.

FRENCH, J.M. AND J.R. KOPLIN. 1977. Distribution, abundance and breeding status of Ospreys in Northwestern California. Pp. 223–240 in J.C. Ogden, ed. *Transactions of the North American Osprey Research Conference.* US Dept. Int., Natl. Park Serv., Trans. and Proc. Ser. No. 2.

GERRARD, J.M., D.A. WHITFIELD AND W. MAHER. 1976. Osprey–Bald Eagle relationship in Sasketchewan. *Blue Jay* **34**: 240–246.

GIECK, C.M., R.G. ECKSTEIN, L. TESKY, S. STUBENVOLL, D. LINDERUD, M.W. MEYER, J. NELSON AND B. ISHMAEL. 1992. Wisconsin Bald Eagle and Osprey surveys 1992. Unpubl. Rep. Wisconsin Dept. Nat. Res. Madison, Wisconsin.

GILMAN, A.P., G.A. FOX, D.B. PEAKALL, S.M. TEEPLE, T.R. CARROLL AND G.T. HAYMES. 1977. Reproductive parameters and egg contaminant levels of Great Lakes Herring Gulls. *J. Wildl. Manage.* **41**: 458–468.

GOVERNMENT OF CANADA. 1991. Toxic Chemicals in the Great Lakes and Associated Effects. Environment Canada, Dept. of Fisheries and Oceans, and Health and Welfare Canada. Ottawa, Ontario.

HARTMAN, W.L. 1988. Historical changes in the major fish resources of the Great Lakes. Pp. 103–131 in M.S. Evans, ed. *Toxic contaminants and ecosystem health: A Great Lakes focus.* J. Wiley and Sons, New York, New York.

HENNY, C.J. 1977. Research, management, and status of the Osprey in North America. Pp. 199–222 in R.D. Chancellor, ed. *Proc. of World Conf. on Birds of Prey, Vienna 1975.* International Council for Bird Preservation, Cambridge, UK.

—. 1983. Distribution and abundance of nesting Ospreys in the USA. Pp. 175–186 in D.M. Bird, ed. *Biology and management of bald eagles and ospreys*. Harpell Press, Ste. Anne de Bellevue, Quebec.

—. 1986. Osprey (*Pandion haliaetus*). Section 4.3.1, US Army Corps of Engineers Technical Rep. EL-86-5. US Army Engineer Waterways Exp. Stn., Vicksburg, Mississippi.

— AND J.L. KAISER. 1996. Osprey population increase along the Willamette River, Oregon and the role of utility structures, 1976–93. Pp. 97–108 in D. Bird, D.E. Varland and J.J. Negro, eds. *Raptors in Human Landscapes*. Academic Press, London.

—, M.M. SMITH AND V.D. STOTTS. 1974. The 1973 distribution and abundance of breeding Ospreys in the Chesapeake Bay. *Chesapeake Sci.* **15**: 125–133.

HOLLA, T.A. AND P. KNOWLES. 1988. Age structure of a virgin white pine, *Pinus strobus*, population. *Can. Field-Nat.* **102**: 221–226.

HUGHES, J. 1990. Distribution, abundance and productivity of Ospreys in interior Alaska. Alaska Dept. of Fish and Game, Division of Wildlife Conservation Nongame Wildlife Program Report. 13 pp. Juneau, AK.

KEVAN, S.D. AND D.V. WESELOH. 1992. A survey on bird predation at Ontario Trout farms. *Ont. Aquaculture Assn. Newsl.* **1**(7): 1–3.

LAWRIE, A.H. AND J.F. RAHRER. 1973. Lake Superior – a case history of the lake and its fisheries. Gt. Lakes Fish. Comm. Tech. Rep. No. 19.

MCILWRAITH, T. 1886. *The Birds of Ontario*. The Hamilton Association: Hamilton, Ont.

MCKEANE, L. AND D.V. WESELOH. 1993. *Bringing back the bald eagle to Lake Erie*. State of the Environment Fact Sheet No. 93-3. Environment Canada, Ottawa, Ontario.

OGDEN. J.C. 1975. Effects of Bald Eagle territoriality on nesting Ospreys. *Wilson Bull.* **87**: 496–505.

OLENDORFF, R.R., A.D. MILLER AND R.N. LEHMAN. 1981. *Suggested practices for raptor protection on power lines – the state of the art in 1982*. Raptor Research Report No. 4. Raptor Research Foundation.

POOLE, A.F. 1981. The effects of human disturbance on Osprey reproductive success. *Col. Waterbirds* **4**: 20–27.

—. 1989. *Ospreys: A natural and unnatural history*. Cambridge Univ. Press, Cambridge and New York.

POSTUPALSKY, S. 1969. The status of the Osprey in Michigan in 1965. Pp. 338–340 in J.J. Hickey, ed. *Peregrine falcon populations, their biology and decline*. Univ. Wisconsin Press, Madison.

—. 1971. *Toxic chemicals and declining Bald Eagles and Cormorants in Ontario*. Can. Wildl. Serv. (Pesticide Section) Ms. Report No. 20, Ottawa, Ontario.

—. 1977a. Status of the Osprey in Michigan. Pp. 153–166 in J.C. Ogden, ed. *Transactions of the North American Osprey Research Conf., Feb. 1972*. US Dept. Int. Natl. Park Serv., Trans. and Proc. Ser. No. 2.

—. 1977b. A critical review of problems of calculating Osprey reproductive success. Pp. 1–12 in J.C. Ogden, ed. *Transactions of the North American Osprey Research Conf., Feb. 1972*. US Dept. Int. Natl. Park Serv., Trans. and Proc. Ser. No. 2.

—. 1978. Artificial nesting platforms for Ospreys and Bald Eagles. Pp. 35–45 in S.A. Temple, ed. *Endangered birds: Management techniques for preserving threatened species*. Univ. Wisconsin Press, Madison.

—. 1988. Studies of Ospreys, Bald Eagles, and other raptors – 1987. Unpubl. rep. to Michigan Dept. Nat. Resources, E. Lansing.

—. 1989. Osprey. Pp. 297–313 in I. Newton, ed. *Lifetime reproduction in birds*. Academic Press, London.

— and S.M. STACKPOLE. 1974. Artificial nesting platforms for Ospreys in Michigan. *Raptor Res. Rep.* **2**: 105–117.

REESE, J. 1977. Reproductive success of Ospreys in central Chesapeake Bay. *Auk* **94**: 202–221.

RICKLEFS, R.E. 1980. *Ecology (2nd Edition)*. T. Nelson and Sons Ltd., England.
SAUNDERS, W.E. AND E.M.S. DALE. 1933. History and list of birds of Middlesex County, Ontario. *Trans. Roy. Can. Inst.* **24**: 241–314.
SCOTT, F. AND C.S. HOUSTON. 1983. Osprey nesting success in west-central Saskatchewan. *Blue Jay* **41**: 27–32.
SLY, P.G. 1991. The effects of land use and cultural development on the Lake Ontario ecosystem since 1970. *Hydrobiologia* **213**: 1–75.
SOKAL, R. R. AND F.J. ROHLF. 1981. *Biometry*. W.H. Freeman and Co., New York.
SPITZER, P.R., R.W. RISEBROUGH, W. WALKER, R. HERNANDEZ, A. POOLE, D. PULESTON AND I.C.T. NISBET. 1978. Productivity of Ospreys in Connecticut-Long Island increases as DDE residues decline. *Science* **202**: 333–335.
STOCEK, R.F. 1983. Distribution and reproductive success of Ospreys in New Brunswick, 1974–1980. Pp. 215–221 in D.M. Bird, ed. *Biology and management of bald eagles and ospreys*. Harpell Press, Ste. Anne de Bellevue, Quebec.
— AND P.A. PEARCE. 1978. *The Bald Eagle and the Osprey in the maritime provinces*. Can. Wildl. Serv., Wildl. Tox. Div. Rep. No. 37.
TODD, W.E.C. 1940. *The Birds of Western Pennsylvania*. Univ. Pittsburgh Press, Pittsburgh.
WEIR, R. 1987. Osprey. Pp. 108–109 in M.D. Cadman, P.F.J. Eagles and F.M. Helleiner, eds. *Atlas of the breeding birds of Ontario*. Univ. Waterloo Press: Waterloo, Ontario.
WESELOH, D.V., S.M. TEEPLE AND M. GILBERTSON. 1983. Double-crested Cormorants of the Great Lakes: egg-laying parameters, reproductive failure and contaminants in eggs, Lake Huron 1972–73. *Can. J. Zool.* **61**: 427–436.
WESTALL, M.A. 1983. An Osprey population aided by nest structures on Sanibel Island, Florida. Pp. 287–291 in D.M. Bird, ed. *Biology and management of bald eagles and ospreys*. Harpell Press, Ste. Anne de Bellevue, Quebec.
WETMORE, S.T. AND D.I. GILLESPIE. 1976. Osprey and Bald Eagle populations in Labrador and northeastern Quebec, 1969–1973. *Can. Field-Nat.* **90**: 330–337.
WHITFIELD, D.W.A., J.M. GERRARD, W.J. MAHER AND D.W. DAVIS. 1974. Bald Eagle nesting habitat, density, and reproduction in central Saskatchewan and Manitoba. *Can. Field-Nat.* **88**: 399–407.

14

The Osprey in Germany: Its Adaptation to Environments Altered by Man

Bernd-Ulrich Meyburg, Otto Manowsky and Christiane Meyburg

Abstract – As a regular breeding species in Central Europe the osprey is presently confined to the Federal States of Mecklenburg-Vorpommern and Brandenburg in eastern Germany, and Pomerania and Mazuria in Poland. This is probably due to human persecution, especially in earlier decades. In Mecklenburg the population reached its lowest level with only 37 pairs in the DDT-period between 1968 and 1972. In Brandenburg, a slow but steady increase has occurred, from ca. 45–50 pairs in the early 1980s to over 120 pairs today. There has probably been a relationship between contamination with pesticides, reproductive success and population development which, however, has been very poorly studied. One limiting factor for the osprey population may have been the scarcity of suitable trees for nesting. The species prefers the tops of isolated old trees or trees on the edge of the forest dominating the adjacent trees. Due to forestry, such trees have become increasingly rare to the point that only a small fraction of the osprey population can nowadays reproduce in the traditional way. Fortunately, ospreys started to breed on power lines as early as 1938. On the pylons the nests are apparently safer than in trees. Nowadays over 75% of ospreys nest on these artificial structures in Germany, although no such breeding is known in Poland. This important adaptation may have helped the species to recover. The breeding success of 258 tree-nests and 366 nests on power-lines was studied. While the tree-nesting population remained rather stable, the pylon nesters strongly increased. On average, pylon-nesting ospreys produced more young than tree-nesting ospreys.

Key words: osprey; Germany; Poland; breeding success; pylon-nesting; tree-nesting.

The osprey occurs in four races virtually throughout the world with the exception of Antarctica. As a breeder, it is absent only from South America except for the extreme north. The northern boundary of its range coincides with the limit of tall trees.

The osprey in Western Europe was persecuted during the nineteenth and early twentieth centuries, resulting in its virtual extirpation apart from a few isolated localities, e.g. in the Balearics, on Corsica and a small area in Portugal. Spontaneous recolonization has taken place in Scotland since 1954, and recently in Central France.

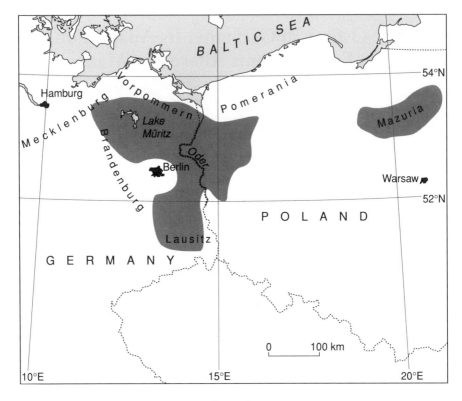

Figure 1.

DISTRIBUTION IN CENTRAL EUROPE

The present distribution of the osprey in Central Europe is very fragmentary, and possibly conditioned by past human persecution. The species has become extinct or eliminated as a regular breeder in several countries: since 1911 in Switzerland, since 1932 in Austria and since the turn of the century in the Czech and Slovak Republics. In the former West Germany it is likewise regarded as extinct but may possibly still breed occasionally.

Regular occurrence in Central Europe today is confined to the lowlands east of the river Elbe. East of the river Oder some 50–60 pairs still breed almost exclusively in the lake districts of Pomerania and Mazuria. In the former East Germany, a relatively significant population was able to survive and is now again on the increase, but with only a slight tendency to spread. Why there are no signs of increase in adjoining Poland, especially in Pomerania and Mazuria, where the species had also sharply declined, remains a mystery.

BREEDING HABITAT AND MANAGEMENT

The osprey needs open and clear water in which to fish. Suitable nesting trees have for decades become increasingly scarce in Germany, so that a lack of nesting trees alone would most likely have led to a decline of the species had not a switch to high tension pylons taken place.

In 1938, for the first time, a nest was found by Rüppel and Rüppel (1938) on a high tension pylon between Angermünde and Templin, north of Berlin. Since then the first German ospreys have taken increasingly to nesting on 100 kW lines (Pehlke 1966, Hemke 1987, Meyburg and Meyburg 1987, Schmidt 1993). Now, around 75% of all pairs do so. Because of the lack of suitable natural trees, this switch to artificial "nest trees" has permitted the osprey population in Germany to maintain itself and increase, all the more so since breeding success among the pylon-nesters is even greater than among tree-nesters (see below). In 1982, for the first time, 20 kW medium voltage pylons were also adopted, but these, due to their configuration, had to be fitted with protective shields so that the birds can safely use the cross-bars as perches. A further chapter opened in 1993, when five nests were built on 380 kW pylons in Brandenburg (Bülow 1994).

The erection of artificial nest-platforms on these pylons has for many years given rise to an additional aspect of management, namely to ensure that special consideration is given to the ospreys during maintenance work on the pylons. Beyond that, any management in the sense of active measures to promote re-colonization of areas where currently the osprey no longer occurs is prevented by the negative attitude of most nature conservationists towards the removal of young birds from nests in the present breeding areas for release elsewhere.

In other parts of the world, ospreys also breed on human-built structures. They readily adopt man-made nests and, where suitable trees or pylons are lacking, will build on platforms fitted to the top of tall poles (Postupalsky 1978). In some regions they also nest on the ground, e.g. Baja California, Mexico and on the Red Sea, and on cliffs, e.g. in Corsica.

In Central Europe today the preference is for well-wooded lakelands. In Mecklenburg-Vorpommern, the Baltic Sea coast, where formerly the Darss Peninsula held the greatest breeding density, has been completely abandoned and so far, apart from one pylon-nesting pair on the island of Rügen (Tusche 1982), has not been recolonized. Nest site and water for fishing can easily be several kilometres apart.

POPULATION DENSITY IN CENTRAL EUROPE

In Central Europe the osprey is capable of forming small, semi-colonial groups. The greatest concentration in Germany was previously on the Darss Peninsula and the Rostock heathlands on the Baltic Coast. In 1925, Peus (1927) counted 15–16 active nests in the 21 000 ha of the Darss. In 1950, 18–20 pairs were known there (Brüll and Kankel in Bijleveld 1974). In 1962, there were still 7–10 pairs and 3 in 1963. By 1970, the species had finally disappeared from this region, once so famous for its birds of prey (Hemke 1984), and also from other parts of the coast.

Presently the osprey breeds in greatest density in the region of Müritz in the federal state of Mecklenburg-Vorpommern and adjacent to the Peitzer fishponds in Lausitz in the federal state of Brandenburg. In terms of districts, the species reaches its highest density in Neustrelitz with 2.3 breeding pairs per 100 km^2. In the districts of Sternberg, Lübz and Uckermünde there are 0.3–0.4 pairs per 100 km^2, and in Röbel, Güstrow and Grimmen 0.1–0.3 (Hauff 1995). In some places, small colonies form and a sequence of several occupied nests can be seen on consecutive pylons. In Brandenburg the overall density is 0.4 pairs per 100 km^2; in its central northern part north of Berlin it reaches 2.8 pairs per 100 km^2 (Sömmer 1994, Ruhle 1994).

In immediately adjacent Poland only 50–60 pairs breed today, predominantly in Mazuria and Pomerania (Mizera and Szymkiewicz 1995). To date there has been no clear sign of a population increase, as in Germany. Also, the ospreys in Pomerania and Mazuria show little inclination to nest on pylons (Mizera 1994).

POPULATION TRENDS

The osprey was formerly widespread in Central Europe (Schmidt 1994, Schmidt and Kapfer 1994). In the former West Germany the species must be now regarded as extinct, or at best a very sporadic breeder (Ringleben 1966, Schäfer 1967, Thielcke 1975, Heller 1984). In the former East Germany, where the species breeds in appreciable numbers only in the new federal states of Mecklenburg-Vorpommern and Brandenburg, the population in 1991 was estimated to be around 170 (± 20) pairs (Nicolai 1993). By now it is well over 200 pairs. At present, there are only a few pairs in the neighbouring federal states: Niedersachsen (Lower Saxony) 1, Sachsen (Saxony) 0–4, Thüringen 2, Sachsen-Anhalt 4.

The population status and increase in Germany have been particularly well documented in Mecklenburg-Vorpommern and rather less so in Brandenburg. The majority of the ospreys are concentrated in the southeastern part of the Mecklenburg/Brandenburg lake district. A further nucleus has come into being in southeast Brandenburg. Here new colonies have developed, especially during the 1970s. In all, Brandenburg, where about half of all the ospreys breed, has shown a marked population increase during the last 25 years, from ca. 45–50 pairs in the early 1980s to over 120 in 1993.

In 1935 there were not more than 25 breeding pairs in Mecklenburg-Vorpommern, of which 14 alone were in the Darss Peninsula on the Baltic. Up to 1957 – probably due to absence of hunting – the number had risen to 70 breeding pairs. Between 1960 and 1969 the Mecklenburg population sharply declined, to reach a low level of 39–41 pairs between 1970 and 1975. It was at that time that the population on the Darss, once the greatest concentration in Germany, was extirpated.

Ospreys breed only irregularly on the Baltic coast, but in 1979 an artificial nest platform on the island of Rügen was occupied, forming a genuine case of resettlement. One cause among others of the steep decline on the Baltic coast may have been water pollution and the annual campaign to control mosquitoes by spraying insecticides from the air (Klafs 1991).

After 1976 the Mecklenburg population slowly increased, with 62 pairs in 1980, 73 in 1986, 90 in 1989 (Klafs 1991) and 94 in 1993 (Köhler 1994). This was also reflected in the former district of Schwerin, on the western edge of the species' range, where the number of pairs rose from 5 in the mid-1970s to 18 in 1986 (Hauff *et al.* 1986, Köhler 1991).

Around 1870 the osprey was evidently a frequent bird in most parts of Brandenburg where it even bred in colonies, e.g. Dubrow with 8–10 pairs, and the Peitzer fishponds with 25–30 pairs. The subsequent rapid decline continued into the 1920s. During the 1930s and '40s there was apparently a remarkable recovery of the population which, however, was not adequately documented. From about 45 pairs around 1980 the population rose to at least 120 pairs in 1992.

PESTICIDES AND BREEDING SUCCESS

During the 1950s and 1960s in the USA a sharp decline in the reproduction rate and hence in the osprey population was brought about by DDT and other organochlorine pollutants, followed by a renewed increase after those pesticides were banned. The osprey's important function as a bioindicator, impressively demonstrated in North America (Ames 1966, Wiemeyer *et al.* 1975, Spitzer *et al.* 1977, 1978), can only be presumed true for Central Europe. During the cold-war years research of this kind could not be pursued. With regard to the use of agricultural chemicals, their negative effect on the environment was denied by communist regimes. Nevertheless, in the former East Germany 19 eggs that had failed to hatch were analysed for organochlorine residues between 1978 and 1981, and yielded an average of 4.5 ppm (maximum 10.8 ppm) of DDT, so that Poole's statement (1989a: 176) that "most European ospreys escaped the trauma of pesticides" is probably not correct, for Germany at least.

The impact of pesticides was also reflected in the brood size, which fell from 2.2 in 1959 to 0.9 in 1966 in Mecklenburg (Moll 1967). Also in Mecklenburg, between 1956 and 1976 more clutches failed to develop and mean brood size was smaller (around 0.8) than during the periods 1932–37 and 1976–90 (Banzhaf 1938, Klafs 1991).

If no similar decline in breeding success was detected in Brandenburg (Feiler in Rutschke 1987, Loew 1981), it was probably because two-thirds of the broods examined were from regular and highly productive pairs (Klafs 1991). Some

Table 1. Terminology of osprey breeding success used.

A = Number of occupied nests with known outcome (= Number of pairs on territory, regardless of whether or not they lay)
B = Number of active nests (in which at least one egg was laid)
C = Nest success (fledged 1, 2, 3 or 4 young)
D = Mean number of young fledged per occupied nest
E = Mean number of young fledged per active nest

individual pairs continued to breed well even during the bad years of the 1950s and '60s so that these pairs should be separated from the rest when calculating the impact of toxic chemicals. Those pairs were probably breeding in areas scarcely affected by pesticides.

OUR OWN STUDY OF BREEDING SUCCESS

Our study covers a relatively long period of time and a large number of broods, thus making it possible to evaluate the correlation between breeding success and population growth on the one hand and comparison between tree- and pylon-nesting on the other.

All observations were carried out from a considerable distance and from the ground, using binoculars or a spotting scope. The nest sites were first inspected in April or early May, to determine whether they were occupied. Following Postupalsky (1977) we considered a territory to be occupied if either one adult bird was lying flat in the nest, or if two adults were present at the nest, or if a nest was clearly in use, e.g. fresh nest material. Breeding success was monitored over several days in July, to determine the number of fledglings. Despite careful observation, it is nevertheless possible that young were overlooked. No attempt was made to climb up to nests. Tree-nesters were monitored by O. Manowsky and pylon-nesters by B.-U and C. Meyburg. See Table 1 for the basic terminology for reproductive success.

Among tree-nesters, there was a relatively stable population in the Schorfheide, an extensive wooded area largely unspoiled and now a biosphere reserve, lying 50 km north from the centre of Berlin. Up to 1988 ospreys nested exclusively on trees. In 1989 new pairs began to nest on pylons. No evidence was found of pairs alternating between trees and pylons.

The increase in the number of annually controlled pylon nests north of Berlin reflects only a part of a general increase in the number of pairs. This is primarily explained by coverage of an ever wider area and more intensive search.

Comparative analysis of the breeding results from the total number of 624 nests monitored (see Tables 2a–3b) provides support to the hypothesis that pylon-nesters are on average more successful than tree-nesters. From this it emerges that the pylon-nests provide greater security from natural enemies, or that both eggs and young risk falling from tree nests due to the swaying of the

Table 2a. The breeding success of ospreys in Germany nesting on trees (1972–83).

Category*	1972	1973	1974	1975	1976	1977	1978	1979	1980	1981	1982	1983
A*	9	9	10	10	10	11	10	12	9	10	9	7
B	9	9	10	10	10	11	10	10	9	10	9	7
C1	2	1	3	2	4	1	0	0	0	6	0	1
2	3	3	2	2	4	5	3	2	5	2	3	2
3	3	4	3	5	0	2	2	0	4	0	1	1
D	1.89	2.11	1.60	2.10	1.20	1.55	1.20	0.33	2.44	1.00	1.00	1.14
E	1.89	2.11	1.60	2.10	1.20	1.55	1.20	0.40	2.44	1.00	1.00	1.14

*See Table 1.

Table 2b. The breeding success of ospreys in Germany nesting on trees (1984–93).

Category*	1984	1985	1986	1987	1988	1989	1990	1991	1992	1993	Total
A*	11	13	12	14	14	17	17	15	15	14	258
B	10	10	9	11	12	15	14	12	14	11	232
C1	4	0	1	5	2	3	3	2	1	1	42
2	1	1	2	3	3	3	5	4	6	3	67
3	1	4	0	1	4	4	3	5	4	4	55
D	0.82	1.08	0.42	1.00	1.43	1.24	1.29	1.67	1.67	1.36	1.32
E	0.90	1.40	0.55	1.27	1.66	1.40	1.57	2.08	1.78	1.73	1.47

*See Table 1.

trees. It had been shown before in North America (Postupalsky 1978, Poole 1989b) that pairs breeding on artificial structures are more successful than tree nesters. However, confirmation had been lacking for Europe where the artificial "nesting trees" were, in contrast to the New World, not specifically built for the birds in most cases.

A higher loss from natural causes was in fact evident among the tree-nesting pairs that we studied. Several young were killed by goshawks and even one case was recorded of an adult bird being struck. Other losses were due to falling from the nest or to the whole nest being blown down.

Confirmation of this hypothesis reveals that adaptation to pylons benefits the osprey in two ways, i.e. by compensating for the lack of natural nest-trees, and leading to an improved reproduction rate.

A comparison of the mean number of young fledged per occupied nest between tree-nesters and pylon-nesters from 1980 to 1993 is shown in Fig. 2. The total number of pylon-nesters (pairs on territory) and tree-nesters under study was 366 and 258, respectively. At least 334 pylon-nesting and 232 tree-nesting pairs proceeded to egg-laying. The number of young fledged per occupied nest was 165 with pylon-nesters and 1.32 with tree-nesters. Brood size was almost similar – 2.22 and 2.08, respectively. Among tree-nesters no successful broods of four were recorded. The number of unproductive nests was 94 in both populations (25.7% vs. 36.4%, respectively) and the number of

Table 3a. The breeding success of ospreys in Germany nesting on power lines (1978–86).

Category*	1978	1980	1981	1982	1983	1984	1985	1986
A*	3	8	12	17	18	19	19	21
B	3	8	10	15	17	17	19	19
C1	2	2	4	0	3	4	3	1
2	1	2	5	6	12	5	8	12
3	0	3	1	5	1	4	5	5
4	0	0	0	0	1	0	1	0
D	1.33	1.87	1.42	1.59	1.89	1.37	2.00	1.90
E	1.33	1.87	1.70	1.80	2.00	1.53	2.00	2.10

*See Table 1.

Table 3b. The breeding success of ospreys in Germany nesting on power lines (1988–93).

Category*	1988	1989	1990	1991	1992	1993	Total
A*	29	33	43	47	42	55	366
B	26	33	37	39	39	52	334
C1	1	5	2	0	5	9	41
2	10	16	16	10	14	20	137
3	10	9	14	9	13	9	88
4	1	0	0	1	0	2	6
D	1.89	1.94	1.77	1.08	1.71	1.53	1.65
E	2.12	1.94	2.05	1.31	1.85	1.62	1.81

*See Table 1.

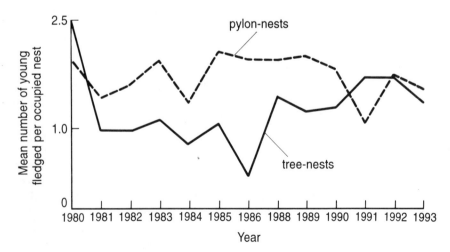

Figure 2. Breeding success of ospreys in Germany (1980–93).

unsuccessful nests was 62 (16.9%) vs. 68 (26.4%). Supporting the hypothesis that nests on pylons offer greater security, 18.5% of clutches and broods were lost among pylon-nesters and 29.3% among tree-nesters.

The tree-nesting population maintained its size at a fairly constant level over 21 years. This must similarly have been the case with tree-nesters in other regions, whereas a notable increase was evident among the pylon-nesters which still continues. This was conditioned by an overall increase in breeding success, probably following the reduction in use of pesticides and adaptation to nesting on different medium and high tension pylons which made more effective management possible. This has ultimately led to an expansion, admittedly only slight to date, of the breeding range westwards and southwards in Germany. Provided that environmental conditions remain unchanged, it may be expected that further areas west of the present breeding range will be gradually recolonized. In adjacent breeding areas in Pomerania and Masuren (present-day Poland), no such adaptation to nesting on pylons has taken place and thus, no increase in the rather small population. The reasons for this difference remain unknown.

REFERENCES

AMES, P.L. 1966. DDT residues in the eggs of the Osprey in the northeastern USA and their relation to nest success. *J. Appl. Ecol.* 3: 87–97.

BANZHAF, W. 1938. Naturdenkmäler aus Pommerns Vogelwelt. 2. Der Fischadler (*Pandion haliaetus* (L.)). *Dohrniana* 17: 74–79.

BIJLEVELD, M. 1974. *Birds of Prey in Europe*. London & Basingstoke: Macmillan Press.

BÜLOW, D. 1994. Schutzmaßnahmen für auf Hochspannungsmasten brütende Fischadler. *Rundbrief Weltarbeitsgr. Greifvögel* 19/20.

HAUFF, P. 1995. Der Fischadler *Pandion haliaetus* in Mecklenburg-Vorpommern. In B.-U. Meyburg and R.D. Chancellor, eds. *Eagle studies*. Berlin, London and Paris: World Working Group on Birds of Prey.

—, W. KÖHLER AND W. MEWES. 1986. Zur Bestandsentwicklung einiger vom Aussterben bedrohter Vogelarten im Bezirk Schwerin. *Naturschutzarb. Mercklenb.* 29: 71–76.

HELLER, M. 1984. Mehrjähriges Auftreten desselben Fischadlers *Pandion haliaetus* mit Paarbildung und Ansiedlungsversuchen im NSG Aalkistensee/Württemberg. *Anz. Orn. ges. Bayern* 23: 225–231.

HEMKE, E. 1984. Geschütze heimische Tiere. Der Fischadler (*Pandion haliaetus*). *Kulturinformationen Kreis Eberswalde* 1/84: 15–18.

—. 1987. Fischadler auf Hochspannungsmasten. *Falke* 34: 256–259.

KLAFS, G. 1991. Die Bestandsentwicklung des Fischadlers *Pandion haliaetus* in Mecklenburg-Vorpommern unter populationsökologischen Gesichtspunkten. *Populationsökol. Greifvogel- u. Eulenarten* 2: 183–192.

—. 1991. Zur Populationsdynamik des Fischadler (*Pandion haliaetus* L.) im Bezirk Schwerin. *Populationsök. Greifvogel- u. Eulenarten* 2: 193–197.

KÖHLER, W. 1994. Der Brutbestand des Fischadlers in Mecklenburg-Vorpommern. Pp. 1–2 in D. Schmidt, ed. *Fischadler in Mitteleuropa*. Intern. Fachtagung. Singen: ILN.

LOEW, M. 1981. Zum Brutbestand und zum Schutz der vom Aussterben bedrohten Adler im Bezirk Potsdam. *Mitt. d. Bezirksarbeitsgr. "Artenschutz"* **2**: 17–24.
MEYBURG, B.-U. AND C. MEYBURG. 1987. Der Fischadler (*Pandion haliaetus*) als Brutvogel in Mitteleuropa. Sitzungsber. *Ges. naturf. Freunde Berlin* (N.F). **27**: 34–41.
MIZERA, T. 1994. Warum ist der Fischadler (*Pandion haliaetus*) ein seltener Brutvogel in Polen? Pp. 19–20 in D. Schmidt, ed. *Fischadler in Mitteleuropa*. Intern. Fachtagung. Singen: ILN.
— AND M. SZYMKIEWICZ. 1995. The present status of the Osprey *Pandion haliaetus* in Poland. In B.-U. Meyburg and R.D. Chancellor, eds. *Eagle studies*. Berlin, London and Paris: World Working Group on Birds of Prey.
MOLL, K.-H. 1967. Der Fischadler. *Falke* **14**: 134–135.
NICOLAI, B. 1993. *Atlas der Brutvögel Ostdeutschlands*. Jena & Stuttgart: G. Fischer Verlag.
PEHLKE, G. 1966. Fischadler auf "eisernen Bäumen". *Naturschutzarb. Mecklenb.* **9**: 42.
PEUS, F. 1927. Vom Fischadler. *Beitr. Fortpflanzungsbiol. Vögel* **3**: 120–122.
POOLE, A.F. 1989a. *Ospreys: A natural and unnatural history*. Cambridge: Cambridge Univ. Press.
—. 1989b. Regulation of Osprey (*Pandion haliaetus*) populations: the role of nest site availability. Pp. 227–234 in B.-U. Meyburg and R.D. Chancellor, eds. *Raptors in the modern world*. Berlin, London and Paris: World Working Group on Birds of Prey.
POSTUPALSKY, S. 1977. A critical review of problems in calculating Osprey reproductive success. Pp. 1–11 in J.C. Ogden, ed. *Transactions of the North American Osprey Research Conference*. US Dept. Int. Nat. Park Serv.
—. 1978. Artificial nesting platforms for Ospreys and Bald Eagles. Pp. 35–45 in S.A. Temple, ed. *Endangered birds: management techniques for preserving endangered species*. Madison: Univ. Wisconsin Press.
RINGLEBEN. H. 1966. Der Fischadler als Brutvogel in Niedersachsen. *Ber. naturh. Ges. Hannover* **110**: 67–76.
RÜPPEL. W. AND L. RÜPPEL. 1938. Fischadlerhorst auf einem eiserenen Gittermast. *Orn. Mber.* **46**: 138–142.
RUHLE, D. 1994. Schutz und Bestandsentwicklung des Fischadlers in der Niederlausitz, Brandenburg. Pp. 3–6 in D. Schmidt, ed. *Fischadler in Mitteleuropa*. Intern. Fachtagung. Singen: ILN.
RUTSCHKE, E. 1987. *Die Vogelwelt Brandenburgs*. 2. Aufl. Jena: G. Fischer Verlag.
SCHÄFER, K. 1967. Der Fischadler wieder Brutvogel in Westfalen? *Falke* **14**: 422–423.
SCHMIDT, D. 1993. *Zur Nisthabitatstruktur des Fischadlers (Pandion haliaetus) in Mittel- und Nordwesteuropa*. Diplomarbeit: Univ. Freiburg.
—. 1994. Zur historischen Brutverbreitung des Fischadlers (*Pandion haliaetus*) in Westdeutschland. Pp. 15–18 in D. Schmidt, ed. *Fischadler in Mitteleuropa*. Intern. Fachtagung. Singen: ILN.
— AND A. KAPFER. 1994. Fischadler in Mitteleuropa – Bericht über eine internationale Fachtagung. *Ber. z. Vogelsch.* **32**: 103–106.
SÖMMER, P. 1993. Zur Situation des Fischadlers (*Pandion haliaetus*) in Brandenburg. Pp. 7–12 in D. Schmidt, ed. *Fischadler in Mitteleuropa*. Intern. Fachtagung. Singen: ILN.
SPITZER, P.R., R.W. RISEBROUGH, J.W. GRIER AND C.R. SINDELAR. 1977. Eggshell thickness – pollutant relationship among North American Ospreys. Pp. 13–19 in J.C. Ogden, ed. *Transactions of the North American Osprey research conference*. US Dept. Int. Nat. Park Serv.
—, —, W. WALKER, R. HERNANDEZ, A. POOLE, D. PULESTON AND I.C.T. NISBET. 1978. Productivity of ospreys in Connecticut-Long Island increases as DDE residues decline. *Science* **202**: 333–335.
THIELCKE, G. 1975. *Das Schicksal der Greifvögel in der Bundesrepublik Deutschland*. Greven: Kilda-Verlag.

TUSCHE, W. 1982. Der Fischadler ist Brutvogel auf Rügen. *Naturschutzarb. Meckl.* **25**: 41–42.
WIEMEYER, S.N., P.R. SPITZER, W.C. KRANTZ, T.G. LAMONT AND E. CROMARTIE. 1975. Effects of environmental pollutants on Connecticut and Maryland Ospreys. *J. Wildl. Manage.* **39**: 124–139.

15

Effectiveness of Artificial Nesting Structures for Ferruginous Hawks in Wyoming

James R. Tigner, Mayo W. Call and
Michael N. Kochert

Abstract – Between fall 1987 and fall of 1992, 71 artificial nesting structures (ANSs) were erected for ferruginous hawks in southcentral Wyoming. Each year between 11 and 41 ANSs were used by nesting hawks. Of pairs laying eggs on the platforms, 80–100% successfully fledged young, producing an average of 1.9 to 3.7 young per laying pair. Nesting success and young fledged per laying pair were significantly higher on the artificial structures than on natural substrates. No difference was observed with brood size at fledging between the two substrates. Management benefits of ANSs for ferruginous hawks include enhanced nesting opportunities and increased productivity.

Key words: artificial nesting structures; ferruginous hawk; Wyoming; nesting success; productivity.

The ferruginous hawk, the largest North American *Buteo*, or soaring hawk, ranges west of a line from southern Manitoba through the Dakotas and southward to Texas (Olendorff 1993). This hawk is the least numerous of all the buteos, with perhaps no more than 10 000 to 15 000 birds in existence (Olendorff 1993). It is a species of high national concern and is currently listed a Category II Species by the US Fish and Wildlife Service (US Fish and Wildl. Serv. 1992). When more definitive information on population status becomes available, it could be listed as either Threatened or Endangered.

Wyoming is the approximate center of the ferruginous hawk breeding range and has one of the largest nesting populations of any state or province (Olendorff 1993). Specifically, the Shamrock Hills, about 13 km west of Rawlins has one of the highest nesting densities of these birds within Wyoming (Call 1988, 1989). Because of the relatively high number of pairs (approximately 35) that annually nest within or near the Shamrock Hills, the US Bureau of Land Management (BLM) designated this area as an Area of Critical Environmental Concern (ACEC).

Some investigators believe that the ferruginous hawk has declined in certain parts of its range (see Olendorff 1993). Lockhart (pers. comm.) felt that the number of nesting ferruginous hawks declined in southern Wyoming and northern Colorado during the late 1980s. The exact reasons for this decline are not known, but, as in other areas of the hawk's range, both habitat alteration and human disturbance are believed to have been major factors (White and Thurow 1985, Olendorff 1993). This hawk is considered to be sensitive to human disturbance and will abandon nests if disturbed, particularly before eggs hatch (Smith and Murphy 1973, Olendorff 1973, 1993, Howard 1975, Powers et al. 1975).

Prior to 1987, BLM biologists noted numerous nesting failures of ferruginous hawks on natural substrates throughout southcentral Wyoming (BLM unpubl. data). These failures resulted from destroyed eggs and missing or dead chicks as a consequence of environmental or human disturbance. Concern about the impacts of increased human disturbance, i.e. recreationists and livestock and energy developments, and adverse natural factors on ferruginous hawk productivity prompted wildlife managers to implement action to enhance this species. Therefore in 1987, BLM staff erected 10 experimental artificial nesting structures (ANSs) 64 km northwest of Rawlins, Wyoming. They built 10 more in 1988, six in 1989, four in 1990, and five in 1991. By 1993, 38 ANSs were available for nesting. Also, as part of the mitigating measures for a proposed coal gasification plant near the Shamrock Hills, Energy International Inc., Pittsburgh, Pennsylvania constructed 30 ANSs (16 in 1987 and 14 in 1988) primarily along the western side of the Shamrock Hills and began a nest monitoring program. In March 1989, the US Air Force also constructed four additional ANSs west of Shamrock Hills.

This paper presents results of efforts to enhance ferruginous hawk nesting habitat in southcentral Wyoming by using artificial nesting platforms. It also compares nesting success and productivity from hawks nesting on the platforms with those nesting on natural substrates and assesses the effectiveness of these platforms as a management tool.

STUDY AREAS AND METHODS

The Great Divide Resource Area (GDRA) of the Rawlins District, BLM is diverse in physiography and vegetation. The area is dominated by big sagebrush, but saltbush (*Atriplex* spp.) and greasewood are common in alkaline areas (BLM unpubl. data). The southern part of the GDRA has extensive quaking aspen stands, but coniferous forests are limited. Rocky outcrops, erosion remnants, and cliffs occur throughout the area and are used by a variety of raptors for nesting (BLM unpubl. data). Ferruginous hawks nest on trees such as junipers (*Juniperus* spp.), large shrubs such as serviceberry (*Amelanchier* spp.), big sagebrush, rock outcrops or pillars, and even on the ground. The Shamrock Hills ACEC lies on the western flank of the Rawlins Uplift which is a large northwest trending anticlinal fold separating the Hanna Basin on the east from the Great Divide Basin on the west (Call 1988, 1989). The sandstones are often resistant to weathering and stand as ridges. These ridges offer good nesting habitat for ferruginous hawks; however, in the surrounding flatlands they nest in large shrubs or on the ground (Call 1988, 1989).

ANSs consisted of a wooden framework approximately 1 m by 1 m attached to the end of a 4 m pressure-treated post set approximately 1 m in the ground (Fig. 1). Tree branches were wired and nailed to the posts to simulate trees that ferruginous hawks commonly use for nesting. Branches were also wired to the nest framework to help strengthen it and extended above the nests to help hold nest material, prevent the young from falling off, and possibly deter aerial predators.

Areas were searched for nests on foot, from vehicles, and from a helicopter

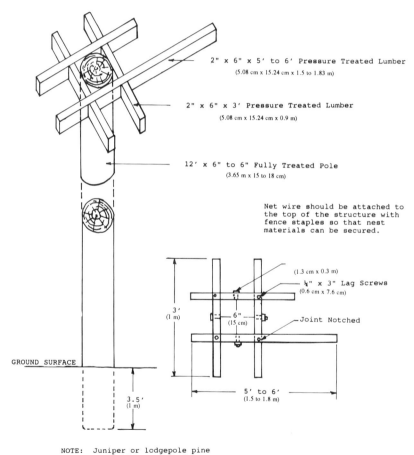

Figure 1. Ferruginous hawk artificial nesting structure.

Table 1. Occupancy, success, and productivity of ferruginous hawks nesting on artificial nesting structures in the Rawlins BLM District, Wyoming, 1988–93.

Year	ANSs[a] (N)	Number of laying pairs	Number successful	Total fledged	Number fledged/ laying pair	Number fledged/ successful pair
1988	25	11	11	38	3.45	3.45
1989	54	34	30	65	1.91	2.16
1990	61	33	31	83	2.51	2.67
1991	65	41	33	83	2.02	2.51
1992	71	37	37	100	2.70	2.70
1993	71	29	26	62	2.14	2.38
All yr	347[b]	185	168	431	2.46	2.64

[a]ANSs – Artificial Nesting Structures.
[b]Total number of structures available.

(only in 1988 for natural substrates). Natural substrate nests in and near the Shamrock Hills were monitored at least three times: early May, early June, and late June or early July. Because of the distance involved (some trips involved 400 km/day), monitoring of natural nests away from the Shamrock Hills usually involved two visits; one in May and a second in late June. ANSs were also monitored at least three times: early May, late May or early June, and late June or early July. First visits were to assess egg-laying, middle visits to age young, and because ferruginous hawks fledge about the first week of July, last visits were to determine productivity. Structures were examined by standing on vehicles top using ladders, or a pole mirror (Steenhof 1987). If possible, nestlings were counted from a distance using binoculars or a spotting scope.

A breeding attempt, i.e. egg-laying, was confirmed if a nest contained an incubating adult or eggs (Steenhof and Kochert 1982). A breeding attempt was considered successful if one or more chicks reached 33 d of age (Steenhof 1987). Moritsch's (1985) photographic guide was used to determine the approximate age of the nestlings at each visit. Although we counted eggs in early May of 1987 and 1988, we avoided lengthy disturbances. In subsequent years we observed incubating birds from a distance and did not disturb them until after hatch.

RESULTS

Between 1988 and 1993, 25–71 ANSs were available for nesting, totaling 347 nesting opportunities (Table 1). Each year between 11 and 41 ferruginous hawk pairs laid eggs on the platforms (using 41–63% of the available platforms). Annual proportion of laying pairs successful on the structures ranged from

Table 2. Success and productivity of ferruginous hawks nesting on natural substrates, Rawlins BLM District, Wyoming 1987–91.

Year	Number of laying pairs	Number successful (%)	Number of young fledged	Number fledged/ laying pair	Number fledged/ successful pair
1987[a]	31	14 (45)	38	1.22	2.71
1988[a]	8	6 (75)	10	1.25	1.67
1989[a]	16	9 (56)	25	1.56	2.78
1990[b]	25	22 (88)	41	1.64	1.86
1991[b]	22	17 (77)	38	1.72	2.23
All yr	102	68 (68)	152	1.49	2.25

[a]Shamrock Hills only.
[b]Shamrock Hills plus adjacent areas.

80.4–100%. Productivity ranged from 1.91–3.45 young fledged per laying pair and 2.16–3.45 young per successful pair (Table 1). During the study 185 breeding attempts were recorded from the 347 nesting opportunities (53%). Of these attempts 171 (91%) successfully produced 431 young (Table 1). We observed almost no nestling mortality on the ANSs. Only three young hawks were known to have died after hatching on three artificial nests.

The number of laying pairs observed on natural substrates ranged from eight to 31 between 1987 and 1991 (Table 2). The annual proportion of successful egg-laying pairs on the natural substrates ranged from 45–88%. Annual productivity ranged from 1.22–1.72 young fledged per laying pair and 1.67–2.78 young per successful pair (Table 2).

Between 1988 and 1991, the mean proportion of successful laying pairs was significantly higher for pairs on the ANSs (90.6%) than those on natural substrates (74.0%; $G^0 = 4.68$; $P = 0.04$). Also, the mean number of young fledged per laying pair was significantly higher in the ANSs (2.4) than in the natural substrates (1.5; Mann-Whitney $U = 0.0001$; $P = 0.02$). In contrast, the mean number of young fledged per successful attempt was similar between substrates (ANS = 2.7 and natural substrates = 2.2; Mann-Whitney $U = 4.000$; $P = 0.24$).

By 1990 when all platforms were in place, 70% of the nesting ferruginous hawks found in the Shamrock Hills were in the ANSs; by 1991 this proportion had risen to 78% (Table 3). Although the total number of nesting pairs found fluctuated greatly over the years, the number using the platforms remained relatively stable, fluctuating between 11 and 14 after 1989. We suspect that some pairs may have been attracted from natural substrate sites to the ANSs. For example, in 1989 at least three ferruginous hawk pairs nested on ANSs within the probable territories of pairs that nested on natural substrate sites in 1988. These ANSs contained the only nesting pairs found in 1989 within one km of the previously used natural substrate sites.

Table 3. Number of ferruginous hawk pairs found nesting on artificial nesting structures and natural substrates in the Shamrock Hills, Wyoming 1987–93.

Year	Number of laying pairs[a]		
	ANSs[a]	Natural	Total
1987	0[b]	31	31
1988	4	8	12
1989	14	16	30
1990	14	6	20
1991	14	4	18
1992	13	6	19
1993	11	8	19

[a] ANS – Artificial Nesting Structure.
[b] No ANS were available during the 1987 nesting season.

DISCUSSION

Olendorff and Stoddard (1974), Schmutz et al. (1984), and Olendorff (1993) suggested that artificial nesting platforms could be used to enhance nesting populations of open-country raptors. The artificial nesting structures we studied provided a more secure nesting substrate that resulted in higher productivity than the natural nesting sites. Schmutz et al. (1984) and Steenhof et al. (1993) also observed increased nesting success and productivity of ferruginous hawks nesting on artificial nesting platforms in Alberta and on transmission towers in Idaho.

A significantly higher proportion of hawk pairs successfully nested on the ANSs than on natural sites, but no difference occurred in brood size at fledging between substrates. Lower productivity at the natural sites probably resulted from complete failures of breeding attempts rather than brood reduction. This suggestion is supported by the low incidence of nestling mortality we observed at the ANSs. Most of the natural substrate nests in the study area appeared to be accessible to terrestrial predators, and predation was suspected for most nest destruction (Call 1988, 1989). Nesting platforms seem to provide protection from those factors which tend to cause total nesting failures, i.e. complete loss of clutches or broods, such as predation, human disturbance, and adverse weather. Ferruginous hawk nesting attempts in nests that are accessible or exposed to these factors tend to fail more often than inaccessible or protected nests (Lokemoen and Dubbert 1976, Schmutz et al. 1984, Roth and Marzluff 1989, Steenhof et al. 1993).

Artificial nesting structures can either provide nesting opportunities where none were available or attract birds from natural substrates (Schmutz et al.

1984, Steenhof *et al.* 1993). The total number of nesting pairs found in the Shamrock Hills did not appear to increase as a result of the structures, and we feel that the structures did attract some of the birds that would have normally nested on natural substrates. Also, the platforms may provide stable nesting substrates. The number of pairs found nesting on the ANSs in the Shamrock Hills did not fluctuate appreciably.

Management benefits of ANSs for ferruginous hawks include enhanced nesting opportunities, increased productivity, and, as reported by Schmutz *et al.* (1984), increased nesting population density. The structures also provide nesting opportunities for other species such as golden eagles, which successfully nested on ANSs during the study (BLM unpubl. data). An important management implication of ANSs is mitigation of various resource developments. ANSs may provide secure and potentially more productive nesting sites which may compensate for any potential losses. Also nests can be moved from disturbances or development projects and placed on ANSs (Postovit and Postovit 1987). Nests in the Rawlins BLM District were moved from oil tanks, windmills, and other hazardous locations (BLM unpubl. data). Unless the status of the ferruginous hawk changes to Threatened or Endangered, the main management application of ANSs will probably be mitigation of development activity.

ACKNOWLEDGMENTS

The study was funded by the US Bureau of Land Management and Energy International, Inc. We thank K. Steenhof for assistance with the statistical analysis. J.K. Schmutz and K. Steenhof made many helpful comments on the manuscript.

REFERENCES

CALL, M.W. 1988. Ferruginous hawk monitoring and construction of artificial nests in the Shamrock Hills of Carbon County, Wyoming, Annu. Rep. Energy International, Inc., Pittsburgh, Pennsylvania, USA.

—. 1989. Raptor monitoring in the Shamrock Hills: a 1989 monitoring study. Prog. Rep. Energy International, Inc., Pittsburgh, Pennsylvania, USA.

HOWARD, R.P. 1975. Breeding ecology of the ferruginous hawk in northern Utah and southern Idaho. M.Sc. thesis, Utah State Univ., Logan, Utah, USA.

LOKEMOEN, J.T. AND H.F. DUEBBERT. 1976. Ferruginous hawk nesting ecology and raptor ecology and raptor populations in northern South Dakota. *Condor* 78: 464–470.

MORITSCH, M.Q. 1985. Photographic guide for aging ferruginous hawks. US Bureau of Land Manage., Boise, Idaho, USA.

OLENDORFF, R.R. 1973. *The ecology of the nesting birds of prey of northeastern Colorado.* Tech. Report No. 211, Grassland Biome, US Inter. Biol. Prog. Fort Collins, Colorado USA.

—. 1993. *Status, biology, and management of ferruginous hawks: a review*. Raptor Res. and Tech. Asst. Cen. Spec. Pub., US Department of the Interior, Bureau of Land Manage., Boise, Idaho, USA.
— AND J.W. STODDART, JR. 1974. The potential for management of raptor populations in western grasslands. Pp. 47–88 in F.M. Hamerstrom, B.E. Harrell and R.R. Olendorff, eds. *Management of raptors*. Raptor Res. Rep. No. 1. Raptor Res. Found., Vermillion, South Dakota, USA.
POSTOVIT, H.R. AND B.C. POSTOVIT. 1987. Impacts and mitigation techniques. Pp. 183–213 in B.A. Giron Pendleton, B.A. Millsap, K.W. Kline and D.M. Bird, eds. *Raptor management techniques manual*. Natl. Wildl. Fed., Washington, D.C. USA.
POWERS, L.R., R. HOWARD AND C.H. TROST. 1975. Population status of the ferruginous hawk in southwestern Idaho and northern Utah. Pp. 153–157 in J.R. Murphy, C.M. White and B.E. Harrell, eds. *Population status of raptors*. Raptor Res. Rep. No. 3, Raptor Res. Found., Vermillion, South Dakota, USA.
ROTH, S.D. JR. AND J.M. MARZLUFF. 1989. Nest placement and productivity of ferruginous hawks in western Kansas. *Trans. Kans. Acad. Sci.* **92**: 132–148.
SCHMUTZ, J.K., R.W. FYFE, D.A. MOOR AND A.R. SMITH. 1984. Artificial nests for ferruginous hawks and Swainson's hawks. *J. Wildl. Manage.* **48**: 1009–1013.
SMITH, D.G. AND J.R. MURPHY. 1973. Breeding ecology of raptors in the eastern Great Basin of Utah. Brigham Young Univ. Sci. Bull., Biol. Series **13**: 1–76. Provo, Utah, USA.
STEENHOF, K. 1987. Assessing raptor reproductive success and productivity. Pp. 157–170 in B.A. Giron Pendleton, B.A. Millsap, K.W. Kline and D.M. Bird, eds. *Raptor management techniques manual*. Natl. Wildl. Fed., Washington, D.C. USA.
— AND M.N. KOCHERT. 1982. An evaluation of methods used to estimate raptor nesting success. *J. Wildl. Manage.* **46**: 885–893.
—, — AND J.A. ROPPE. 1993. Nesting by raptors and ravens on electrical transmission line towers. *J. Wildl. Manage.* **57**: 271–281.
US FISH AND WILDLIFE SERVICE. 1992. Endangered and threatened wildlife and plants: Notice of finding on petition to list the ferruginous hawk. Federal Register 57(161): 37507–37513. Washington, D.C. USA.
WHITE, C.M. AND T.L. THUROW. 1985. Reproduction of ferruginous hawks exposed to controlled disturbance. *Condor* **87**: 14–22.

16

Peregrine Falcons: Power Plant Nest Structures and Shoreline Movements

Gregory A. Septon, John Bielefeldt,
Tim Ellestad, Jim B. Marks and
Robert N. Rosenfield

Abstract – Along the western shoreline of Lake Michigan, nest boxes were erected on man-made structures, including power plants, as potential nest sites for peregrine falcons. Successful nesting has occurred at power plants and other man-made sites. Nest boxes were likewise erected on similar man-made structures along the Upper Mississippi River. With power plants relatively frequent along shorelines that were historically and are currently used for nesting, migration and other movements, these shoreline structures may provide additional opportunities for nesting peregrines in Wisconsin and elsewhere in the Upper Midwest.

Key words: peregrine falcon; nest box; power plants; shoreline; Midwest.

In Wisconsin, peregrine falcons traditionally migrated along two autumnal routes: the Mississippi River (Cade 1982) and the western shoreline of Lake Michigan (Mueller *et al.* 1988). Historically, peregrines also nested along the Mississippi River (Berger and Mueller 1969) and at least one record exists of peregrines nesting along the western shore of Lake Michigan at Racine, Wisconsin (Kumlien and Hollister 1903). Probably because of habitat changes due to fire suppression and predation by great horned owls (P. Redig and H. Tordoff unpubl. data), reintroduced peregrines have not successfully reoccupied historical nest sites along the Upper Mississippi River.

Located along or near these two shorelines are power plants (electric generating stations) with nest boxes installed for peregrines. Although the use of power plants and smoke stacks as peregrine nesting sites is a relatively new phenomenon, it is well known that peregrines currently nest on other man-made structures, buildings and bridges (Cade *et al.* Chapter 1).

The increase of peregrine populations in the Upper Midwest may be limited mostly by the availability of suitable nest sites. The use of power plants and smoke stacks for erecting nest boxes in proximity to historical nest sites offers another means of addressing nest site availability.

METHODS

In 1991, four nest boxes were installed in Wisconsin along the western shore of Lake Michigan (Fig. 1): three on smoke stacks and one on the corner of a power plant roof. These were attached to the sides of smoke stacks on galvanized steel mountings near existing catwalks about 100 m above the ground. The nest box attached to the power plant roof was secured to a southeast corner parapet at a height of approximately 80 m.

The installation of these four nest boxes complemented the erection of additional nest boxes on buildings and grain elevators along the same route. As of 1993, 10 of 16 nest boxes in Wisconsin were along the western shore of Lake Michigan and three were along the Upper Mississippi River.

In addition to the lakeshore nest boxes in Wisconsin, additional nest boxes were installed between 1988 and 1993 at electric power plants elsewhere in the Upper Midwest (Fig. 1). Table 1 shows the status of each site by state in 1993.

RESULTS AND DISCUSSION

In 1993, there were 48 known occupied peregrine territories in the Upper Midwest (P. Redig pers. comm.). Redig indicated that 32 (66%) of these nests were successful. Five of the successful nests (16%) were at power plants, where 15 young were produced. A total of 82 young fledged in this region in 1993 and 15 (18%) of these were produced at power plants.

There were 21 power plants with nest boxes in the Upper Midwest in 1993: 18 had peregrine sightings; nine had one or more peregrines present during the nesting season; and six of these were occupied by pairs, five of which nested successfully. In 1993 in Minnesota, all five of the nest boxes at power plants were occupied by pairs and four successfully fledged young (R. Anderson pers. comm.). Considering the lack of success peregrines have experienced attempting to reoccupy most of their historical nest sites in the Upper Midwest (P. Redig and H. Tordoff unpubl. data), power plants could add significantly to the number of man-made structures available as nest sites within this region and possibly serve to increase the regional peregrine population as well.

Some individuals have expressed concern about the use of power plants as nest sites for peregrines because of potential problems with pollution from sulfur dioxide, nitrous oxide and particulates. Although there are power plants where these issues may be a concern, much effort has been undertaken in Wisconsin to alleviate emissions.

The Wisconsin Electric Power Company, which owns the plants where three of the lakefront nest boxes are installed, has, over the past 20 yr, reduced its particulate emissions at coal powered plants from 80% to 99.95% (E. Neckar pers. comm.). This company has also switched to burning low sulfur coal, has installed state-of-the-art precipitators and has full stack opacity monitors on each of their smoke stacks (E. Neckar pers. comm.).

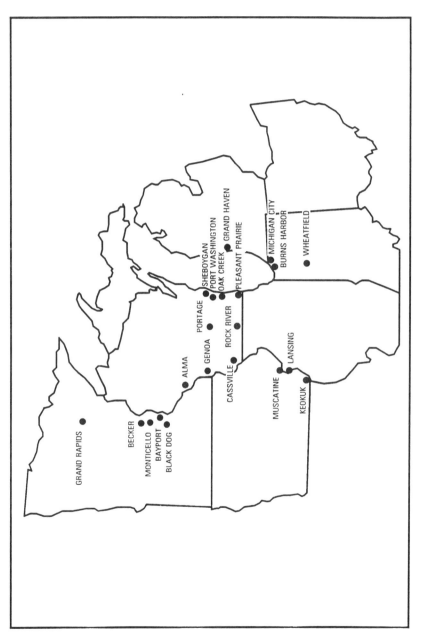

Figure 1. Upper Midwest power plants with nest boxes, 1993.

Table 1. Upper Midwest power plants with nest boxes for peregrine falcons, 1993.

State	Location/site	Status
Wisconsin	Genoa	N
	Alma	T
	Cassville	N
	Sheboygan	A
	Port Washington	O
	Oak Creek	N
	Pleasant Prairie	T
	Portage	N
	Rock River	N
Minnesota	Bayport	A
	Monticello	A
	Becker	A
	Cohassett	A
	Black Dog	A
Indiana	Michigan City	T
	Burns Harbor	N
	Wheatfield	S
Iowa	Lansing	N
	Muscatine	N
	Keokuk	N
Michigan	Grand Haven	N

A = Active nest, T = Territorial bird(s), N = No activity, O = Other i.e. great horned owl use, S = Installation scheduled.

Benefits of power plant nest sites may include fewer window collisions and vehicle kills and reduced human disturbance. Such possible benefits at power plants are perhaps attributable to power plant design, i.e. very few windows, and limits on traffic and access for security reasons.

Across the Upper Midwest in Michigan, Wisconsin, Ohio, Indiana, Illinois and Iowa there are > 150 electric power plants. Many of these are along the shorelines of rivers and lakes which are important sources of cooling water. In Wisconsin, the two main regions utilized are along the Mississippi River and the western shore of Lake Michigan, both of which are also important migration routes for peregrines (Cade 1982, Mueller et al. 1988).

Along the western shoreline of Lake Michigan where we have concentrated our observations, numerous examples of shoreline movements have been recorded. These include two wild-produced young that traveled from their Milwaukee nest site and were later trapped in separate years at the Cedar Grove Ornithological Research Station about 65 km north of Milwaukee (W. Robichaud pers. comm.). In 1992, five juvenile peregrines from other sites were observed interacting with the falcons released at the Kenosha, Wisconsin hack site (see Sherrod et al. [1987] for a description of hacking). These interactions included playful "dogfighting" and feeding on quail provided for the hacked young. The visiting falcons included one 1992 Milwaukee wild-produced young, (leg band number 77R77) and four young released from the 1992 Racine, Wisconsin hack site (71Z71, 84Z84, 25N25 and 27N27) (R. Donnelly and D. Gunther unpubl. data).

Peregrines have also dispersed along this shoreline and successfully nested. In Sheboygan, Wisconsin, a captive-produced female peregrine (42V42) nested successfully in 1992. She had been hacked in Zion, Illinois, in 1989 and had nested successfully in the Chicago area in 1990 and in 1991. Two males nesting in 1994 in Sheboygan (70T70) and Milwaukee (74T74) were hacked from Madison, Wisconsin, in 1990. In 1994, a captive-produced female peregrine (black/red color band = b/r; b/r 2/8) augmented at the Sheboygan nest site in 1993, paired with a male peregrine at Irving Park (north of Chicago). Also in 1994, a 1993 hatch female peregrine (b/r 5/3) from the Wacker Drive nest site in Chicago spent a week with a 1992 Kenosha hacked male (b/r C/D) at a Milwaukee grain elevator site. Also in 1994, at Milwaukee's newest nest site (Landmark on the Lake), another male (76T76) hacked at Madison in 1990 paired with an unbanded adult female and nested successfully.

It is interesting to note that three of the males hacked at Madison in 1990 have all paired and are nesting along the western Lake Michigan shoreline where most of the nest boxes in Wisconsin are concentrated. In each of these cases, movement between sites was < 190 km. If more man-made structures with suitable nest ledges or nest boxes were available within this region, the mean dispersal distance might be reduced.

With nest boxes now installed at many power plants, peregrines following these shorelines may find and utilize these readily visible structures. Indeed, at the Sheboygan, Wisconsin site referred to earlier, the nest box was occupied by a successful nesting pair in 1992, less than one year after it was installed. At other sites across the Upper Midwest, peregrines have also arrived shortly (1–3 yr) after the installation of nest boxes (J. Castrale and R. Anderson pers. comm.). Relatively rapid occupancies indicate that the Upper Midwest peregrine population may be limited by the number of adequate nest sites. With the serious problems at historical cliff nest sites, an increase in the number of nest boxes on appropriate man-made structures such as buildings, grain elevators, water towers, power plants and smoke stacks could provide peregrines with more nest sites and possibly increase the regional population.

Clustered around cities and along major waterways, power plants could effectively serve to expand nesting opportunities for peregrines throughout the Upper Midwest. Although cities also offer nest sites, the number of potential sites is related to the number of buildings and their proximity to each other. Thus, even though a city may have many tall buildings for nesting, the number of actual nest sites will be limited by the territory dynamics of the pair(s) present. In the Twin Cities region of Minnesota, two pairs of peregrines successfully nested approximately 6.5 km apart (P. Redig pers. comm.). Two pairs in Milwaukee in 1994 nested 1.6 km apart. Prey density may also play an important role in determining territoriality. If territoriality limits nesting densities of peregrines

Table 2. Dispersal distances (km) for 11 peregrine falcons from six nest sites in the Twin Cities area in 1993.

Dispersal distance	Males ($N = 6$)	Females ($N = 5$)
Mean	63	140
Median	64	40
Range	0–121	0–362

Dispersal distances (km) for nine falcons from five nest sites on the southern Lake Michigan shoreline in 1993.

Dispersal distance	Males ($N = 4$)	Females ($N = 5$)
Mean	135	349
Median	145	241
Range	0–249	193–547

within cities, then nest boxes at power plants (located relatively close to each other yet far enough apart not to trigger territorial skirmishes) may offer new nesting opportunities that could result in an increase in local peregrine numbers and the possible formation of regional populations or demes. Observations of shoreline movements and dispersal indicate that this may already be occurring along the southern Lake Michigan shoreline in Wisconsin, Illinois and Indiana and also possibly in the Twin Cities area of Minnesota.

Temple (1988) offered the following comment in support of this theory.

> As well all know, *Falco peregrinus* literally means the wandering falcon, and one is generally given the impression in the literature that Peregrines are capable of migrating and dispersing over great distances and typically do so. Despite this reputation, the reports given at this symposium have revealed a bird that, at least in terms of dispersal, is something of a "homebody", showing consistently strong philopatry to natal areas. Although this tendency for Peregrines to breed as close as possible to their natal areas has probably always been the case, the pattern becomes particularly noticeable in closely monitored, small, isolated populations.

Dispersal distances (the distances from hatch or hack sites to eventual territories or nest sites) of peregrines nesting in the Twin Cities area and the southern Lake Michigan shoreline in 1993 may be an important early indicator of the formation of regional populations in the Upper Midwest (Table 2). The greater mean dispersal distances in the southern Lake Michigan shoreline (Table 2) may reflect differences in proximity (and possibly clustering) of nest sites on man-made structures within these regions. Tracing a route from Sheboygan, Wisconsin south and east to Michigan City, Indiana, the southern Lake Michigan shoreline covers a distance of approximately 320 km. Spaced along this route are 16 man-made structures with peregrine nest boxes or trays. (Peregrines have also nested on a building in Chicago but no nest box or tray has been installed at this site.) Six (38%) of these nest boxes are on power plants. With the exception of the Milwaukee area, none of the nest box sites along the southern Lake Michigan shoreline are clustered to the same degree as sites within the Twin Cities area where all are within 50 km of each other (M. Martell pers. comm.). Of the six nest box sites located within the Twin Cities area, two are at power plants.

As indicated earlier, there have been noticeable movements by peregrines between sites along the Lake Michigan shoreline. This is also occurring in the Twin Cities region where peregrines from one nest or hack site have also been observed at one or more nearby sites (R. Anderson, P. Redig and H. Tordoff pers. comm.).

Thus, young falcons moving away from their hack or hatch sites, which we will refer to as their site of origin, stopped at least temporarily at nearby tall man-made structures, hack sites or nest sites. If suitable sites had been available closer to their site of origin, they would probably have stopped there as well. Dispersal, at least in part, may therefore be determined by the proximity of other sites. C. White (pers. comm.) suggested that peregrines in a specific region will often occupy all suitable nest sites within that region before any significant dispersal and colonization takes place.

Many Upper Midwest peregrines have nested in or occupied territories relatively close to their sites of origin (Table 2). Peregrines whose sites of origin lack other nearby suitable nest sites however, may need to move great distances to find nest sites. A case in point is a female peregrine (20V20) hacked at Isle Royale in 1988. She travelled about 650 km to Milwaukee where, at the time, the only nest box in Wisconsin existed. She has nested there successfully every year since. While there are exceptions, it appears that many peregrines in the Upper Midwest nest relatively close to their sites of origin.

From data on released and wild-produced peregrines in the Upper Midwest between 1987 and 1993, their band numbers and the locations of hack or nest sites (P. Redig and H. Tordoff unpubl. data), we were able to track 70 peregrines from their sites of origin to eventual nest sites during this period. Only two (3%) of these birds originated outside the Upper Midwest: a male from West Virginia (J42J) that nested unsuccessfully in LaCrosse, Wisconsin in 1993 and a female Canadian peregrine (red band, X/H) that nested at Cohasset, Minnesota in 1993. Long-distance dispersers seem to be the exception in what, for the most part has been a regionally influenced population.

We further determined that 75% (29) of marked female peregrines originating at hack or nest sites in the Upper Midwest between 1987 and 1993 nested $\leq$ 355 km from their sites of origin. For marked males during the same period, 75% (24) nested $\leq$ 170 km from their sites of origin.

In summary, the relatively large number of power plants in the Upper Midwest might provide additional nest sites for peregrines in this region. Historically (pre-DDT), about 30 pairs of peregrines nested along a 320 km stretch of the upper Mississippi River (Redig and Tordoff 1988). Perhaps it was the proximity of these cliff nest sites to each other that accounted for the population of peregrines that formerly existed in this region. A similar situation may have occurred in the southcentral Wisconsin River area where at least five historical eyries existed close together (Stoddard 1969, N. Braker pers. comm.). It would be revealing to know the extent to which gene flow existed within and between these regions in the past. Genetic analysis of blood samples from wild-produced peregrines as part of a cooperative study of the recovering Upper Midwest population of peregrines (P. Redig and S. Moen pers. comm.) and further documentation of dispersal distances may indicate the extent of regional populations in the Upper Midwest. If regional populations do arise, power plants may play an important role in their expansion by augmenting the supply of available nest sites. Likewise, along shorelines where peregrines historically nested and especially where habitat changes have occurred, it is also possible that power plants and other man-made structures could replace historical eyries as preferred nest sites for the recovering population.

ACKNOWLEDGMENTS

We would like to thank Mike D'Allessandro and Ed Neckar of the Wisconsin Electric Power Company for their enthusiasm and hard work with nest box installations and for their input pertaining to the company's pollution abatement programs. Joanne Thiel, Ken Koele and the employees of the Wisconsin Power & Light Company's Edgewater Generating Station in Sheboygan, Wisconsin are to be thanked for their heartfelt cooperation. We also thank Mark Washburn in Iowa, Robert Anderson in Minnesota, John Castrale in Indiana and John Will in Michigan for their data on nest boxes at power plants. Mark Martell, Patrick Redig and Harrison Tordoff in Minnesota provided information regarding the Twin Cities peregrines.

We would also like to thank Clayton White for his input pertaining to population saturation and dispersal and Laurie Goodrich, Frances Hamerstrom, Mark Martell, and Dan Varland for reviewing this manuscript. Lastly, we thank Jackie Oldham of BXG, Inc. of Boulder, Colorado for the Midwest portion of their map of coal deposits and power plants across the US.

REFERENCES

BERGER, D. AND H.C. MUELLER. 1969. Nesting peregrine falcons in Wisconsin and adjacent areas. Pp. 115–122 in J.J. Hickey, ed. *Peregrine falcon populations: their biology and decline.* Univ. of Wisconsin Press, Madison, Wisconsin USA.

CADE, T.J. 1982. *The falcons of the world.* Comstock/Cornell Univ. Press, Ithaca, New York USA.

—, M. MARTELL, P.T. REDIG, G.A. SEPTON AND H.B. TORDOFF. 1996. Peregrine falcons in urban North America. Pp. 3–13 in D.M. Bird, D.E. Varland and J.J. Negro, eds. *Raptors in human landscapes.* Academic Press, London, UK.

KUMLIEN, L. AND N. HOLLISTER. 1903. *The birds of Wisconsin.* Bulletin of the Wisconsin Natural History Society, Vol. 2, Nos. 1, 2 and 3. Milwaukee Public Museum, Milwaukee, Wisconsin USA.

MUELLER, H.C., D.D. BERGER and G. ALLEZ. 1988. Population trends in migrating peregrines at Cedar Grove, Wisconsin, 1936–1985. Pp. 497–506 in T.J. Cade, J.H. Enderson, C.G. Thelander and C.M. White, eds. *Peregrine falcon populations: their management and recovery*. The Peregrine Fund Inc., Boise, Idaho USA.

REDIG, P.T. AND H.B. TORDOFF. 1988. Peregrine falcon reintroduction in the Upper Mississippi Valley and Western Great Lakes. Pp. 559–563 in T.J. Cade, J.H. Enderson, C.G. Thelander and C.M. White, eds. *Peregrine falcon populations: their management and recovery*. The Peregrine Fund Inc., Boise, Idaho, USA.

SHERROD, S.K., W.R. HEINRICH, W.A. BURNHAM, J.H. BARCLAY AND T.J. CADE. 1987. *Hacking: a method for releasing peregrine falcons and other birds of prey*. The Peregrine Fund Inc., Boise, Idaho USA.

STODDARD. H.L. 1969. *Memoirs of a naturalist*. Univ. of Oklahoma Press, Norman, Oklahoma USA.

TEMPLE, S.A. 1988. Future goals and needs for the management and conservation of the peregrine falcon. Pp. 843–848 in T.J. Cade, J.H. Enderson, C.G. Thelander and C.M. White, eds. *Peregrine falcon populations: their management and recovery*. The Peregrine Fund Inc., Boise, Idaho USA.

17

Competition for Nest Boxes Between American Kestrels and European Starlings in an Agricultural Area of Southern Idaho

Marc J. Bechard and Joseph M. Bechard

Abstract – Occupancy of nest boxes by American kestrels and European starlings was monitored from 1987 to 1993 in an agricultural area of southern Idaho where the native habitat had been eliminated by livestock grazing and the cultivation of alfalfa for hay production. Combined occupancy of boxes by both species averaged over 99%. Occupancy by starlings averaged 33% from 1987 through 1989, but decreased to 0% thereafter. Since 1990, all but 1 of the 60 available nest boxes were occupied by kestrels with nest productivity averaging 3.5 young per breeding pair. This study showed that kestrels can successfully outcompete starlings for limited numbers of nest cavities and ultimately exclude them from farmland in Idaho where livestock grazing and hay production are the major forms of land use. Therefore, we suggest that future nest box programs targeted at augmenting kestrels populations in Idaho be concentrated in agricultural areas supporting large amounts of hayfields and pasturelands.

Key words: kestrel; starling; nest box; agriculture; competition; Idaho.

The American kestrel is a widely distributed, ecologically versatile species that occurs from sea-level up to about 3700 m in the Rocky Mountains of North America and 4300 m in the Andes Mountains of South America (Cade 1982). Overall, it favors open habitats with few trees, but forest edges near openings are frequently used as nesting habitats (Johnsgard 1990). It is an obligate cavity nester and typically uses old woodpecker holes or natural cavities in trees, cliffs and earthen banks for its nest sites (Roest 1957, Smith et al. 1972, Cade 1982, Palmer 1988, Johnsgard 1990). In addition to an available supply of suitable nest cavities, an abundance of elevated perches and open terrain from which kestrels can hunt insects and small mammals are habitat requisites (Balgooyen 1976).

The kestrel also occurs in a variety of human-altered habitats such as farmlands, vacant building sites in cities and towns, airfields, athletic fields, cemeteries, city parks, highway and railway rights-of-way, electricity transmission lines, and reclaimed surface mines (Stahlecker 1979, Cade 1982, Palmer 1988, Varland and Loughin 1993). Due to the elimination of potential nest holes in dead trees through the urbanization of habitat and the spread of agriculture, nest cavities increasingly have become the principal factor limiting the density of breeding populations (Hamerstrom et al. 1973). It is not surprising, therefore, that pairs nesting in human-altered habitats are often found using a variety of man-made structures such as farm buildings (Bent 1938, Cruz 1976, Scott et al. 1977, Craig and Trost 1979, Sutton and Tyler 1979). As a result, nest box programs have been advocated as a means of maintaining or increasing densities of nesting populations in these areas (Hamerstrom et al. 1973, McArthur 1977, Craig and Trost 1979, Stahlecker 1979, Jones 1982, Bloom and Hawks 1983, Wilmers 1987).

Nest cavities have become even more limiting in human-altered habitats due to the spread of another obligate cavity nester, the European starling. Introduced into Central Park, New York in 1890, the North American starling population grew to an estimated 200 000 000 birds by the early 1980s (Feare 1984). Tolerant of people, most starlings occur in areas that are affected heavily by human disturbance and it is under these circumstances that the two species compete. Initiating breeding earlier than the kestrel, the starling not only begins using cavities first, but because it can raise two or more broods in a single nesting season, it can totally preclude the use of cavities by kestrels. Starlings have, in fact, undermined nest box programs undertaken to increase populations of kestrels by using an average 62% of nest boxes erected in human-altered habitats of New York, Idaho, West Virginia, and Colorado (Lincer and Sherburne 1974, Craig and Trost 1979, Stahlecker 1979, Wilmers 1987).

Few studies have attempted to address the issue of interspecific competition between starlings and kestrels for nest cavities, especially in areas that are heavily impacted by humans. It is not known, for instance, if starlings are always the dominant species or if kestrels are capable of outcompeting starlings for nest sites where food is abundant and only nest cavities limit population size. Our study was undertaken to document the outcome of competition in such an area in southern Idaho. The area had all of the habitat requisites of kestrels and starlings except that it lacked an abundant supply of both natural and unnatural nest cavities. By placing nest boxes in this area, it was hoped that the outcome of interspecific competition between these two species could be documented.

STUDY AREA AND METHODS

The study was conducted in southwestern Idaho in the Camas Prairie (Fig. 1). The area was chosen because it forms a natural basin in the foothills of the Sawtooth Mountains. The basin measures 80×30 km and is closed at the west end by mountains. It opens into the Snake River plain to the east. Elevation averages 2000 m. Historically, the basin supported shrub-steppe vegetation dominated by sagebrush, rabbitbrush and bunch grasses such as *Agropyron spicatum* and *Elymus cinereus*. There were very few cliffs with natural cavities and, except for small willows (*Salix* sp.) that grew along creeks and marshes, it

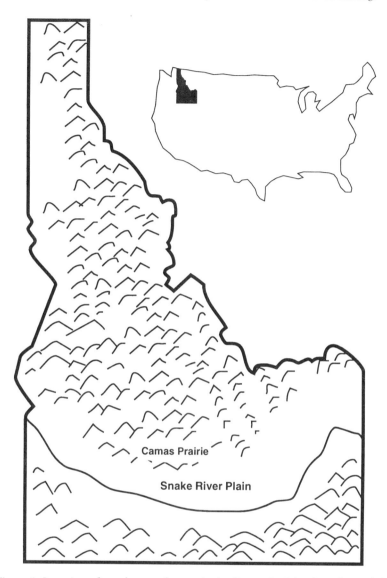

Figure 1. Location of nest box study area in the Camas Prairie of southern Idaho.

was treeless. The area was first settled in the 1850s. Most native habitat has been eliminated. The area is only moderately grazed by cattle and sheep, but it has been extensively cultivated. Alfalfa, timothy and clover are the major crops, however there is also some cereal crop production. Clean farming practices have eliminated most of the native willow habitat in marshy areas and irrigation practices and drainage of marshy areas have eliminated many of the marshes and creeks. Dead and dying trees are commonly burned to prevent the spread of weeds. Most of the remaining trees are cottonwoods (*Populus* spp.) that have been planted around farm sites.

The area supports a variety of wildlife. Birds including upland species, waterfowl, raptors, and starlings are abundant. It supports an abundance of potential prey species, including deer mice, voles and grasshoppers, and there are numerous utility structures suitable for hunting perches. Indeed, other than its apparent lack of natural cavities, it seems to be an ideal area for nesting kestrels. There were very few kestrels observed in the area when it was first selected for study.

In 1986, 60 wooden nest boxes measuring 25 × 25 × 40 cm with entrance holes measuring 8 cm in diameter were installed on poles approximately 5 m above the ground. All box openings faced south and the tops were hinged to allow access. They were spaced approximately 0.8 km apart in a grid pattern measuring roughly 100 km^2. Initially, sawdust was used to line the bottoms of boxes but, after 1988, it was replaced with dried straw which was not blown out by wind so easily. Boxes needing repair and those with accumulated debris and fecal matter were visited and repaired and cleaned each winter.

Each year, all boxes were checked for occupancy in mid- to late May after egg laying was completed and adult kestrels and starlings were incubating. Incubating adult kestrels were captured in boxes and banded with US Fish and Wildlife Service aluminum leg bands. If they were already banded, their band numbers were recorded. Clutch size, i.e. the number of eggs being incubated, was recorded, and any prey items in boxes were identified and counted. All boxes occupied by kestrels were revisited in late June or early July, at which time young were sexed and banded, and prey remains were noted. The productivity of kestrel pairs was also recorded at this time based on the number of young that had reached legal banding age (Steenhof 1987). Incubating and brooding starlings were left undisturbed and neither adult nor young starlings were banded. No attempt was made to determine the nesting success of starlings.

RESULTS

Combined occupancy of nest boxes by both species averaged over 99% between 1987 and 1993 (Fig. 2). In 1987, an equal number of kestrels and starlings occupied boxes. Thereafter, occupancy by starlings decreased annually (Fig. 2). Beginning in 1990, not a single box was occupied by starlings and, except for 1993 when one box was empty, occupancy by kestrels averaged nearly 100%. After starlings were totally excluded by kestrels in 1990, there was a 4-yr average annual population density of 60 breeding pairs of kestrels per 100 km^2.

When the two species first occupied the boxes in 1987, they were almost evenly distributed throughout the study area, but starlings tended to be more concentrated toward the center (Fig. 3). As kestrels began to dominate in 1988

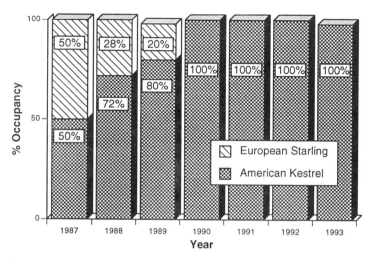

Figure 2. Percent occupancy of nest boxes by American kestrels and European starlings from 1987 to 1993.

and 1989, they increasingly replaced starlings in the center, forcing them to use boxes located on the periphery of the study area. By 1989, the original occupancy pattern was reversed, with kestrels dominating the center and starlings restricted to boxes located at the periphery.

Only two nest boxes were occupied by starlings all three years prior to their exclusion from the boxes. An additional four nest boxes were occupied by starlings for two of the three years, but the remainder of the boxes were occupied by starlings for only one year. In all but four cases, once a box was taken over by a kestrel it was never again reused by a starling.

Numbers of eggs and nestlings of starlings were recorded when boxes were first visited in May, but these boxes were not revisited to determine overall starling reproductive success. Starling clutches found during incubation averaged 4.3 ($N = 23$, $SE = 0.34$) eggs per nest. Fewer nests were found to contain young starlings, but broods averaged 4.1 ($N = 12$, $SE = 0.54$) young per nest, indicating a high degree of egg hatchability. Kestrel clutches averaged 4.4 ($N = 369$, $SE = 0.23$) eggs per nest. Kestrels were found incubating three mixed clutches containing four or five kestrel eggs and one or two starling eggs in 1987 and 1988. All starlings eggs in these clutches were removed and found to be infertile. In all three cases, all of the kestrel eggs in mixed clutches hatched, and all young successfully fledged. Between 1987 and 1993, kestrel brood size and fledging success averaged 3.7 ($SE = 0.36$) and 3.6 ($SE = 0.28$) young per nest, respectively.

A total of 154 prey items was observed in nest boxes occupied by kestrels. Of these, 48% (75) were deer mice and 44% (67) were grasshoppers. Prey in the remaining 8% of the sample consisted of voles (1), horned lizards (7), horned larks (1), and beetles (2).

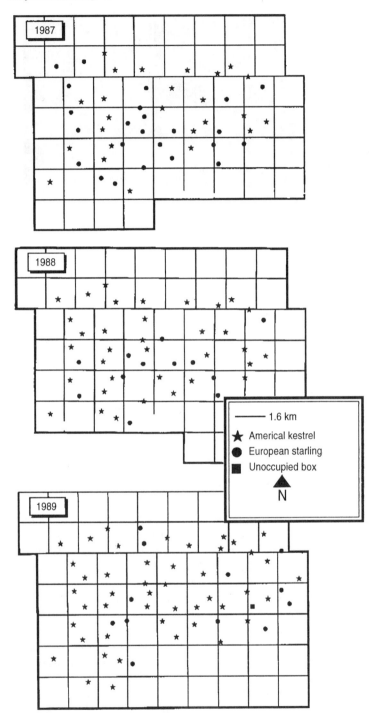

Figure 3. Pattern of nest box occupancy by American kestrels and European starlings from 1987 through 1989.

DISCUSSION

The 120 European starlings that were introduced into the United States in 1890 increased a millionfold in 50 years (Davis 1950). Today the species occurs throughout much of North America. The introduction and spread of the starling in North America undoubtedly brought competition over an increasingly limited number of natural nest cavities to a peak. It has unquestionably forced the starling and kestrel into a new and intensely competitive relationship. Cade (1982) suggested that the kestrel may be the loser in competitive encounters and as a result may have been competitively excluded in some areas. Indeed, there are many reports of kestrels abandoning areas after starlings began to nest in them (Pearson 1936, Hamerstrom et al. 1973, Wilmers 1987).

Balgooyen (1976), however, has suggested that kestrels can outcompete starlings for nest cavities under certain circumstances. When food and hunting perches are plentiful and only nest sites are limited, it appears that kestrels may be the dominant species. This study provides convincing evidence that kestrels can successfully outcompete starlings for nest cavities. They began to use nest boxes in equal numbers, but within four years, kestrels totally excluded starlings from using the nest boxes. Not only did they usurp nest boxes from starlings but they prevented starlings from reoccupying them after taking them over. No estimate of the number of breeding pairs of kestrels was made prior to the construction of the boxes, but there were very few kestrels observed in the study area during the initial stages of the study. With a density of 0.6 pairs of breeding kestrels per km^2, this population is one of the densest reported (Craighead and Craighead 1956, Enderson 1960, Smith et al. 1972, Cruz 1976). Nagy (1963) and Hamerstrom et al. (1973) have also reported considerable increases in nesting kestrel densities through the installation of nest boxes in suitable agricultural areas of Pennsylvania and Wisconsin.

It is plausible that something other than kestrel competition was responsible for the reduction of starling use of nest boxes. Lacking any evidence for a decline in the availability of starling prey, outbreaks of starling parasites, or increasing numbers of starling predators, none of these alternatives seemed likely.

Results of this study also indicate that, in selecting locations for nest box programs aimed at reestablishing or increasing kestrel populations, a few essential habitat features must be met. First, an abundance of hunting perches is important since kestrels frequently use a sit-and-wait style of hunting during the breeding season to hunt small mammals. Second, locations must be selected on the basis of the availability of prey. Areas supporting dense populations of deer mice or other species of small mammals are good, but areas with the greatest potential for success must also have large numbers of grasshoppers during the summer months. The availability of insects is particularly important in ensuring that broods of young kestrels survive to fledging. Conditions such as this typically exist in agricultural areas of the western United States where the primary land uses are livestock grazing and hay production, rather than intensive crop cultivation.

REFERENCES

BALGOOYEN, T.G. 1976. Behavior and ecology of the American Kestrel (*Falco sparverius* L.) in the Sierra Nevada of California. *Univ. Calif. Publ. Zool.* **103**: 1–83.

BENT, A.C. 1938. Life histories of North American birds of prey. *US Natl. Mus. Bull.* 170.

BLOOM, P.H. AND S.B. HAWKS. 1983. Nest box use and reproductive biology of the American Kestrel in Lassen County, CA. *Raptor Res.* **17**: 9–14.

CADE, T.J. 1982. *The falcons of the world.* Cornell Univ. Press, Ithaca, New York USA.

CRAIG, T.H. AND C.H. TROST. 1979. The biology and nesting density of American Kestrels and Long-eared Owls in the Big Lost River, southeastern Idaho. *Wilson Bull.* **91**: 51–61.

CRAIGHEAD, J.J. AND F.C. CRAIGHEAD. 1956. *Hawks, owls, and wildlife.* Stackpole Co., Philadelphia, USA.

CRUZ, A. 1976. Food and foraging ecology of the American kestrel in Jamaica. *Condor* **78**: 409–412.

DAVIS, D.E. 1950. The growth of starling, *Sturnus vulgaris*, populations. *Auk* **67**: 460–465.

ENDERSON, J.H. 1960. A study population of the sparrow hawk in east-central Illinois. *Wilson Bull.* **72**: 222–231.

FEARE, C. 1984. *The starling.* Oxford Univ. Press, Oxford, UK.

HAMERSTROM, F., F.N. HAMERSTROM AND J. HART. 1973. Nest boxes: an effective management tool for kestrels. *J. Wildl. Manage.* **42**: 400–403.

JOHNSGARD, P.A. 1990. *Hawks, eagles, and falcons of North America: biology and natural history.* Smithsonian Institution Press, Washington, DC USA.

JONES, R. 1982. The 1982 nesting year. *Kestrel Karetaker News* **3**: 1.

LINCER, J.L. AND J.A. SHERBURNE. 1974. Organochlorines in kestrel prey: a north–south dichotomy. *J. Wildl. Manage.* **38**: 427–434.

MCARTHUR, L.B. 1977. Utilization of nest boxes by birds in three vegetational communities with special reference to the American Kestrel (*Falco sparverius*). M.Sc. thesis, Brigham Young Univ. Provo, Utah USA.

NAGY, A.C. 1963. Population density of kestrels in eastern Pennsylvania. *Wilson Bull.* **75**: 93.

PALMER. R.S., ed. 1988. *Handbook of North American birds, Volume 5, Diurnal raptors, Part 2.* Yale University Press, New Haven, Connecticut USA.

PEARSON, T.G., chief ed. 1936. *Birds of America.* Garden City Publ. Co., Garden City, New York USA.

ROEST, A.I. 1957. Notes on the American sparrow hawk. *Auk* **74**: 1–19.

SCOTT. V.E., K.E. EVANS, D.R. PATTON AND C.P. STONE. 1977. Cavity-nesting birds of North American forests. *US Dept. Agric. Handbook* 511.

SMITH, D.G., C.R. WILSON AND H.H. FROST. 1972. The biology of the American Kestrel in central Utah. *Southwestern Nat.* **17**: 73–83.

STAHLECKER, D.W. 1979. Raptor use of nest boxes and platforms. *Wildl. Soc. Bull.* **7**: 59–62.

STEENHOF, K. 1987. Assessing raptor reproductive success and productivity. Pp. 157–170 in B.A. Giron Pendleton, B.A. Millsap, K.W. Cline and D.M. Bird, eds. *Raptor management techniques manual.* Nat. Wildl. Fed. Washington, DC, USA.

SUTTON, G.M. AND J.B. TYLER. 1979. On the behavior of American Kestrels nesting in towns. *Okla. Ornith. Soc.* **4**: 25–28.

VARLAND, D.E. AND T.M. LOUGHIN. 1993. Reproductive success of American kestrels nesting along an interstate highway in central Iowa. *Wilson Bull.* **105**: 465–474.

WILMERS, T.J. 1987. Competition between starlings and kestrels for nest boxes: A review. Pp. 156–159 in D.M. Bird and R. Bowman, eds. *The ancestral kestrel.* Raptor Res. Found., Inc. and MacDonald Raptor Res. Centre of McGill Univ., Ste. Anne de Bellevue, Quebec Canada.

Raptors in Cultivated Landscapes

18

White-tailed Kite Movement and Nesting Patterns in an Agricultural Landscape

Andrea L. Erichsen, Shawn K. Smallwood, A. Marc Commandatore, Barry W. Wilson and Michael D. Fry

Abstract – We examined patterns of white-tailed kite habitat use and nesting success in the agricultural landscape of the Sacramento Valley, California. Road surveys covering 320 km were conducted from 1990–93. The number of kites observed per km of transect decreased through the years. White-tailed kites preferred areas of natural vegetation, which constituted a small percentage of the landscape along the transect. Seasonal and local changes in land use also affected kite habitat use. Kites preferred rice stubble in spring as the few remaining fields were being tilled. A nest survey covering 625 km^2 was conducted in 1993. Nine of 22 nests successfully fledged young. Nesting habitat was characterized using Atlas^{TM} Geographic Information System. Significant differences in land uses were found using 0.8 km radii around successful and failed nests. Successful nests contained more natural vegetation and more human development (abandoned farms, vacant lots, a cemetery), but none were in urban areas. Differences in land uses were not significant when radii were increased to 1.6 km around successful and failed nests. Forty-one percent ($N = 22$) of nests were located in riparian corridors, but only one of these nests was successful. The successful nests shared in common: adjacent low-lying medium density natural vegetation (>50 m × 30 m), water <1.5 km from nest, placement of nest on edge of the vegetation, nest located >100 m from roads.

Key words: white-tailed kites; Sacramento Valley California; Geographic Information System; habitat use analysis.

The white-tailed kite was nearly extinct in California early in this century. Whereas the successful recovery of the species has demonstrated its resilience in adapting to human-caused disturbances (Pruett-Jones *et al.* 1980), land uses within its range are always changing, and could present future threats to its existence.

The white-tailed kite is a small rodentivore raptor found in California, Arizona, southern Texas, and the Gulf Coast, with recent expansions occurring into Oregon and Washington (Eisenman 1971, Henny and Annear 1978, Harrington-Tweit 1980, Larson, 1980). In California the white-tailed kite is found in the Central Valley, the north coast of Del Norte County, and south to

San Diego County (Fry 1966, Eisenman 1971, Pruett-Jones et al. 1980). Within its California range the kite is primarily non-migratory, inhabiting moist lowland areas, grasslands, and cultivated fields (Eisenman 1971, Warner and Rudd 1974, Pruett-Jones et al. 1980). The white-tailed kite's range in the Central Valley includes lands dominated by intensive agriculture with heavy use of pesticides.

The spatial scale of our research reflects the importance of conserving raptors as an assemblage, not as individual species, in a human-modified ecosystem. Conservation of the kite, and other raptors in California, depends on understanding how these raptors use the present landscape, especially those elements that are managed intensively. Land-use practices that stress wildlife species inhabiting agricultural landscapes include annual till, chemical use, introduction of exotic species (both plant and animal), and loss of natural habitat, especially riparian forests. Today, less than 8% of the original natural wetlands and riparian forests of the Sacramento Valley remain (Katibah 1984, Griggs 1993). Raptors must use agricultural lands for space and food resources, especially when densities are highest during annual migration (Bloom 1985, Smallwood et al. 1996).

In this paper, we incorporate results from a 4-yr road survey (1990–94) and a nesting success survey (1993). The goal of our research is to elucidate temporal and spatial patterns of white-tailed kite habitat use and nesting success in an agricultural landscape, as part of a study of the actions of multiple stresses on wildlife.

STUDY AREA AND METHODS

The study site (15–25 m elevation) was located in the southern portion of the Sacramento Valley 20 km west of Sacramento, California (Fig. 1). The climate of the region is mediterranean with annual precipitation averaging 43.5 cm (Koppen and Geiger 1954).

Road Survey

One of the authors (KSS) initiated wildlife surveys in January 1990 along a 58-km road transect through Sacramento Valley farmland. The white-tailed kite was added to the survey on 11 March 1990. The transect was enlarged in late 1990 to 200 km along a 320-km loop (Fig. 1). The survey was conducted from a car travelling 55 km hr^{-1} with one passenger who served as principal observer. The frequency of surveys varied from once per week to once per month. Observations of birds and mammals were tape-recorded, and included number, location, activity and association with a landscape element.

White-tailed kite observations were analyzed for association with landscape elements, using typical percentages of landscape elements along the transect

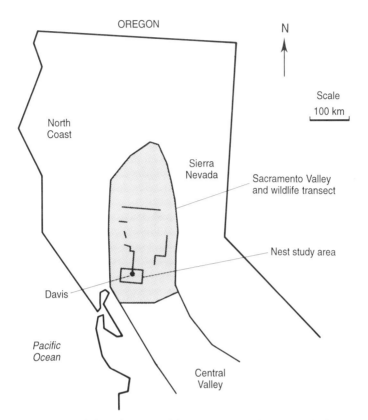

Figure 1. Locations of the white-tailed kite nesting study and wildlife survey in the Sacramento Valley, California.

during winter (November–February) and during spring (March–July). The majority of kite observations occurred in these two seasons.

A measure of association between kites and land-use elements was derived by dividing the land-use percentage value into the proportion of the wildlife observations that occurred on or above that landscape element. The resulting ratio is a measure of association, which is interpreted as the number of species A at landscape element i as a multiple of that to be expected by chance. The parameters of this ratio can be rearranged to equal the observed divided by expected values of χ^2 analysis (Smallwood 1993). Hence, the observed and expected values can be rearranged again to obtain a probability value for a χ^2 test. Landscape associations and number of observations were compared between cumulative distances of transect, years and seasons.

Nesting Habitat

Nesting white-tailed kites were studied in a 625-km² area surrounding Davis, California (38° 33′N, 121° 45′W), during the breeding season of 1993 (January–September).

Pairs of white-tailed kites were located during road surveys. Nest sites were located by observing breeding displays such as soaring, flutter flight, twig carrying, nest building and copulation (Palmer 1988). Courting pairs were first observed during late January. The last observed nesting pair was courting in mid-August.

Nesting pairs were monitored at least twice weekly to locate nests and to record the nesting stage (courtship, incubation, post-hatch). To minimize the risk of nest abandonment, we climbed to nests and measured habitat only after chicks hatched (Fyfe and Olendorff 1976). Pole-mounted mirrors were used for eight nests which were in the inaccessible top meter of the tree canopies.

Nest habitat land use was characterized within 0.8 and 1.6 km radii from the nest. Hawbecker (1940) reported that breeding male kites (which feed the females during incubation) rarely hunt more than 1 km from the nest. Our nest observations indicated that nesting kites frequently used hunting areas immediately adjacent to the nest (<0.5 km) and out to approximately 1.6 km of the nest. The immediate nest territory was the area where all aggressive interactions between white-tailed kites and other raptors occurred. Therefore, we generated both 0.8 and 1.6 km radii to include immediate nest surroundings and adjacent land uses which might affect nest success. Land-use maps, obtained from the California Department of Environmental Quality, were ground-truthed against current land uses (Division of Water Resources Sacramento, California). Subsequently, the maps were digitized into Atlas GIS^{TM}, a Geographic Information System (Strategic Mapping, Inc., Santa Clara, California). With Atlas GIS, 0.8 and 1.6 km radii buffers were generated around all nests (Fig. 2). A "buffer" is a circular area generated by the program around a central point (in this case the nest). Land-use variables analyzed within each buffer included field crops (sugar beet, safflower, corn, beans), grains (wheat, barley), natural and idle grass vegetation, riparian vegetation, pasture (alfalfa), tomato, water, human development (including rural homes, abandoned buildings, vacant lots, livestock facilities, cemeteries, cities, and highways), vineyard, and orchard. Areas of land uses and their diversities (Simpson's Diversity Index) were measured and compared within the 0.8 km and 1.6 km radii for differences between nest successes and failures using logistic regression and the Mann-Whitney Test (Afifi and Clark 1984, Daniel 1990).

RESULTS

Road Survey

The number of white-tailed kite observations per km during winter declined from 1990/91 to 1992/93 (Fig. 3). Nearly all winter-time observations of white-tailed kites were of single birds. Only 9 of 213 observations included 2 individuals. The numbers of kites observed during spring also declined, but not

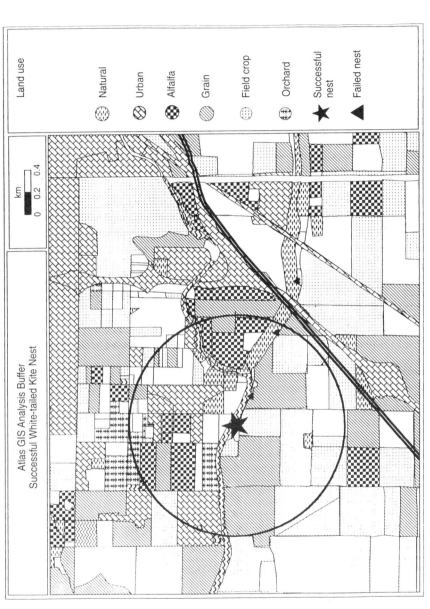

Figure 2. A simplified GIS map demonstrating a 1.6 km "buffer" used to analyze land-use composition around a successful white-tailed kite nest. The legend displays selected land uses. Note the three failed nests which are in close proximity to each other and the successful nest on this riparian corridor. The closeness of these nests was atypical.

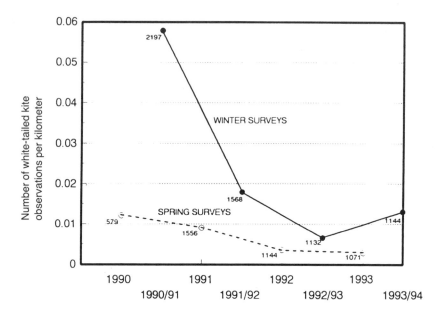

Figure 3. Winter- and spring-time trends in total number of white-tailed kite observations divided by the total distances of road survey in the Sacramento Valley from 1990–93. Cumulative survey distances are indicated at each data point.

as dramatically. The population of white-tailed kites was less concentrated during spring than during winter (Fig. 3).

Based on cumulative survey distance, use of the landscape was non-homogeneous during both seasons ($P < 0.001$ for all χ^2 tests, d.f. = 12, Fig. 4). For example, white-tailed kites occurred in sugarbeets 7.7 times more than expected by chance, which was the cumulative number of kite observations on the transect multiplied by 0.5%, the percentage of sugarbeets on the transect. White-tailed kites also preferentially used natural vegetation, which included riparian, wetland, and upland habitats. Land-use preferences changed with season. In spring kites maintained a strong preference for riparian habitat but increased preference for rice, alfalfa, dry pasture and orchards. Small areas of rice stubble, which have not been burned in the winter, are disked in early spring. Plowed fields, rural areas and irrigated pastures were avoided by kites throughout the survey.

The perch types used by white-tailed kites varied with season. Fewer white-tailed kites were seen on the ground during winter, and they avoided trees and utility poles during spring and fall ($\chi^2 = 18.2$, d.f. = 6, $P = 0.006$). Overall, 39 kites (19%) were on trees, 12 (6%) were observed on utility poles, 11 (5%) on the ground, three (1.4%) were on artificial hawk poles, and one (0.5%) on a cattail (*Typha* spp.).

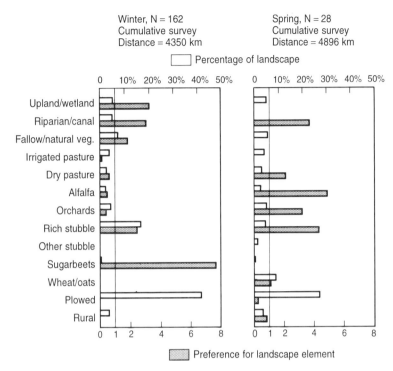

Figure 4. The character of the landscape along the survey transect and white-tailed kite preferences for landscape elements. The top horizontal axes correspond with incidence of landscape elements, and the lower axes correspond with the measure of ecological association, M_i, or selection "preference". The solid vertical lines at $M_i = 1.0$ correspond with the selection to be expected based on chance.

On road surveys 68% ($N = 128$) of sightings were of kites flying, with nearly half of these (34%, $N = 128$) being hovering flight. The proportion of kites seen hovering did not change with season ($\chi^2 = 3.28$, d.f. $= 2$, $P = 0.19$).

Nesting Habitat

Twenty-two nests were located within the 625-km² study area between late February and mid-September 1993. Nine of 22 pairs (40.9%) successfully fledged young. The 13 nest failures occurred prior to or during incubation by the female.

Six of the 13 white-tailed kite nest failures (46%) were due to displacement by Swainson's hawks in mid-March and early April. All of the kite nest territories displaced by Swainson's hawks were located along riparian corridors. Aggressive interactions, in which physical contact was made when white-tailed kites stooped on Swainson's hawks, were common. Red-tailed hawks and northern harriers and other kites were also chased by white-tailed kites in nesting territories. One kite nest was successful along a riparian corridor and was unusually close to three failed nests (Fig. 2). All were within a 1.6-km stretch of the riparian corridor. Two nests, within 1 km of each other, failed when one of the mates was killed. The remaining five nests (38%) failed for undetermined reasons.

Features common to successful nest sites were:

(1) nests were in rows (hedgerows between fields) or patches of trees (including riparian corridors, and unmanaged vegetation around buildings) overlooking low-lying natural vegetation, fallow fields, wet pasture and alfalfa;
(2) low-lying vegetation contiguous with the nest tree was at least 50 m × 30 m;
(3) nests were <1.5 km from water, whether man-made or natural;
(4) nests were >100 m from roads.

Eight failed nests (66%) located along riparian corridors shared the above characteristics. Five failed nests were in single trees ($N = 2$), in patches less than 50 m × 30 m, or >1.5 km from water.

Successful nests had more natural, fallow, and riparian vegetation as well as more human development within 0.8 km buffers than failed nests (One-sided Mann-Whitney $T = 77$, $P < 0.025$). The human development was characteristically low use, abandoned, or maintained private property. Analysis of 1.6 km buffers also showed that successful nests tended to have more natural vegetation, although not significantly so (Mann-Whitney $T = 70$, $0.10 > P > 0.05$). No significant differences were found between 0.8 and 1.6 km buffers using logistic regression.

Successful nests were distributed as follows: 2 in *Cupressacaea* spp.; 3 in *Quercus* spp.; 1 in *Eucalyptus* spp.; 1 in *Pinus* spp.; 1 in *Populus* spp.; and 1 in *Podocarpus* spp. Average nest height was 14 m, located within 1 m of the crown. Snags were consistently part of the nest tree or adjacent to the nest and were favored as perches by adults and offspring.

DISCUSSION

In their analysis of Christmas Bird Counts, Pruett-Jones et al. (1980) stated that it is "critical" to monitor California's white-tailed kite population. They noted that the kite population of the Central Valley, which was the center of California's kite population, had declined significantly since 1975 (Eisenman 1971, Larson 1980). Other than Christmas Bird Count analysis, there are few records of kite populations and historical nest locations in the Sacramento Valley.

The decreasing frequency of white-tailed kite observations along the survey transect probably indicates a decreasing population trend rather than a shift in location. The white-tailed kite population is probably decreasing in the Sacramento Valley as suggested by the decrease occurring on every stretch of the transect, which sampled populations in a large area of the valley. The population decrease might be attributed to land-use changes such as urban development of agricultural lands, loss of natural habitats, drought, prey cycles, and shifts in agricultural crops grown in the area. Competition between the white-tailed kite and Swainson's hawk and other tree nesting raptors is also a potential factor.

Seasonal changes in agricultural activities have direct impacts on white-tailed kites. Changing agricultural practices in the middle part of the century drastically increased rodent abundance and altered species composition, particularly in irrigated crops such as sugarbeets and alfalfa (Warner and Rudd 1974, Pruett-Jones et al. 1980). These changes in prey availability had a positive effect on recovering white-tailed kite populations (Stendell and Meyers 1973; Warner and Rudd 1974). The expansion of irrigation, along with federal and state protection, also assisted the kites' recovery (Eisenman 1971).

Association analysis values support findings of Stendell and Meyers (1973) and Warner and Rudd (1974) that kites selectively use sugarbeet and alfalfa crops. Rice stubble is another important foraging habitat for the white-tailed kite, especially in the spring. Most of the rice stubble is burned during the winter, and in spring, the remainder is plowed under. Subsequently, white-tailed kites begin to increase foraging in alfalfa and orchards.

The California vole is the primary prey of the white-tailed kite (Stendell and Meyers 1973, Mendelson and Jaksic 1989). This vole depends upon natural vegetation and standing water for breeding and dispersal (Stendell and Meyers 1973, Lidicker 1980). In general, California vole densities are low and their distribution is patchy in agricultural habitats (Ostfeld and Klosterman 1986, Estep 1989). In terms of agricultural crops, sugarbeets and alfalfa generally contain the highest densities of voles (Lidicker 1980, Ostfeld and Klosterman 1986, Estep 1989). The hunting success of white-tailed kites is directly proportional to the abundance and composition of prey species (Stendell and Meyers 1973, Warner and Rudd 1974). Therefore, it is likely that transect populations and habitat selection measurements reflect, in a large part, the abundance of California voles.

We found Atlas GIS to be a user-friendly, effective tool for analyzing the spatial use of habitat by kites. However, there were limitations. The land-use maps, which serve as templates for GIS-based analyses, are out-dated and lack resolution. Specifically, the land-use coverages do not show small fallow strips of vegetation between fields, or along roads. These linear elements are important hunting areas for raptors in the Sacramento Valley (Smallwood et al. 1996).

In general, our analyses of nesting habitats demonstrate that kites in the study area exploit a highly heterogeneous landscape and select nest territories that are in patches and corridors of natural vegetation amidst agricultural fields and human development. Analysis using the 0.8-km buffers produced significant results whereas the 1.6-km buffers did not. Successful nests had more natural vegetation and human development within 0.8 km of the nest than unsuccessful nests. This suggests that the immediate nesting habitat might be a more important factor for nest success than the larger breeding territory. Furthermore, the aggressive interactions between kites and Swainson's hawks occurred in the areas within 0.8 km of the nest.

Indeed, land-use analysis of kite nesting habitat is perhaps confounded by the loss of territories to Swainson's hawks, which strongly prefer riparian corridors. Therefore, the use of fallow developed areas by kites may be secondary if Swainson's hawks avoid these areas. The selection of riparian corridors, by 41% of the pairs, one of which was successful, implies that nest success is influenced by factors other than nest habitat availability. Additionally, 88% of the successful nests were located in varied classes of human development such as unmanaged vegetation adjacent to developed lands, rural homes, ranches, abandoned buildings, and construction sites (One tailed Mann-Whitney $T = 77$, $P < 0.025$). White-tailed kites might be able to adapt to the pressures of competition for primary nest sites, such as riparian corridors, if suitable alternatives are available. Still, the availability of alternatives is waning as urban development continues spreading across the Sacramento Valley. The habitats of two successful 1993 nests, that were located in urban coverages, are not available this year due to the commencement of construction and removal of existing trees and fallow vegetation.

In the 1993 breeding season, 41% of white-tailed kites in our study area (625 km^2) had one or more fledglings per pair. White-tailed kites occupied nesting territories from late January through late September 1993 and there is potential for kites to lay second clutches, or to establish a second nesting territory, if they fail in one location. However, we do not know whether failed breeding pairs renested. A telemetry study is underway to track changes in the nesting, hunting, and communal roosting habitats of a breeding pair of white-tailed kites. As of 1 May 1994 the pair had remained together within the 1.6 km buffer area along a riparian corridor throughout the winter, renested in the riparian area once, failed in late March (aggression by kites was frequently observed towards a pair Swainson's hawks <100 m to the west), and renested in riparian habitat within 0.2 km of the first nest. We found no references to past levels of reproductive success for local populations of this species.

Almost half of the failed nests were caused by kites losing nest territories to other hawks, namely Swainson's hawks. Swainson's hawks have also been documented to usurp the high-quality territories of red-tailed hawks (Janes 1994). At this time Swainson's hawks are threatened in California. Research on the Swainson's hawk is focusing on interspecific interactions for nest territories between tree-nesting raptors in the Central Valley (P. Moreno pers. comm.). The primary nesting and foraging habitats that remain in California for the Swainson's hawk are riparian corridors (Estep 1989, Risebrough et al. 1989). We propose that the same trend applies to white-tailed kites and that future studies will be necessary on the comparative habitat uses of both species.

The prey of the two species are similar as well. Swainson's hawks also prefer California voles and hunt in crops that are associated with vole abundance such as fallow fields, alfalfa, sugarbeet, and tomato fields. Estep (1989) reported that all Swainson's hawk nests were adjacent to primary foraging areas. It is possible that white-tailed kites and Swainson's hawks in our study area preferentially hunt the same prey and thus select very similar foraging and nesting territories.

Cooperative efforts between researchers and farmers are underway to create farm habitat for wildlife. Native grasses and trees have been planted along roadsides, ditches, and fields. In contrast, conventional farming practices involve maintaining roadsides, ditches, and field edges as bare dirt strips using herbicides (J. Anderson pers. comm.). This practice is expensive and provides no habitat for wildlife dispersal, hunting, or breeding.

The white-tailed kite, like the Swainson's hawk and other raptors inhabiting the Sacramento Valley, needs habitats which are rapidly disappearing from the landscape. Certainly the white-tailed kite has adapted in many ways to human-altered landscapes. It is unlikely, however, that its population will remain stable if riparian habitats, natural and fallow lands continue to be fragmented and developed.

ACKNOWLEDGMENTS

This work was supported by the United States Environmental Protection Agency (US EPA) Center for Ecological Health Research at the University of California, Davis (Grant Number R819658) and the US Department of Agriculture, National Research Initiative Competitive Grants Program. Although the information in this document has been funded wholly or in part by the US EPA, it may not necessarily reflect the views of the Agency and no official endorsement should be inferred. Special thanks are also extended to Brenda Nakamoto, Nancy Ottum, Robin Rhoten and the Olmo Ranch for their valuable assistance. Additional gratitude is extended to Drs. R. Titman and J. Bustamante for their critical assistance in preparing this manuscript.

REFERENCES

AFIFI, A.A. AND V. CLARK. 1984. *Computer-aided multivariate analysis*. Wadsworth Press: Belmont, California USA.

BLOOM, P.H. 1985. Raptor Movements in California. *Proc. Hawk Migration Conference* IV: 313–324. Hawk Migration Association of North America.

DANIEL, W.W. 1990. *Applied non-parametric statistics*. PWS-Kent Publishing Co., Boston, Massachusetts USA.

EISENMAN, E. 1971. Range expansion and population increase in North and Middle America of the white-tailed kite. *Amer. Birds* **25**(3): 529–536.

ESTEP, J.A. 1989. *Biology, movements, and habitat relationships of the Swainson's hawk in the Central Valley of California, 1986–87*. Calif. Dep. Fish and Game, Nongame Bird and Mammal Sec. Rep.

FRY, D.H. JR. 1966. Recovery of the white-tailed kite. *Pac. Disc.* **19**: 27–30.

FYFE, R.W. AND R.R. OLENDORFF. 1976. *Minimizing the dangers of nesting studies to raptors and other sensitive species*. Can. Wildl. Serv. Occas. Paper No. 23.

GRIGGS, F.T. 1993. Protecting biological diversity through partnerships: The Sacramento River Project. Pp. 235–237 in J.E. Keely, ed. *Interface between ecology and land development in California*. Southern California Academy of Sciences, Los Angeles, California USA.

HARRINGTON-TWEIT, B. 1980. First records of the white-tailed kite in Washington. *West. Birds* **11**: 151–153.

HAWBECKER, A.C. 1940. The nesting of the white-tailed kite in southern Santa Cruz County, California. *Condor* **42**: 106–111.

HENNY, C. AND J. ANNEAR. 1978. A white-tailed kite breeding record for Oregon. *West. Birds* **9**: 130–131.

JANES, S.W. 1994. Partial loss of red-tailed hawk territories to Swainson's hawks: Relations to habitat. *Condor* **96**: 52–57.

KATIBAH, E.F. 1984. A brief history of riparian forests in the Central Valley, California. Pp. 23–36 in R.E. Warner and K.M. Hendrix, eds. *California riparian systems*. University of California Press, Berkeley, California USA.

KOPPEN, W. AND R. GEIGER. 1954. Climate of the Earth (map). A.J. Nystrom and Co., Chicago, Illinois USA.

LARSON, D. 1980. Increase of White-tailed kite populations in California and Texas: 1944–1978. *Amer. Birds* **34**: 689–699.

LIDICKER, W. JR. 1980. The social biology of the California vole. *The Biologist* **62**: 46–55.

MENDELSON, J.M. AND F.M. JAKSIC. 1989. Hunting behavior of black-shouldered kites in the Americas, Europe, Africa, and Australia. *Ostrich* **60**: 1–75.

OSTFELD, R.S. AND L.L. KLOSTERMAN. 1986. Demographic substructure in a California vole population inhabiting a patchy environment. *J. Mammal.* **67**: 693–704.

PALMER, R.S., ed. 1988. Black-shouldered kite. Pp. 132–147 in *Handbook of North American birds*. Yale University Press, New Haven, Connecticut USA.

PRUETT-JONES, S.G., M.A. PRUETT-JONES AND RICHARD KNIGHT. 1980. The White-tailed kite in North and Middle America: Current status and recent population changes. *Amer. Birds* **34**: 682–690.

RISEBROUGH, R.W., R.W. SCHLORFF, P.H. BLOOM AND E.E. LITTRELL. 1989. Investigations of the decline of Swainson's hawk populations in California. *J Raptor Res.* **23**: 63–71.

SMALLWOOD, S.K. 1993. Understanding ecological pattern and process by association and order. *Acta Oecologica* **14**: 443–462.

—, F.J. NAKAMOTO AND S. GENG. 1996. Association analysis of raptors in a farming landscape. Pp. 177–190 in D.M. Bird, D.E. Varland and J.J. Negro, eds. *Raptors in human landscapes*. Academic Press, London UK.

STENDELL, R.C. AND P. MEYERS. 1973. White-tailed kite predation on a fluctuating vole population. *Condor* **75**: 359–360.

WARNER, J.S. AND R.L. RUDD. 1974. Hunting by the white-tailed kite. *Condor* **77**: 226–230.

19

Association Analysis of Raptors on a Farming Landscape

Shawn K. Smallwood, Brenda J. Nakamoto and Shu Geng

Abstract – We developed an extensive sampling program for a landscape study of raptors on the farming landscape of the Sacramento Valley, California. By September 1993, after 47 months and 97 surveys along a 200-km road transect, we mapped 5662 observations of 11 species of Accipitridae, 1505 of 3 species of Falconidae and 986 of turkey vulture. With each observation, we recorded activity or perch used, and association with landscape elements and other species. Pooled observations of Accipitridae and Falconidae demonstrated a strong migratory cycle and aggregation in the landscape. They preferentially selected riparian, wetland and upland vegetation, alfalfa fields, and rice stubble and other crop debris, all of which occurred rarely. They avoided human settlements, plowed fields, and most grain and row crops. Accipitridae preferred trees with snags and open views, and utility poles with multiple horizontal beams. The associations we identified will help agriculturalists develop effective strategies for increasing raptor populations on farmland. Strategies under study include farmscaping, changes in cultural practices, and landscape engineering, all of which will benefit raptor conservation, farming efficiency and the aesthetic value of the landscape.

Key words: Accipitridae; agriculture; association analysis; conservation; Falconidae; landscape ecology; perch preference; road survey; turkey vulture.

Agriculture is conducted on lands that used to support the most diverse and abundant flora and fauna in the corresponding regions (Pimentel *et al.* 1992). Cropland and grassland/pasture comprise about 100 000 000 km^2 of the earth's terrestrial surface area, and most of the cropland was added during the last 300 yr (Meyer and Turner 1992). Amidst such rapid change, many of the native species still exist on these managed lands (Banaszak 1992, Kromp and Steinberger 1992, Smallwood in press), often within riparian corridors (Naiman *et al.* 1993), nature reserves, or fragments of natural habitat. But administrative boundaries of nature reserves should not be expected to translate into real biological boundaries (Schonewald-Cox and Bayless 1986). Agriculture impairs inter-reserve movement and intra-reserve survival of native mammal species (Smallwood in press), thus reducing the food supply of raptors across most of the landscape.

Prey-bearing habitat on the farmland of the Sacramento Valley, California, might limit raptor populations regionally, because many raptors from western North America aggregate in the Valley during one phase of their annual migration (Bloom 1985), and because raptor populations respond to their prey population dynamics (Grant et al. 1991). Therefore, raptor populations can be conserved, and even augmented, by increasing the availability of prey-bearing habitat in the Sacramento Valley with changes in landscape structure and cultural practices. Such changes might be made most efficiently by learning how raptors use the existing landscape, given its structure, perches and management.

Previous studies have shown that populations of bird species benefit from greater structural complexity on farms (O'Connor and Shrubb 1986) and roadside edges (Ferris 1979, van der Zande et al. 1980, Arnold and Weeldenburg 1990). Insectivorous birds best infiltrate corn fields that are adjacent to woodlots (Best et al. 1990). Existing landscape engineering efforts in the Sacramento Valley (Griggs 1993, Yolo County Farmland and Open Space Conference, May 8, 1993) are focused on increasing the spatial extent of natural vegetation, mainly as a network (*sensu* Forman 1981) of restored riparian habitat and strips of natural vegetation between agricultural fields. These 'corridors' are where the richest and most diverse assemblages of wildlife species occur in the Valley (Griggs 1993). However, the commodity type and cultural practices in agricultural fields can significantly affect the food base of raptors that would use the adjacent or nearby corridor network. Vertebrate pest control programs in some agricultural fields can seriously reduce the food supply of hawks. Certainly, raptor conservation on the farming landscape requires more than open space; it also requires a suite of crops and cultural practices (or lack thereof) that provide resources for wildlife and that are non-destructive. Our goal is to provide these types of information by combining landscape ecology with ecosystem ecology (Forman 1981, Risser 1985) and with conservation biology (Turner 1989).

We report on patterns of seasonal distribution, activity and perch use of functional/taxonomic raptor assemblages among landscape elements of the Sacramento Valley. A landscape is a spatially heterogeneous area in which its patch and corridor elements are *structured* by type, size, shape and configuration, and in which they *function* to store and channel the flow of energy, material and organisms (Foreman 1981, Turner 1989). Landscape ecology can help identify how raptors interact with landscape elements, and with landscape structure, so long as the *extent* of the study is at a spatial scale large enough to recognize structure (Turner 1989). To achieve a sufficiently large spatial scale for a landscape study of raptors, we used non-conventional, high-speed surveys, which provided large sample sizes across large areas, but also new types of error (Smallwood and Geng in press). Most other road surveys for raptors occurred only during winter in smaller areas with simple categories of landscape elements (Wilkinson and Debban 1980, Bauer 1982, Craig et al. 1986, Andersen and Rongstad 1989).

METHODS

We conducted wildlife surveys along 200 km of road transect, which was in 7 parts along a 320-km loop in the Sacramento Valley (Fig. 1). One person drove a car at 84 kph, while the other searched for wildlife and recorded observations into a tape-recorder. Each survey was begun between 0800 h and 1000 h and ended by 1400 h. Surveys were conducted during all weather conditions, except during the rare winter storms when visibility was severely limited. Along this transect, we mapped crops, natural vegetation, the Sacramento River (encountered 3 times), creeks, canals, roads, houses, roadside vegetation height and width, and potential perches such as trees, utility poles, transmission towers, fences, standpipes and artificial hawk perches. Landscape changes and cultural

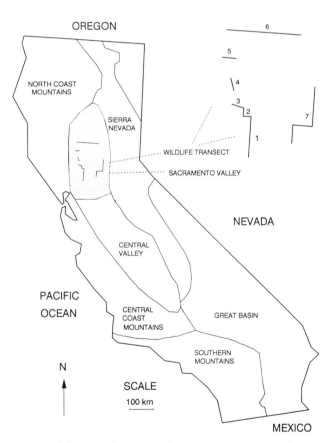

Figure 1. Locations of the 7 road transects for raptor survey in the Sacramento Valley, California. Dominant landscape elements were riparian and upland/wetland along transect 1, dry and irrigated pasture along transect 7, rice along transects 5 and 6, and field crops along transects 2 and 3. Transect 4 was a mix of pasture, riparian, field crops and rural.

practices were monitored and mapped. For example, we recorded dates of planting, harvest, irrigation, plowing, burning of debris, and spraying of pesticides when possible. We surveyed the first 58 km of the transect 97 times no more than once per week for 47 months. We recorded the abundance, locations, and activities and types of perches of 82 vertebrate species. Locations included odometer reading to the nearest 0.16 km, side of road, distance from road, and the associated landscape element. All animal observations of questionable taxonomic identification were not recorded. Species were added to the survey as developing expertise allowed accurate identification.

Similar to Marion and Ryder (1975) and Preston (1990), we identified associations between raptors and landscape elements by comparing the incidence of the species population on a landscape element (percentage "use") with the incidence of that element on the sampled landscape during the sampled time (percentage "availability"). Association of a species with a landscape element can be expressed by the ratio:

$$M_i = \frac{r_i}{p_i} = \frac{n_i}{N * p_i} = \frac{\text{Observed}}{\text{Expected}},$$

where n_i is the number of individuals counted in the ith landscape element, N is the number of individuals counted in the entire sample, r_i is n_i/N (as used by Marion and Ryder 1975), and p_i is the incidence of the ith landscape element in the sample (Smallwood 1993). M_i interprets as the species occurrence in the ith landscape element as the multiple of that to be expected from a uniform pattern of the species on the landscape. Values <1 express avoidance of the ith landscape element, and values >1 express selection preference. Observed and expected values were also rearranged for testing the hypothesis of homogeneity:

$$\chi^2 = \sum \frac{(O - E)^2}{E},$$

where O is observed and E is expected values, and the degrees of freedom is $i - 1$ landscape elements.

RESULTS

Populations of Accipitridae, Falconidae and turkey vulture varied annually (Fig. 2), seasonally (Fig. 3) and among sites (Table 1). Accipitridae and Falconidae were most numerous in the valley during winter, although the Swainson's hawk was most numerous during summer. Vultures were most numerous during spring and summer. As these raptor populations varied, they changed their use of landscape elements and cultural practices, both of which also varied spatially and temporally (Fig. 4).

Accipitridae

Hawks were most dense along transect 2, and least dense along transect 5 (Table 1, Fig. 3). The species distribution of Accipitridae among transects was heterogeneous (groups compared: ferruginous hawk, red-tailed hawk, Swainson's hawk, rough-legged hawk, red-shouldered hawk, northern harrier, white-tailed kite, and sharp-shinned and Cooper's hawks combined, $\chi^2 = 183.1$, df = 35, $P < 0.001$). More than the expected number of ferruginous hawks and Swainson's hawks occurred on transects 2 and 3, rough-legged hawks on transects 1 and 6, northern harriers on transects 3 and 6, white-tailed kites on transect 6, and accipiters on transect 1.

Hawks preferred alfalfa and riparian, upland, and wetland vegetation throughout the year, but alfalfa was most preferred during summer (Fig. 4). Rice stubble was also preferred, but unavailable during summer. Burned and unburned rice stubble were equally preferred during fall and winter. Grain crops other than rice were preferred during fall, and annual vegetable crops were preferred during winter, when they were rare. Dry pasture, rice, plowed fields and rural areas were avoided throughout the year.

The incidence of potential perches on the landscape has not been quantified, so an association analysis of perch use could not be completed. Of the hawks observed, 30% were in flight, 29% on utility poles, 15% on telephone wires, 15% on trees, 5% on the ground, 4% on the rarely occurring transmission towers, and 2% on miscellaneous perches such as irrigation standpipes and pumps, fences, houses/buildings, tractors, street lights, signs and bridge rails. Hawks were observed on artificial hawk perches (narrow dowel on thin pole) only three times. Hawks used major types of perches/activity seasonally ($\chi^2 = 351$, df = 15, $P < 0.001$); they were observed more often than expected by chance in flight during spring and summer, on trees during winter, on utility poles during fall, and on transmission towers during summer. Of those observed in flight, more than the expected number of hawks were soaring during spring and fall, and in flutter-flight during winter and summer ($\chi^2 = 75.5$, df = 6, $P < 0.001$). Hawks preferred utility poles with multiple crossbeams or no crossbeams (Fig. 5). When possible, utility poles were used in such a way as to keep some structure above the hawk. Of the hawks observed in trees, 31% were on willows (*Salix* spp.), 30% on cottonwoods (*Populus fremontii*), 17% on oaks (*Quercus* spp.), 13% on American elm, 5% on eucalyptus (*Eucalyptus globulus*), and 4% on walnuts (*Juglans* spp.).

Falconidae

Falcons preferentially selected natural vegetation throughout the year (Fig. 4). They preferred annual vegetable crops during winter; rice stubble, annual vegetables and irrigated pasture during spring; alfalfa, fallow and plowed fields during summer; and grain fields and burned rice stubble during fall. Falcons avoided rural areas, rice, asparagus and dry pasture throughout the year.

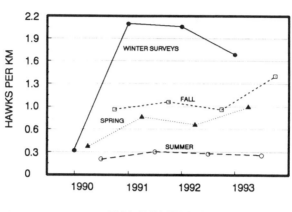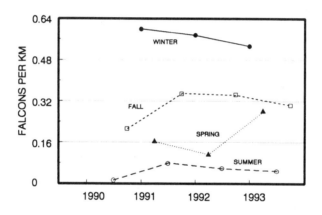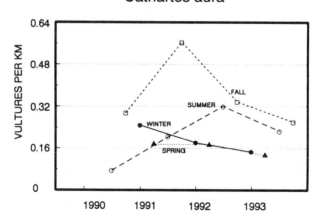

Of the falcons observed, 55% were on telephone wires, 23% were on utility poles, 14% were in flight, 4% were in trees, 1% were on the ground, and 3% were on transmission towers and miscellaneous perches. Of the falcons in flight, they tended to be in flutter-flight more often during spring and summer, and less often during fall ($\chi^2 = 7.27$, df = 3, $P = 0.064$).

Turkey Vulture

This species preferentially selected upland/wetland vegetation throughout the year (Fig. 4). It also preferred riparian vegetation and orchards during all seasons except fall, and irrigated pasture during all seasons but spring. Winter vegetables were strongly preferred, as were rice stubble and debris from summer vegetables. Dry pastures, asparagus, rice, plowed fields and rural areas were avoided throughout the year.

Of the vultures observed, 85% were in flight, of which 53% were soaring. Twelve percent were on the ground (usually at carrion), 1% were on trees, and the rest of the available perches were scarcely used.

DISCUSSION

We only analyzed seasonal variation in selection of landscape elements, but the substantial locational variation in density (Table 1) suggests the need for very large samples of raptor observations in time and space. Future analysis for raptor associations with landscape elements needs to include seasonal variation at each transect. The heterogeneity of species distribution among transects suggests that future analysis also needs to be conducted for each species separately. As sample sizes increase, and the association analysis is further split in time, space, and by species, we will learn much more about how raptors interact with landscape elements, landscape structural properties and agricultural practices.

Repeat, high-speed surveys for raptors along extensive road transects can provide the precision and heterogeneous distribution that are needed for pattern analysis (Smallwood and Geng in press). Predictions from models of distributional dynamics can be tested by surveying for raptors along extensions of the transect into nearby areas. The measure of association and related statistics are easy to use and free of the unrealistic assumptions required of surveys for

---

Figure 2. Population dynamics of the three species groups studied during the road surveys. Densities were calculated by summing among transects the numbers observed and distances traveled during each season of each year.

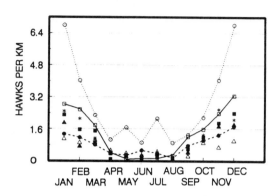

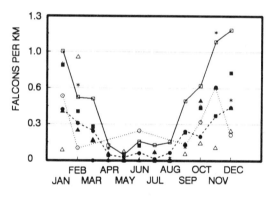

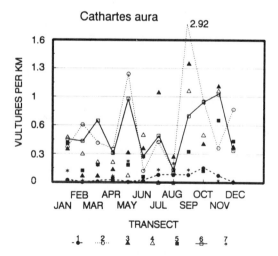

population estimation (Seber 1973, Eberhardt 1978, Burnham et al. 1980, Mikol 1980). However, sightability varies among species, landscape elements and season (perch use and flight patterns change seasonally), and needs to be assessed with long-term observation of individual hawks, e.g. Diesel (1989), Preston (1990), and other methods.

Riparian, upland and wetland vegetation were preferred by all raptors during most seasons, yet they comprised a minority of the landscape. A landscape engineering effort for restoring and extending riparian and strip corridors would likely increase raptor populations in the Sacramento Valley. Most raptor activity was concentrated at or nearby the Sacramento River and major streams with riparian habitat. Adjacency to riparian, upland and wetland vegetation accounted for many of the raptor associations with other landscape elements, e.g. orchards and pasture. Craighead and Craighead (1959) also observed many hawks off their preferred hunting grounds, but nearby.

Burned rice stubble attracted falcons during fall, but contributed only a short-term pulse of food. Unburned rice stubble provided a longer-lasting food source, and was preferred by raptors, especially as it grew more rare during the spring. Other crop debris, when they occurred, were also preferred by hunting raptors. Such strong preference for fields with crop debris suggests that food-bearing land is a limiting resource for wintering raptors on this farming landscape; there is little of it when it is most needed.

Plowed fields comprised a large portion of the landscape along our survey transects, but they were avoided by most raptors most of the time. Many falcons were recorded at plowed fields, but they were always near fallow strips of vegetation along canals, road verges or fence lines. We did not include these strips of vegetation in the association analysis. A widespread delay in plowing crop residue until just before the next planting would probably support substantially larger winter aggregations of raptors in the Sacramento Valley, which would translate into increased regional populations of raptors. Raptor conservation could be served by addressing this and other cultural practices. Farmers have practical and economic reasons for their farm management, but there usually are alternatives that would benefit wildlife without risking economic loss to the farmer.

Consistent avoidance of rural or developed areas by raptors demonstrated the need for preventing extensive housing and industrial development in the Sacramento Valley. Again, regional populations of raptors would be severely affected. Farming *is* less destructive to wildlife than is urban and industrial expansion. Many farmers do not wish to sell their farms to developers, but often yield to pressure from city and county governments and economic necessity.

Figure 3. Seasonal dynamics of the three species groups. Densities per month were calculated by summing among years the numbers observed and distances surveyed in each month.

Table 1. Number of observations of raptor species during the Sacramento Valley farmland survey from January 1990 until September 1993.

Species Common names	Transect							Total
	1	2	3	4	5	6	7	
Cumulative one-way survey distance (km)	5148	364	835	498	610	2174	2119	11748
Accipitridae	2026	522	320	135	299	1431	929	5662
Hawks, unidentified	299	116	44	50	40	148	133	830
Red-tailed hawk	497	226	99	43	227	340	306	1738
Swainson's hawk	78	5	11	2	8	17	9	130
Red-shouldered hawk	5	7	3	1	4	9	5	34
Rough-legged hawk	4	1	0	0	0	5	0	10
Ferruginous hawk	15	0	2	0	1	1	1	20
Osprey	0	0	1	0	0	0	0	1
Accipiters, unidentified	2	0	0	0	0	2	2	6
Sharp-shinned hawk	1	1	0	0	0	0	0	2
Cooper's hawk	2	4	0	0	0	3	0	9
Kites, unidentified	12	2	0	3	4	28	8	57
Northern harrier	138	11	46	4	30	180	100	509
White-tailed kite	46	29	6	3	8	79	34	205
Golden eagle	1	0	0	0	1	1	0	3
Unidentified	423	87	24	8	24	125	37	728
Falconidae	484	33	85	21	94	494	294	1505
Falcons, unidentified	405	25	60	17	79	405	244	1235
American kestrel	71	6	22	2	13	81	44	239
Merlin	5	0	0	1	0	1	3	10
Prairie falcon	3	2	3	1	2	7	3	21
Cathartidae								
Turkey vulture	86	101	141	81	61	422	94	986

Now is the time to manage farmland for raptor conservation, because the loss of farmland is a loss to wildlife, to our food base and to the farmer's way of life. A new farming landscape, with more raptors and supporting habitat, would strengthen its aesthetic value and its prospects for protection from development and over-intense use.

Patterns of perch use in the Sacramento Valley suggest that farmscaping could successfully attract raptors to fields where prey is available. Perches were not continuously present along our survey transects. Some alfalfa fields were not bordered by trees or utility poles; these fields were visited by few falcons and hawks. Many other fields were bordered by trees that raptors did not use, or by utility poles that lacked the structure preferred by hawks. Preferred tree species could be added to field borders, or a non-utility crossbeam could be added 0.5 m beneath the utility beam on many poles. This additional lower crossbeam would not only attract hawks to field borders, but also increase the distance between hawks and the dangerous aspect of the pole. It would reduce the risk of electrocution to hawks, especially in combination with other protective measures (Olendorff 1981, Ferrer *et al.* 1991). Multiple crossbeams possibly offer perched hawks protection from aggressive approaches by other hawks.

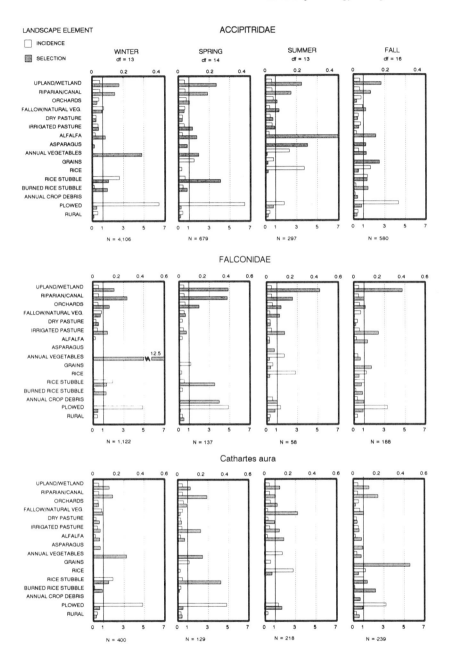

Figure 4. Seasonal variation in selection of landscape elements by species of Accipitridae, Falconidae and turkey vulture (*Cathartes aura*). The top horizontal axes correspond with incidence of landscape elements, and the lower axes correspond with the measure of ecological association, M_i, or "selection". The solid vertical lines at $M_i = 1.0$ corresponds with the selection to be expected based on chance. Use of the landscape was nonhomogeneous for all three taxonomic groups during each season ($P < 0.001$ for all χ^2 tests).

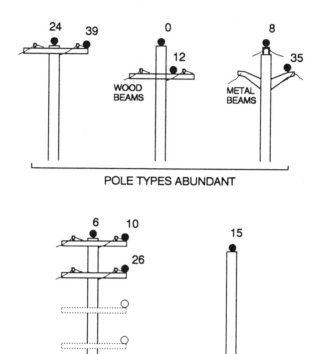

Figure 5. Occurrence frequency of Accipitridae on different types and aspects of utility poles during the latest nine months of the survey.

ACKNOWLEDGMENTS

Our study was partly funded by a grant from the USDA's National Research Initiative Competitive Grants Program. We thank J. Rodriguez for his assistance, and R. A. Ryder and J. A. Donazar for their review of our manuscript.

REFERENCES

ANDERSEN, D.E. AND O.J. RONGSTAD. 1989. Surveys for wintering birds of prey in southeastern Colorado: 1983–1988. *J. Raptor Res.* 23: 152–156.

ARNOLD, G.W. AND J.R. WEELDENBURG. 1990. Factors determining the number and species of birds in road verges in the wheatbelt of Western Australia. *Biol. Conserv.* 53: 295–315.

BANASZAK, J. 1992. Strategy for conservation of wild bees in an agricultural landscape. *Agric., Ecosyst., Environ.* **40**: 179–192.
BAUER, E.N. 1982. Winter roadside raptor survey in El Paso County, Colorado, 1962–1979. *Raptor Res.* **16**: 10–13.
BEST, L.B., R.C. WHITMORE AND G.M. BOOTH. 1990. Use of cornfields by birds during the breeding season: the importance of edge habitat. *Am. Midl. Nat.* **123**: 84–99.
BLOOM, P.H. 1985. Raptor movements in California. *Proc. Hawk Migration Conference* **IV**: 313–324. Hawk Migration Association of North America.
BURNHAM, K.P., D.R. ANDERSON AND J.L. LAAKE. 1980. Estimation of density from line transect sampling of biological populations. *Wildl. Monogr.* **72**: 1–202.
CRAIG, E.H., T.H. CRAIG AND L.R. POWERS. 1986. Habitat use by wintering golden eagles and rough-legged hawks in southeastern Idaho. *Raptor Res.* **20**: 69–71.
CRAIGHEAD, J.J. AND F.C. CRAIGHEAD. JR. 1959. *Hawks, owls, and wildlife.* Dover Publications, Inc., New York.
DIESEL, D.A. 1984. Evaluation of the road survey technique in determining flight activity of red-tailed hawks. *Wilson Bull.* **96**: 315–318.
EBERHARDT, L.L. 1978. Transect methods for population studies. *J. Wildl. Manage.* **42**: 1–31.
FERRER, M., M. DE LA RIVA AND J. CASTROVIEJO. 1991. Electrocution of raptors on power lines in southwestern Spain. *J. Field Ornithol.* **62**: 181–190.
FERRIS, C.R. 1979. Effects of Interstate 95 on breeding birds in Northern Maine. *J. Wildl. Manage.* **43**: 421–427.
FORMAN, R.T.T. 1981. Interaction among landscape elements: a core of landscape ecology. Pp. 35–48 in *Proc. Int. Congr. Neth. Soc. Landscape Ecol., Veldhoven.* Pudoc, Wageningen.
GRANT, C.V., B.B. STEELE AND R.L. BAYN, JR. 1991. Raptor population dynamics in Utah's Uinta Basin: the importance of food resource. *Southwest. Nat.* **36**: 265–280.
GRIGGS, F.T. 1993. Protecting biological diversity through partnerships: the Sacramento River Project. Pp. 235–237 in J.E. Keeley, ed. *Interface between ecology and land development in California.* Southern California Academy of Sciences, Los Angeles.
KROMP, B. AND K.-H. STEINBERGER. 1992. Grassy field margins and arthropod diversity: a case study on ground beetles and spiders in eastern Austria. (*Coleoptera: Carabidae, Arachnida: Aranei, Opiliones*). *Agric., Ecosyst., Environ.* **40**: 71–93.
MARION, W.R. AND R.A. RYDER. 1975. Perch-site preferences of four diurnal raptors in northeastern Colorado. *Condor* **77**: 350–352.
MIKOL, S.A. 1980. *Field guidelines for using transects to sample nongame bird populations.* US Fish Wildl. Serv., Office of Biological Service -80/58, Washington, D.C.
MEYER, W.B. AND B.L. TURNER, III. 1992. Human population growth and global land-use/cover change. *Ann. Rev. Ecol. Syst.* **23**: 39–61.
NAIMAN, R.J., H. DECAMPS AND M. POLLOCK. 1993. The role of riparian corridors in maintaining regional biodiversity. *Ecol. Applic.* **3**: 209–212.
O'CONNOR, R.J. AND M. SHRUBB. 1986. *Farming and birds.* Cambridge University Press, New York.
OLENDORFF, R.R. 1981. *Suggested practices for raptor protection on power lines.* Raptor Res. Rep. no. 4.
PIMENTEL, D., H. ACQUAY, M. BILTONEN 1992. Conserving biological diversity in agricultural/forestry systems. *Bioscience* **42**: 354–362.
PRESTON, C.R. 1990. Distribution of raptor foraging in relation to prey biomass and habitat structure. *Condor* **92**: 107–112.
RISSER, P.G. 1985. Toward a holistic management perspective. *Bioscience* **35**: 414–418.
SCHONEWALD-COX, C.M. AND J.W. BAYLESS. 1986. The boundary model: a

geographical analysis of design and conservation of nature reserves. *Biol. Conserv.* **38**: 305–322.

SEBER, G.A.F. 1973. *The estimation of animal abundance.* Hafner Press, New York.

SMALLWOOD, S.K. 1993. Understanding ecological pattern and process by association and order. *Acta Oecologica* **14**: 443–62.

—. In press. Site invasibility by exotic birds and mammals. *Biol. Conserv.*

— AND S. GENG. In press. Landscape strategies for biological control and IPM. In *Proc. Integrated Resource Management for Sustainable Agriculture.* Beijing, China.

TURNER, M.J. 1989. Landscape ecology: the effect of pattern on process. *Ann. Rev. Ecol. Syst.* **20**: 171–197.

WILKINSON, G.S. AND K.R. DEBBAN. 1980. Habitat preferences of wintering diurnal raptors in the Sacramento Valley. *Western Birds* **11**: 25–34.

VAN DER ZANDE, A.N., W.J. TER KEURS AND W.J. VAN DER WEIDEN. 1980. The impact of roads on the densities of four bird species in an open field habitat: evidence of a long-distance effect. *Biol. Cons.* **18**: 299–321.

20

Sparrowhawks in Conifer Plantations

I. Newton

Abstract — Sparrowhawks in Britain prefer to nest in conifer plantations rather than in native broadleaved woods, despite lower prey densities in conifers. In Eskdale, south Scotland, sparrowhawks did not use the available plantations at random for nesting, but preferred those where nest success was demonstrably highest. They occupied plantations for nesting after trees had been thinned for the first time, at around 20 yr of age. Occupancy and nest success were then high for the next 10 yr or so, but declined thereafter, as the plantation matured. Owing to rotational management, in which a proportion of plantations was felled and replanted each year, the age structure of available plantations remained fairly stable through time. Sparrowhawk breeding numbers were also fairly stable, but each year some old plantations were abandoned and young ones occupied.

Key words: sparrowhawk; conifer plantation; territory quality; habitat quality; habitat succession; population regulation.

In this paper I shall describe the breeding of sparrowhawks in conifer plantations, examining nest success in relation to spatial and temporal variation in plantation habitat. The paper is based on data collected over a 20 yr period (1972–91) from a 200 km^2 study area centred on Eskdale (55g 16h N 3g 5h W), a mixed farmland–plantation landscape in south Scotland. Within this area, I searched all suitable plantations every year to find nests and count the numbers of young raised. I also banded the young, and as many of the breeders each year as I could catch. As in the rest of Britain, sparrowhawks were resident in Eskdale throughout the year, feeding on other birds, especially songbirds. Within their plantation nesting habitat, the hawks tended to nest in the same restricted localities year after year. Other aspects of the study have been described in detail elsewhere (Newton 1986, 1991).

THE PLANTATION HABITAT

Some 2000 yr ago, as the pollen record reveals, most of Britain was blanketed by broadleaved deciduous forest. But by the fourteenth century, an expanding human population had reduced the forest cover to only 10% of its original area (Rackham 1986). Since then further reductions and modifications have occurred, so that today no natural old-growth forest remains anywhere in Britain. In the last 70 yr however, substantial areas of new conifer plantation have been added to the remaining fragments of modified broadleaved woodland. These new forests consist of Norway and Sitka spruces, Scots and Lodgepole pines and larches (*Larix* spp.), mostly introduced to Britain from other parts of the world (only Scots pine is native). The young trees are planted close together (usually 1.5 × 2 m) in monocultural plantations varying from a few hectares to several thousand hectares in area. After about 20 yr, the resulting impenetrable thickets are thinned to produce a more open structure. Thinning is then repeated every 5 yr or so until the trees reach 40–50 yr of age, when they are clear-felled. A fresh tree crop is then planted on the cleared ground.

Compared with diverse broadleaved woodland, these conifer monocultures – with their uniform structure – tend to be poor in birds. In south Scotland, counts revealed songbird densities of about 200–600 pairs per km^2 in 20–46 yr plantations, compared with up to 500–1800 pairs per km^2 in nearby semi-natural broadleaved or mixed woods of similar age (Moss 1978, Moss *et al.* 1979). The difference in avian biomass was even greater, for most of the birds in plantations weighed less than 20 g, giving an overall biomass of 5–12 kg per km^2, whereas many larger birds occurred in broadleaved and mixed woods, giving an overall biomass of 20–70 kg per km^2. However, sparrowhawks nest commonly in the new plantations, preferring them to broadleaved areas (Newton 1986). They hunt partly in the plantations and partly in other woodland or open areas nearby (Marquiss and Newton 1982). Compared with the presumed original situation, sparrowhawks in Britain now nest mostly in conifers, and only to a small extent in broadleaved trees, and obtain more of their food from open areas. Their diet must also have changed greatly, because some of the open-country species now taken commonly as prey would have been scarce or absent in the original forest.

Within the Eskdale study area, almost all woodland was in the form of coniferous plantation, managed on a rotation basis. Each year some areas of mature plantation were felled and replanted, so that at any one time the total plantation area consisted of a mosaic of patches of different ages, giving a fairly uniform age-structure over time. The plantations were scattered among farmland and open sheep pasture, covering a total of about 20 km^2.

Despite continual management of the nesting habitat, sparrowhawk nest numbers remained fairly constant. Over the 20-yr period, the average number of nests found per year was 34, and the numbers in particular years were always within 15% of this mean (Newton 1991). Because of changes resulting from forest management, one analysis given below was based on data from only 10 yr and another from only 15 yr of the 20-yr study. This curtailment was necessary to avoid short-term spatial patterns in occupancy and nest success being obscured by long-term temporal changes.

SPATIAL VARIATION IN HABITAT QUALITY

For present purposes the quality of plantation habitat was measured as the number of young sparrowhawks produced from particular nesting places during a 10-yr period (1977–86). In each year a sparrowhawk pair could raise up to 6 young in the single brood, so in theory each nesting place could have produced 60 young in the 10-yr period. In practice however, the individual nesting places varied greatly in the numbers of young they produced (range 0–33), and none of them yielded anywhere near the maximum possible (Fig. 1). Productive and unproductive nesting places were not concentrated in different parts of the study area, but were interspersed. Some of the most productive places were adjacent to wholly unproductive ones. This spatial variation in productivity was 'natural' in that it was not directly due to human interference.

Much of the overall variation in productivity resulted, as expected, from variation in the number of years a place was used for nesting. Although all patches of plantation habitat in Fig. 1 were available for the whole 10-yr period, not all were used every year. Some places were used only once in that time. In general then, total production of young over 10 yr increased with the number of years a nesting place was occupied (Fig. 2). The variation was considerable however, and one plantation was occupied in 8 of the 10 yr but still produced no young.

In no year in the study area did sparrowhawks occupy all available plantation nesting places (if available was defined on the basis of age, structure and spacing). Nor over a period of years did they occupy the available places at random. Rather they showed marked preferences for certain places, using them more than expected by chance at the population levels found, and avoiding others which they used less than expected by chance (Newton 1991). In other words, the hawks had favoured patches of plantation habitat.

To examine whether such local habitat preferences were related to nest success, I took the data for a 15-yr period and graded all the nesting places used in that time according to the number of years they were occupied. A grade 1 place was used 1–3 yr (not necessarily successive years) in the 15-yr period, a grade 2 place in 4–6 yr, grade 3 in 7–9 yr, grade 4 in 10–12 yr and grade 5 in 13–15 yr (Table 1). Places available for fewer than 15 yr (because of timber felling) were graded according to the proportion of the available years they were occupied.

In general, the proportion of nests that produced young increased with grade of nesting place (= frequency of usage). Evidently sparrowhawks nested most often in those places where their chances of raising young were greatest. Over the five grades, there was a two-fold variation in the average number of young produced per nest.

To summarize so far, plantation nesting habitat varied in quality, as measured by the number of young produced over a period of years. Certain nesting places were used more often than others, and in these favoured places individual nest success was greatest.

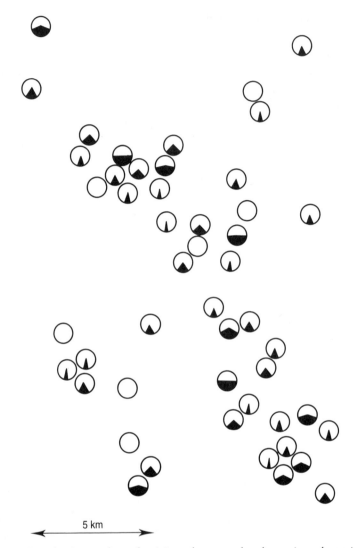

Figure. 1. Distribution and productivity of sparrowhawk nesting places in conifer plantations over a 10-yr period, Eskdale, south Scotland. Circles indicate all plantation nesting places which remained continuously available over a 10-yr period, and were used at least once in that time. Shading within the circles depicts the number of young produced as a proportion of 60, the maximum possible. Thus a half-filled circle indicates that 30 young were produced in the 10-yr period, a quarter-filled circle that 15 were produced, and an open circle that none were produced. From Newton (1991).

The high productivity of certain plantation nesting places could not be attributed to the same 'high quality' birds using these places over long periods. On average, most sparrowhawks remained on particular nesting places for only one year, although others stayed up to six years (mean 1.5 yr; Fig. 3). This high turnover in occupancy was due partly to mortality, and partly to movement, for

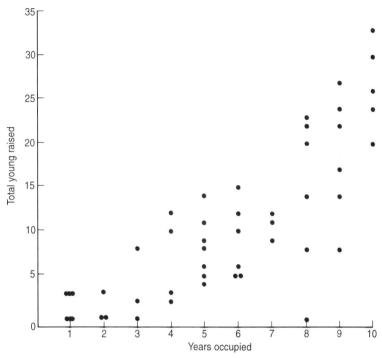

Figure 2. Relationship between the total number of young produced by sparrowhawks on particular nesting places over 10 yr and the number of years those places were occupied. The nesting places are the same as those in Fig. 1.

Table 1. Frequency of use and nest success of sparrowhawks in plantation nesting places over a 15-yr period, Eskdale, south Scotland.

Grade of nesting place	Number of years occupied	Total number of nests	Percentage of nests successful	Mean number of young per nest
1	1–3	26	31	1.19
2	4–6	54	48	1.57
3	7–9	161	47	1.60
4	10–12	150	60	2.11
5	13–15	123	71	2.63

Significance of variation among grades of nesting place in the proportion of nests that were successful: $\chi^2_4 = 26.87, P < 0.001$; and in the mean number of young raised per nest: $r_s = 1.00, P < 0.01$. Frequencies of different brood-sizes among nests were not normally distributed, so no standard deviations were calculated for the right-hand column.

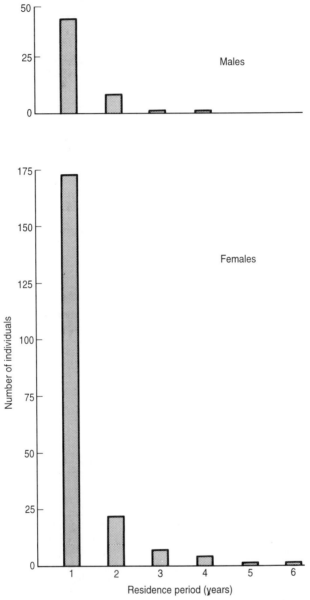

Figure 3. Residence periods of individual sparrowhawks in particular nesting places.

many individuals used different territories in different years (Newton and Wyllie 1992). Continued occupancy of certain nesting places was therefore due to different sparrowhawks occupying them in rapid succession but most staying only 1–2 yr.

There was, however, a significant tendency for high-grade nesting places to be occupied by older (more experienced) birds. In particular, first-year birds, which often have low nest success (Newton 1986), formed a greater proportion of breeders on low-grade places than of those on high-grade places. However, when first-year and older birds were examined separately, nest success within each age group was still higher on the high-grade places (Newton 1991). Hence, the higher performance recorded on high-grade nesting places was partly a function of place quality and partly of bird quality. Despite high turnover, good places tended to be occupied repeatedly by 'good birds'.

TEMPORAL VARIATION IN HABITAT QUALITY

Over a period of 10–20 yr, plantations changed greatly in internal structure. When first occupied, at around 20 yr of age, the trees had been thinned for the first time, enabling sparrowhawks to fly below the canopy, among the trunks and branches. The trees were still small (less than 10 m high) at that stage and close together (less than 4 m apart). Then, as the trees grew, and with further thinning every five yr or so, they gradually became larger and farther apart. As light penetrated, the bare ground became covered with herbaceous vegetation, and eventually supported small trees and shrubs under the main canopy. At 50 yr, the trees themselves had reached more than 30 m in height, and through thinning were more than 6 m apart. A similar change would presumably have occurred in natural forest, e.g. regrowing after fire, but in managed plantations the periodic removal of a proportion of trees by foresters speeded the process.

The question of interest was whether plantations in which sparrowhawks performed well over short periods of 5–10 yr remained good over longer periods of 10–30 yr. The short answer to this question was no, because plantation habitat deteriorated over time.

For the first 5–10 yr after first occupation at around 20 yr, plantation nesting places were occupied frequently and nest success was high (Fig. 4). Then, as the plantation aged, places were occupied less frequently and nest success declined. In consequence, most nesting occurred in plantations of 20–30 yr of age, less in plantations of 30–40 yr, and hardly any in older plantations. So the temporal stability of both population and nest success in the study area as a whole was associated with a system of rotational plantation management, which ensured a continuing availability of stands in the favoured age classes. The numbers of breeding pairs remained approximately constant over the years (see above), but their distribution within the area changed continually, as birds gave up using older plantations and occupied younger ones. Some banded individuals made the switch during their lifetimes. In conclusion then, both spatial and temporal variations in the physical structure of plantation nesting habitat was reflected in the breeding distribution and nest success of the hawks themselves.

DISCUSSION

Although these findings were from plantation forests, there is no reason to suppose that sparrowhawks would behave any differently in natural forest.

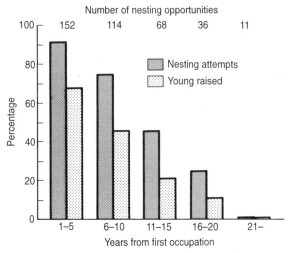

Figure 4. Decline in occupancy and nest success of sparrowhawks, according to years from first occupation of a nesting place (reflecting age of trees). The values along the top show the number of opportunities for nesting in plantations at each stage, dependent in turn on the number of plantations and the numbers of years they remained available. First occupation is defined by the first nest recorded during the growth of a plantation.

Throughout their wide Eurasian range, sparrowhawks mainly occupy young dense forest for nesting, being replaced in older, more open forest by a larger congener, the goshawk (Hald-Mortensen 1974, Newton 1986). Indeed, the preference of sparrowhawks for dense forest may be partly due to reduced opportunities for predation by the larger goshawk, which cannot easily penetrate the thicker areas that sparrowhawks prefer. However, predation avoidance may not be the only reason why sparrowhawks breed best in young forest. There were no goshawks in my study area, and the main causes of sparrowhawk nest failure in older plantations seemed to result from food shortage, namely non-laying, clutch desertion, and chick starvation (Newton 1991). Perhaps, therefore, sparrowhawks hunted less successfully in older, more open plantations, than in young dense ones. Although they hunted partly outside the plantations, most of their hunting in the first half of the breeding season (when most nest failures occurred) took place within 600 m of the nest. Those radio-tagged males that had young plantation within this radius showed a strong preference for hunting in it, and used it much more than expected from its proportion in their total hunting range (Marquiss and Newton 1982). No marked decline in songbird densities was detected in plantations of increasing age between 20 and 50 yr (surveyed in the same calendar years, Moss 1978, Moss et al. 1979), but the opportunities for sparrowhawks to catch such birds may have declined in the older, more open woods. Whatever the basis of the preference for young plantations, in the study area the age structure of the nesting habitat had much more influence on occupancy and nest success than did variations in the surrounding habitats.

It is less easy to explain why sparrowhawks prefer conifer over broadleaved trees for nesting. This preference is probably related to the structure of the trees themselves, because in broadleaved woods that contained only a few conifer trees, sparrowhawks almost invariably chose the conifers for nesting. The structure of conifers, in which several branches emerge at the same level, provides a sound platform for the nest, and the fact that successive whorls of branches emerge every metre or so up the trunk means that the nest is concealed from both above and below. This advantage is especially evident in early spring before foliage develops on broadleaved trees. Moreover, the dead twigs of conifers, being flexible yet breakable, are readily available as nest material. In the study area, larch twigs were especially favoured, the old leaf nodules ensuring that such twigs locked together to form a resilient structure. These may not be the only reasons why conifers were favoured for nesting, but they were the most obvious. Finally, I should add that, although conifers are new to sparrowhawks in most of lowland Britain, Scots pine had always been naturally available to sparrowhawks in certain mountainous areas of Britain, and that conifers are the commonest trees over most of the bird's geographical range in Eurasia. They have therefore been important in the evolutionary history of the species over its total range, if not in that small fraction of its range that falls within lowland Britain.

ACKNOWLEDGMENTS

I am grateful to the Buccleugh Estates and other landowners for permission to work on their land, and to Robert N. Rosenfield and an anonymous referee for helpful comments on the manuscript.

REFERENCES

HALD-MORTENSEN, P. 1974. Nest and nest site of the Sparrowhawk *Accipiter nisus* – and some comparisons with the Goshawk *A. gentilis. Dansk Orn. For. Tidss.* 68: 91–115.

MARQUISS, M. AND I. NEWTON. 1982. A radio-tracking study of the ranging behaviour and dispersion of European Sparrowhawks *Accipiter nisus. J. Anim. Ecol.* 51: 111–133.

MOSS, D. 1978. Song-bird populations in forestry plantations. *Quart J. For.* 72: 5–14.

—, P. N. TAYLOR AND N. EASTERBEE. 1979. The effects on song-bird populations of upland afforestation with spruce. *Forestry* 52: 129–150.

NEWTON, I. 1986. *The Sparrowhawk*. T. & A.D. Poyser, Calton, UK.

—. 1991. Habitat variation and population regulation in Sparrowhawks. *Ibis* 133 (supplement): 76–88.

— AND I. WYLLIE. 1992. Fidelity to nesting territory among European Sparrowhawks in three areas. *J. Raptor Res.* 26: 108–114.

RACKHAM, O. 1986. *The history of the countryside*. Dent, London, UK.

21

Adaptations of Raptors to Man-made Spruce Forests in the Uplands of Britain

Steve J. Petty

Abstract – A brief historical account is provided to show the major influence that deforestation followed by persecution had on raptor populations in Britain, and the subsequent recovery of many species over the last 20 yr. The review then concentrates on how raptors have adapted to large spruce forests that have been created in the British uplands over the last 75 yr. Distinct raptor guilds and potential prey associated with four forest successional stages are described. Many forests were established over short periods of time, resulting in similar raptor guilds over extensive areas, which then dramatically changed as forests passed from one seral stage to the next. At the end of the first rotation, opportunities occur to alter the spatial structure of these extensive even-aged forests, by clear-felling patches much smaller than the initial planting and varying rotation length, to create a mosaic comprising patches of different seral stages. Current research suggests that such an approach will result in forests with more diverse and sustainable raptor populations. Sixteen species of raptor occur in the British uplands; most have either benefited or been little affected by the present level of afforestation. The hen harrier and golden eagle are two exceptions, both dependent on open moorland for foraging. Their future in forested areas will depend on how successfully forests and open habitats can be integrated.

Key words: succession; persecution; afforestation; deforestation; clear-felling; owls; guilds; communities.

Over the last 75 yr the area of forest in Britain has doubled from 5% of the land area to around 10% (Petty and Avery 1990). This has largely been achieved by creating large conifer forests on marginal agricultural ground in upland areas. Until recently, better quality land in the lowlands was unavailable for forestry.

Prior to afforestation, many upland areas in Britain had been virtually treeless for centuries and had developed wildlife communities of high national and international value (Thompson *et al.* 1988). So, it was not surprising that planting non-native conifers generated concern about the future of such communities. However, little consideration was given to the possible role of these forests in diversifying wildlife communities in the uplands (Avery and Leslie 1990, Petty and Avery 1990).

The older plantations are now reaching the end of their first rotation, and opportunities exist to alter spatial and structural composition to mimic characteristics of natural forests (Ratcliffe and Petty 1986, Peterken 1987, 1992).

Raptors are at the top of many forest food-chains, and so provide indicators of the value of these new forests for wildlife. In this paper I review first how raptors adapted to deforestation and subsequently to the creation of extensive spruce forests.

THE DEMISE OF NATIVE FORESTS

Forest cover was at its greatest extent in Britain around 5000 yr ago, covering most of the land surface apart from the highest mountain tops, cliffs and bogs and fens (Godwin 1975). They were mainly broadleaved forests apart from an extensive area of Scots pine in northeast Scotland with smaller pine forests elsewhere. Forest clearance was begun by Neolithic people, and continued at an increasing pace throughout the Bronze Age, Iron Age and into historical times (Dimbleby 1984). It has been estimated that by 1700 forest cover had been reduced to around 12% of the total land surface of Britain, and still further to 5% by 1900 (Holmes 1975). Although climatic changes during this period also influenced forest cover, anthropogenic factors had the greatest impact. Since the 1800s, the British uplands were used largely for rearing sheep, deer (mainly *Cervus elephus*) and red grouse (Ratcliffe 1990).

Such a dramatic loss of forest had a major influence on raptors (Newton 1972). The breeding ranges of forest species such as the honey buzzard, northern goshawk, Eurasian sparrowhawk, and tawny owl were greatly reduced. In contrast, some species such as the golden eagle, common buzzard, common kestrel and merlin adapted to open moorlands by nesting on cliff ledges or on the ground rather than in trees, and by switching their diet to sheep and deer carrion or open-country species.

The final blow for many raptors was the rise in persecution during the nineteenth century (Newton 1972, 1979). This had the most pronounced effect on large species with low breeding densities and/or low reproductive rates. White-tailed eagle, northern goshawk and osprey ceased to breed in Britain, while the breeding ranges of honey buzzard, hen harrier and red kite were reduced to small pockets (Sharrock 1976, Watson 1977, Davis 1993). So, by the early part of this century the composition of raptor guilds had dramatically changed, initially by deforestation and subsequently through intense persecution that reduced the breeding range of some species while eliminating others.

RAPTORS OCCURRING IN THE BRITISH UPLANDS

Sixteen species of raptor breed in the British uplands (Table 1), and some have substantial populations in the lowlands too. Populations range from fewer than 10 pairs of the recently reintroduced white-tailed eagle to 50 000–100 000 pairs for the ubiquitous tawny owl.

Table 1. Breeding populations of raptors in Britain, with an indication of recent populations trends and the effects of afforestation. Only species that breed in the uplands are included, although the number of breeding pairs is for the whole of Britain.

Species	Breeding Pairs			Recent population trend (source)	Effects of afforestation (a) (source)
	Number	Year	Source		
Honey buzzard	10–30	1988–90	1,2	slight increase (1,2)	+
Red kite	97	1992	22	increasing (1,2,22)	= (21,22)
White-tailed eagle	<10	1990	1	increasing (1)	=
Hen harrier	Scotland 570±150 (95%CI) England and Wales 60	1988–89	3	varies regionally (3)	− (3,23,24)
Northern goshawk	400–500	1992–93	4	increasing (1,4,11)	+ (25)
Eurasian sparrowhawk	ca. 25 000	1984	5	increasing (12,13)	+ (12,26)
Common buzzard	12 000–17 000	1983	6	increasing (6)	+ (27)
Golden eagle	minimum 424	1982	7	static (7)	− (28)
Osprey	62	1990	1	increasing (1,2)	=
Common kestrel	up to 80 000	1984	5	fluctuates (5,14)	= (14)
Merlin	550–650	1983–84	8	increasing (15,16,17)	= (13,29,31)
Peregrine falcon	1200	1991	9	increasing (9)	=
Barn owl	4500–9000*	1968–72	10	decreasing/fluctuates (10,18,19,36)	= (19,30)
Tawny owl	5000–100 000	1968–72	10	increasing (10,20,36)	+ (20)
Long-eared owl	2200–7200*	1988–91	36	fluctuates (10,36)	= (32,33)
Short-eared owl	minimum 1000*	1988–91	36	fluctuates (10,36)	− (34,35)

(a): + beneficial, − detrimental, = neutral
* estimated breeding population in Britain and Ireland

References: 1 = Spencer *et al.* 1993, 2 = Batten *et al.* 1990, 3 = Bibby and Etheridge 1993, 4 = Petty unpubl., 5 = Newton 1984, 6 = Taylor, K. *et al.* 1988, 7 = Dennis *et al.* 1984, 8 = Bibby and Nattrass 1986, 9 = Ratcliffe 1993, 10 = Sharrock 1976, 11 = Marquiss 1981, 12 = Newton 1986, 13 = Newton and Haas 1984, 14 = Village 1990, 15 = Haworth and Fielding 1988, 16 = Ellis and Okill 1990, 17 = Little and Davison 1992, 18 = Shawyer 1987, 19 = Taylor *et al.* 1988, 20 = Petty 1992, 21 = Newton *et al.* 1981, 22 = Newton *et al.* 1993a, 23 = Watson 1977, 24 = O'Flynn 1983, 25 = Petty 1989a, 26 = Newton 1991, 27 = Newton *et al.* 1982, 28 = Watson 1992, 29 = Bibby 1986, 30 = Taylor *et al.* 1992, 31 = Parr 1994, 32 = Village 1981, 33 = Village 1992, 34 = Lockie 1955, 35 = Village 1987, 36 = Gibbons *et al.* 1993.

Since the 1970s, population trends have shown an increase for 10 out of the 16 species (63%), four (25%) have populations that fluctuate due mainly to annual variations in rodent density, one (6%) has a static population, and just the barn owl may be declining, although its population now appears to have stabilized (Table 1). The increases have resulted from numerous factors. A reduction in illegal poisoning has helped the red kite and common buzzard (Taylor K. et al. 1988, Batten et al. 1990); white-tailed eagle, red kite and northern goshawk have been reintroduced into areas from which they had become extinct (Marquiss and Newton 1981a, Love 1983, Batten et al. 1990); osprey have naturally recolonized Scotland (Dennis 1987); reducing levels of organochlorines in food-chains has greatly helped the recovery of the bird-eating Eurasian sparrowhawk and peregrine falcon (Newton and Haas 1984, Newton et al. 1993b, Ratcliffe 1993), and the recent improvements in merlin populations may also be attributable to this factor; and the increasing area of forest has allowed the northern goshawk, Eurasian sparrowhawk, common buzzard and tawny owl to extend their range. So, prospects for raptors in Britain are at present encouraging.

Three raptors have a mainly lowland distribution in Britain. These are marsh harrier, Eurasian hobby and little owl (Sharrock 1976, Gibbons et al. 1993). Most of the British population of Montagu's harrier breeds in cereal fields in the lowlands, although in the past they have bred on upland moors and in young conifer plantations, a habit which could occur again if the population increases (Brown 1976, Batten et al. 1990). These four species are not discussed further.

In Britain, all raptors are legally protected throughout the year under the Wildlife and Countryside Act 1981, and some species and their habitats are given additional protection under the European Community Wild Birds Directive. The habitats of raptors are less easy to protect legally because of low breeding densities and the relatively small size of most nature conservation areas. Therefore, it is essential to evaluate how raptors adapt to any extensive land-use change, such as afforestation.

THE NEW FORESTS

Afforestation commenced in earnest with the formation of the state-run Forestry Commission in 1919, although in the late eighteenth and during the nineteenth centuries there was some localised tree planting on private estates (Holmes 1975). At the last Forestry Commission Census of Woodland and Trees in 1979–82 there were 2.1 million ha under trees, 9.4% of the surface area of Great Britain (Locke 1987). About 47% was managed by the state-run Forestry Commission while the rest was in private ownership. So, in a 75 yr period the area of forest in Britain had almost doubled.

Much of this new planting was done in upland regions that had been tree-less for centuries, by creating large forests of introduced conifers (Petty and Avery 1990). It was not uncommon for 2000–3000 ha to be planted in a 2–3 yr period, and some of the largest forests are now over 50 000 ha. Sitka spruce from northwest America was the most important tree because it was easy to establish in a wide range of soil types and climates, and yielded a higher volume of usable timber than other conifers (Henderson and Faulkner 1987). More forest is now planted with Sitka spruce than with any other tree species (Petty and Avery 1990).

The creation of spruce forests generated much controversy, partly because they were seen as a threat to unique wildlife communities that had adapted to the open nature of many upland areas following the felling of native forests (Thompson *et al.* 1988, Turner and Housden 1992). Of particular concern were Britain's internationally important populations of golden eagle and peregrine falcon, and nationally important populations of red kite, hen harrier, merlin and short-eared owl (Batten *et al.* 1990).

One advantage of planting such extensive areas was that within such forests the illegal killing of raptors was reduced or ceased entirely, particularly in state-run forests. Persecution, although decreasing, is still a major problem in Britain. It is thought to limit the breeding distribution of raptors that kill gamebirds or lambs, particularly the golden eagle (Dennis *et al.* 1984), hen harrier (Bibby and Etheridge 1993), northern goshawk (Marquiss and Newton 1981a, SP unpubl. data) and peregrine falcon, but also red kite and common buzzard which may feed on illegally placed poisoned baits (Cadbury 1992). Thus forests provide safe refugia within otherwise hostile environments, and are large enough to sustain viable populations of many raptors.

RESPONSES OF RAPTORS TO MAN-MADE FORESTS

Forest Succession

One striking feature of man-made conifer forests is their speed of growth. Because of favourable climate and soils, Britain has some of the fastest growth rates for conifers in Europe. Sitka spruce in Britain is usually grown on a rotation of 45–55 yr, after which stands are clear-felled and replanted (restocked) (Henderson and Faulkner 1987).

I have used four serial stages to describe the development of a spruce plantation following afforestation (Table 2), and related these to changes in composition of raptor guilds and their food-supplies. Food supply appears ultimately to regulate the populations of many birds including raptors (Newton 1979), particularly when other habitat requirements are not limited, such as nest sites (see later). The main food source for 14 of the 16 raptors are birds, mammals or carrion (carcasses of sheep, deer or goats) or a combination of these foods (Table 3), so it is pertinent to explore the abundances of these foods.

(1) **Pre-thicket.** Tree heights are between 0–3 m. Domestic stock are excluded by fencing, the ground is cultivated with trees planted at 2–2.2 m spacing in the

Table 2. Main breeding and foraging habitats of 16 species of raptor in upland forests (adapted from Petty 1988).

Species	Successional stages from moorland to old-forest				
	Moorland	Pre-thicket	Thicket	Tall-forest	Old-forest
Honey buzzard	■	■		□■	□■
Red kite	■	■		□	□■
White-tailed eagle	□■	□■			□
Hen harrier	□■	□■			
Northern goshawk	■	■		□■	□■
Eurasian sparrowhawk	■	■	□■	□■	□■
Common buzzard	□■	□■		□■	□■
Golden eagle	□■	□■			□■
Osprey				□	□
Common kestrel	□■	□■	□	□	□■
Merlin	□■	□■	□	□	□■
Peregrine falcon	□■	□■			
Barn owl	□■	□■	□		□■
Tawny owl		■	□	□■	□■
Long-eared owl	■	■	□	□■	□■
Short-eared owl	□■	□■			

□ = breeding habitat, ■ = foraging habitat

Table 3. The main nest-sites and foods of raptors in upland forests in Britain (adapted from Petty 1988).

Species	Nest-sites	Foods
Honey buzzard	*tree	invertebrates
Red kite	*tree	carrion, mammals, birds
White-tailed eagle	*tree, cliff ledge	carrion, mammals, birds, fish
Hen harrier	*ground	birds, mammals
Northern goshawk	*tree	birds, mammals
Eurasian sparrowhawk	*tree	birds
Common buzzard	*tree, cliff ledge	carrion, mammals, birds
Golden eagle	*cliff ledge, tree	carrion, mammals, birds
Osprey	*tree	fish
Common kestrel	+ tree nest, cliff ledge, building	small mammals, birds
Merlin	+ tree nest, cliff ledge, ground	birds
Peregrine falcon	+ cliff ledge	birds
Barn owl	+ large cavity in tree/cliff, building	small mammals
Tawny owl	+ tree nest, tree cavity, building, ground	small mammals, birds, amphibians
Long-eared owl	+ tree nest	small mammals, birds
Short-eared owl	+ ground	small mammals

* = species which build nests of twigs and branches; + = species which do not built nests but form a scrape for their eggs in some existing substrate.

vegetation-free strips (Petty and Avery 1990). The reduction in grazing results in a dramatic recovery of the existing ground vegetation, which peaks when the planted trees are 1–2 m in height. Towards the end of this stage, tree growth begins to shade out the ground vegetation (Hill 1979, 1986).

The removal of domestic stock, most often sheep, removes a source of carrion for the golden eagle, common buzzard and red kite. However, in the short term this loss may be more than balanced by other prey which rapidly increase in response to changes in the ground vegetation. Charles (1981) gave field vole densities of 10 per ha on grazed moorland compared with 50–130 per ha on newly afforested areas. A number of studies have also shown that once released from heavy grazing, field voles not only reach high densities but also develop cycles of abundance at 3–5 yr intervals (Petty 1992, Taylor et al. 1992), similar to the periodicity of microtine cycles in northern Fennoscandia (Hanski et al. 1991, Korpimäki 1992). Common buzzard, red kite, common kestrel, barn owl and long-eared owl all exploit this plentiful food supply (Davis and Davis 1981, Village 1981, 1990, Taylor et al. 1988, Taylor et al. 1992).

Songbird populations also increase, often by a factor of 2–3, initially with growth of populations present prior to afforestation but latterly by colonization of scrub-dwelling species (Moss 1979, Moss et al. 1979). Black grouse can also increase substantially when the ground vegetation comprises partly ericaceous plants (Cayford 1993). The hen harrier is one of the most characteristic raptors to exploit this abundance of bird prey, but only in some areas (Watson 1977, Bibby and Etheridge 1993). Sometimes it is reluctant to colonize young forests. This may be related to the relative abundance of birds on moorlands compared with forest habitats. For instance, if bird prey are scarce on moorlands due to over-grazing, then harriers should preferentially nest and hunt in prey-rich forests, but when there are few differences in prey abundance between habitats, they may be less inclined to switch from their traditional moorland habitats. Both merlin and peregrine falcon hunt birds over these areas too. So, the raptors that exploit these young forests are largely those present prior to afforestation (Table 2). Higher breeding densities and greater reproductive output are also recorded for some raptors in young forests, the short-eared owl is the most notable (Lockie 1955, Village 1987).

(2) **Thicket.** Tree heights are >2–10 m. Canopy-closure occurs and results in the progressive loss of most ground vegetation. This may be a slow process in some areas where tree growth is patchy or poor (Hill 1979).

Most thicket crops are difficult for humans to walk through let alone for raptors to fly through. Small mammal densities decline rapidly around the time of canopy-closure (Charles 1981), as do those raptors dependent on this food supply. However, Village (1992) has shown that vole specialists such as long-eared owls and common kestrels can persist during the thicket stage providing some open ground remains, but at a lower density compared with pre-thicket crops. Songbird density continues to increase, but mainly with woodland species (Moss 1979, Moss et al. 1979).

Songbirds are relatively secure from avian predators due to the dense structure of the forest. However, Eurasian sparrowhawks colonize in the late-thicket, where they nest in faster-growing patches of trees with sufficient space below the tree canopy (Newton 1986). Male Eurasian sparrowhawks also use thicket stage plantations for both roosting and foraging (Marquiss and Newton 1981b). During severe weather conditions these dense crops also provide sheltered roosting places for other raptors (Petty 1992).

Carrion crows start to nest in early-thicket stage crops, often near edges where they build a new nest each year. These nests persist for 2–5 yr and provide suitable nest sites for common kestrel, long-eared owl, tawny owl and merlin (Village 1981, 1990, Parr 1991, 1994, Little and Davison 1992, Petty 1992), providing there are suitable foraging habitats nearby.

(3) **Tall-forest.** Tree heights are >10 m until clear-felling usually at 45–55 yr of age. Stands on sheltered, well-drained sites may have trees progressively removed (thinned) from the late-thicket stage onwards. Crops on more exposed and poorly drained sites often remain unthinned but may be more susceptible to windthrow. Thinning and windthrow allow light to penetrate to the forest floor, thereby encouraging the reestablishment of some ground vegetation. Small mammals are present but are much less abundant than in pre-thicket crops, squirrels reach high densities in good cone years (SJP unpubl. data) and bird densities continue to increase (Moss 1979, Moss et al. 1979).

Space between the trees gives access to the medium-sized common buzzard and northern goshawk for both breeding and foraging. The latter is better adapted than the Eurasian sparrowhawk to kill larger birds such as the carrion crow, Eurasian jay, woodpigeon, and also grey and red squirrels (Opdam 1975, Marquiss and Newton 1981a, Petty 1989a). Tawny owls have often colonized by this stage, but unless small mammals are abundant in adjacent habitats, their breeding density is low and productivity poor (Petty 1992).

(4) **Old-forest.** Crops that are retained well beyond the normal commercial rotation lengths (tall-forest stage), usually for conservation reasons (Ratcliffe and Petty 1986, Peterken 1992). Gaps in the tree canopy allow light to penetrate, and ground vegetation, broadleaved trees/shrubs and conifers to regenerate, provided that deer browsing is not excessive. The greatest diversity of birds is found in these stands (Currie and Bamford 1982), although small mammal density is lower than in the pre-thicket stage. Large-diameter dead or wind-broken snags provide breeding sites for cavity nesting raptors, and large tree crowns provide ample opportunities for some of the larger raptors to build their substantial nests of sticks.

By using these growth-stages, it is possible to show the range of habitats used for breeding and foraging by the different raptor species (Table 2). The thicket stage supports the least number of raptors mainly because access is so restricted, even though potential prey such as songbirds are abundant. This has important implications when planning habitat improvements for raptors. The other three growth stages provide suitable breeding and foraging habitat for a wide range of raptors (Table 2).

Few species appear to be limited by a lack of nest sites in spruce forests, even many of the owls and falcons that do not build their own nests. This is largely because of the abundance of old stick nests of corvids, squirrels and other raptors (Table 3). The barn owl is one striking exception (Petty *et al.* 1994). It is dependent on large cavities in trees and cliffs or buildings in which to breed. The lack of suitable natural nest sites can be overcome by providing nest boxes; these can lead to spectacular increases in abundance so long as there is a nearby source population to provide recruits (Shaw and Dowell 1990, Petty *et al.* 1994). Towards the end of the first rotation it is important to plan for the retention of patches of old-forest. Correctly sited, these provide nesting opportunities for some of the larger raptors. Both golden eagles and white-tailed eagles have used such retentions, where the former species has also bred on artificial nests (SJP unpubl. data).

Second Generation Forests

Much research on raptor populations in the British uplands has concentrated on either evaluating the impact of forestry expansion on open-country species, or on how raptors have adapted to the pre-thicket stages following the initial planting, which at times can have spectacularly rich raptor assemblages. In contrast, relatively little research has been done on how raptor guilds change as forests mature (but see Newton 1991 and Village 1992), and even less on the impact of clear-felling.

Many spruce forests in upland Britain are grown on rotations of 45–55 yr and harvested by a clear-felling system. Clear-felling is used instead of selective logging because unstable site conditions and the frequency of gales can result in catastrophic wind damage if trees are allowed to become too tall. Clear-felling has been underway for about 20 yr in some of the first forests to be established. So, it is only now becoming possible to assess whether there are differences in how raptors utilize the pre-thicket stages of first and second rotation crops.

My first approach has been to explore whether there are differences in raptor food supply between first and second generation crops. Three studies have compared songbird density and species richness (Currie and Bamford 1981, Leslie 1981, Patterson and Ollason 1991, 1992). All have found either greater bird abundance on restocked than afforested sites or little difference between habitats. There has been no direct comparison of small mammal populations, but Petty (1992) showed that field vole populations can exhibit multi-annual cycles of abundance on restocked sites, at a similar periodicity to those on afforested areas (Taylor *et al.* 1992). Thus, the present evidence suggests that there may be a similar food supply (birds and small mammals) present on both first and second generation pre-thicket crops.

The main differences between afforested and restocked sites relates to scale. Clear-felling provides an opportunity to alter spatial patterns within forests. This is being achieved by increasing and decreasing rotation lengths and clear-felling areas much smaller than the initial planting to create a finer-grained patchwork of different-aged stands of trees (Hibberd 1985, Ratcliffe and Petty 1986, Petty and Avery 1990).

This has important implications because it should allow a greater diversity of raptors to occur than on a similar area of afforested ground. On a rotation length of 45–55 yr, the three successional stages (pre-thicket, thicket and tall-forest) each occupy about one-third of the rotation, and each stage has a different raptor guild associated with it (Table 2). One criticism of first-generation plantations is that in a very short period (<10 yr) there can be dramatic habitat changes over vast areas, particularly as crops pass from the rich pre-thicket stage into the less interesting thicket stage. Redesigning forests at the end of the first rotation should lessen the effects of such changes.

Unlike many newly-afforested areas, clear-fells are often surrounded by older forest. This enables tree-nesting common buzzards, common kestrels and tawny owls, together with short-eared owls which also nest on larger clear-fells, to exploit foods such as field voles when these are abundant on clear-fells. More edge habitat is also created. The tawny owl provides a useful example of how the size and spatial arrangement of clear-fells can influence raptor distribution. Tawny owls need thicket and pole-stage crops in which to roost and nest, but obtain most of their food (small mammals) from clear-fells. A fine-grained patchwork of clear-fells, with a high degree of temporal separation between them, results in a higher breeding density of tawny owls than a coarser-grained patchwork (Petty 1989b, 1992).

Most of the raptors present in pre-thicket, first generation forests also utilize the same stage after clear-felling. Currently, hen harriers appear to be one exception. They greatly benefit from newly afforested areas, but show little indication of colonizing clear-fells (Petty and Anderson 1986, Bibby and Etheridge 1993). The reason for this is unclear but does not appear to be related to food supply. So, in the long term, forests may lead to a loss of habitat for this species.

CONCLUDING REMARKS

I have attempted to assess whether man-made forests in the uplands have had a beneficial (+), detrimental (−) or neutral (=) effect on raptors, based on what is available in the literature (Table 1) and the assumption that second generation forests will contain substantial amounts of vole-rich habitat. This indicates that five species have largely benefited, eight species appear little affected, while just three species may have been adversely affected (Table 1). In this latter group, I have already covered the hen harrier. Little evidence is available to make an assessment for short-eared owl. In the long term, forestry may have little effect on this species because at any one time up to a quarter of forest area should be in the vole-rich pre-thicket stage, where field vole densities can be 5–10 times higher than on grazed moorland (Charles 1981). Concern about golden eagles highlights the importance of integrating forestry with areas of open land.

Forestry has allowed golden eagles to colonize some areas, such as Galloway in southwest Scotland, where previously persecution was the limiting factor (Marquiss et al. 1985). However, unless enough open land is available for hunting, populations may subsequently decline, as there is little evidence of eagles using forests after the pre-thicket stage following afforestation (Marquiss et al. 1985, Watson 1992). Fortunately, most areas with eagles also have extensive open areas at too high an altitude for economic forestry.

So, many raptors have generally adapted well to man-made spruce forests (Tables 1 and 2). Forest size and the lack of persecution has allowed viable populations of raptors to develop, sometimes within otherwise hostile environments. First generation forests, often planted over few years, exhibit dramatic habitat changes around the time of canopy-closure, with a corresponding impact on the structure of raptor guilds. At the end of the first rotation, clear-felling allows spatial patterns within forests to be altered, to create a patchwork of different-aged stands. This should provide a more diverse and sustainable habitat for many of the species considered in this review. Hen harrier and golden eagle may not be well catered for by this strategy, and research is underway to understand why, so that forests can be better designed for them in the future.

ACKNOWLEDGMENTS

I thank Robert Kenward, Mick Marquiss, Juan Jose Negro, Andy Village and David Jardine for their helpful comments on an earlier draft of this paper, and Diane Chadwick for word-processing the text.

REFERENCES

AVERY, M. AND R. LESLIE. 1990. *Birds and forestry*. Poyser, London, UK.
BATTEN, L.A., C.J. BIBBY, P. CLEMENTS, G.D. ELLIOTT AND R.F. PORTER. 1990 *Red data birds in Britain*. Poyser, London, UK.
BIBBY, C.J. 1986. Merlins in Wales: Site occupancy in relation to vegetation. *J. Appl. Ecol.* 23: 1–12.
— AND B. ETHERIDGE. 1993. Status of the Hen Harrier *Circus cyaneus* in Scotland in 1988–1989. *Bird Study* 40: 1–11.
— AND M. NATTRASS. 1986. Breeding status of the Merlin in Britain. *Brit. Birds* 79: 170–185.
BROWN, L. 1976. *British birds of prey*. Collins, London, UK.
CADBURY, J. 1992. The illegal killing must stop: A review of bird of prey persecution and poison abuse. *RSPB Conserv. Rev.* 6: 28–33.
CAYFORD, J. 1993. *Black grouse and forestry*. Forestry Commission Technical Paper 1. Forestry Commission, Edinburgh, UK.
CHARLES, W.N. 1981. Abundance of the Field Vole (*Microtus agrestis*) in conifer plantations. Pp. 135–137 in F.T. Last and A.S. Gardiner, eds., *Forest and woodland ecology*. Inst. Terrestr. Ecol., Cambridge, UK.

CURRIE, F.A. AND R. BAMFORD. 1981. Bird populations of sample pre-thicket forest plantations. *Quart. J. Forestry* **75**: 75–82.
— AND —. 1982. The values to wildlife of retaining small conifer stands beyond normal felling age within forests. *Quart. J. Forestry* **76**: 153–160.
DAVIS, P. 1993. The Red Kite in Wales: setting the record straight. *Brit. Birds* **86**: 295–298.
DAVIS, P.E. AND J.E. DAVIS. 1981. The food of the Red Kite in Wales. *Bird Study* **28**: 33–40.
DENNIS, R.H. 1987. Osprey recolonisation. *RSPB Conserv. Rev.* **1**: 88–90.
—, P.M. ELLIS, R.A. BROAD AND D.R. LANGSLOW. 1984. The status of the Golden Eagle in Britain. *Brit. Birds* **77**: 592–607.
DIMBLEBY, G.W. 1984. Anthropogenic changes from the Neolithic through to Medieval times. *New Phytologist* **98**: 57–72.
ELLIS, P.M. AND I.D. OKILL. 1990. Breeding ecology of the Merlin *Falco columbarius* in Shetland. *Bird Study* **37**: 101–110.
GIBBONS, D.W., J.B. REID AND R.A. CHAPMAN. 1993. *The new atlas of breeding birds in Britain and Ireland: 1988–1991*. Poyser, London, UK.
GODWIN, H. 1975. *The history of the British flora*. Cambridge Univ. Press, Cambridge, UK.
HANSKI, I., L. HANSSON AND H. HENTTONEN. 1991. Specialist predators, generalist predators and the microtine rodent cycle. *J. Anim. Ecol.* **60**: 353–367.
HAWORTH, P.F. AND A. FIELDING. 1988. Conservation and management implications of habitat selection in the Merlin *Falco columbarius* L. in the south Pennines, UK. *Biol. Conserv.* **46**: 247–260.
HENDERSON, D.M. AND R. FAULKNER, eds. 1987. Sitka spruce. *Proc. Royal Soc. Edinburgh* **93B**: 1–234.
HIBBERD, B.G. 1985. Restructuring of plantations in Kielder Forest District. *Forestry* **58**: 119–129.
HILL, M.O. 1979. The development of a flora in even-aged plantations. Pp. 175–192 in E.D. Ford, D.C. Malcolm and J. Atterson, eds. *The ecology of even-aged plantations*. Inst. Terrestr. Ecol., Cambridge, UK.
—. 1986. Ground flora and succession in commercial forests. Pp. 71–78 in D. Jenkins, ed. *Trees and wildlife in the Scottish uplands*. Inst. Terrestr. Ecol., Abbots Ripton, UK.
HOLMES, G.D. 1975. History of forestry and forest management. *Phil. Trans. Royal Soc. London B* **271**: 69–80.
KORPIMÄKI, E. 1992. Population dynamics of Fennoscandian owls in relation to wintering conditions and between-year fluctuations of food. Pp. 1–10 in C.A. Galbraith, I.R. Taylor and S. Percival, eds. *The ecology and conservation of European owls*. Joint Nature Conservation Committee, Peterborough, UK.
LESLIE, R. 1981. Birds of the north-east England forests. *Quart. J. Forestry* **75**: 153–158.
LITTLE, B. AND M. DAVISON. 1992. Merlins *Falco columbarius* using crow nests in Kielder Forest, Northumberland. *Bird Study* **39**: 13–16.
LOCKE, G.M.L. 1987. Census of woodlands and trees 1979–82. Forestry Comm. Bull. 63. HMSO, London, UK.
LOCKIE, J. 1955. The breeding habits of Short-eared Owls after a vole plague. *Bird Study* **2**: 53–69.
LOVE, J.A. 1983. *The return of the Sea Eagle*. Cambridge Univ. Press, Cambridge, UK.
MARQUISS, M. 1981. The Goshawk in Britain – its provenance and current status. Pp. 43–56 in R.E. Kenward and I. Lindsay, eds. *Understanding the Goshawk*. International Association for Falconry and Conservation of Birds of Prey, Oxford, UK.
— AND I. NEWTON. 1981a. The Goshawk in Britain. *Brit. Birds* **75**: 243–260.
— AND —. 1981b. A radio-tracking study of the ranging behaviour and dispersal of European Sparrowhawks *Accipiter nisus*. *J. Appl. Ecol.* **51**: 111–133.
—, D.A. RATCLIFFE AND R. ROXBURGH. 1985. The numbers, breeding success and

diet of Golden Eagles in southern Scotland in relation to land use. *Biol. Conserv.* 33: 1–17.
MOSS, D. 1979. Even-aged plantations as a habitat for birds. Pp. 413–427 in E.D. Ford, D.C. Malcolm and J. Atterson, eds. *The ecology of even-aged plantations.* Inst. Terrestr. Ecol., Cambridge, UK.
—, P. N. TAYLOR AND N. EASTERBEE. 1979. The effects on songbird populations of upland afforestation with spruce. *Forestry* 52: 129–147.
NEWTON, I. 1972. Birds of prey in Scotland: some conservation problems. *Scottish Birds* 7: 5–23.
—. 1979. *Population ecology of raptors.* Poyser, Berkhamstead, UK.
—. 1984. Raptors in Britain – a review of the last 150 years. *BTO News* 131: 6–7.
—. 1986. *The Sparrowhawk.* Poyser, Calton, UK.
—. 1991. Habitat variation and population regulation in Sparrowhawks. *Ibis* 133 Suppl. 1: 76–88.
—, P.E. DAVIS AND J.E. DAVIS. 1982. Ravens and Buzzards in relation to sheep farms and forests in Wales. *J. Appl. Ecol.* 19: 681–706.
—, — AND D. MOSS. 1981. Distribution and breeding of Red Kites in relation to land-use in Wales. *J. Appl. Ecol.* 18: 173–186.
— AND M.B. HAAS. 1984. The return of the Sparrowhawk. *Brit. Birds* 77: 47–70.
—, D. MOSS AND P.E. DAVIS. 1993a. *Red Kites and land-use in Wales.* Report to the Forest Enterprise. Inst. Terrestr. Ecol., Huntingdon, UK.
—, I. WYLLIE, AND A. ASHER. 1993b. Long-term trends in organochlorine and mercury residues in some predatory birds in Britain. *Environ. Pollution* 79: 143–151.
O'FLYNN, W.J. 1983. Population changes of the Hen Harrier in Ireland. *Irish Birds* 2: 337–343.
OPDAM P. 1975. Inter- and intraspecific differentiation with respect to feeding ecology in two sympatric species of the genus *Accipiter. Ardea* 63: 30–54.
PARR, S.J. 1991. Occupation of new conifer plantations by Merlins in Wales. *Bird Study* 38: 103–111.
—. 1994. Changes in the population size and nest sites of Merlins *Falco columbarius* in Wales between 1970 and 1991. *Bird Study* 41: 42–47.
PATTERSON, I.J. AND J.G. OLLASON. 1991. Modelling songbird/habitat relationships in spruce forests in the northern uplands of Britain. Unpublished First Annual Report to the Forestry Authority. Univ. of Aberdeen, UK.
— AND —. 1992. Modelling songbird/habitat relationships in spruce forests in the northern uplands of Britain. Unpublished Final Report to the Forestry Authority. Univ. of Aberdeen, UK.
PETERKEN, G.F. 1987. Natural features in the management of upland conifer forests. *Proc. Royal Soc. Edinburgh* 93B: 223–234.
—. 1992. Old-growth conservation within British upland conifer plantations. *Forestry* 65: 127–144.
PETTY, S.J. 1988. The management of raptors in upland forests. Pp. 7–23 in D.C. Jardine, ed. *Wildlife management in forests.* Institute of Chartered Foresters, Edinburgh, UK.
—. 1989a. *Goshawks: Their status, requirements and management.* Forestry Commission Bulletin 81. HMSO, London, UK.
—. 1989b. Productivity and density of Tawny Owls *Strix aluco* in relation to the structure of a spruce forest in Britain. *Ann. Zool. Fenn.* 26: 227–233.
—. 1992. Ecology of the Tawny Owl *Strix aluco* in the spruce forests of Northumberland and Argyll. Ph.D. thesis. The Open Univ., Milton Keynes, UK.
— AND D. ANDERSON. 1986. Breeding by Hen Harriers *Circus cyaneus* on restocked sites in upland forests. *Bird Study* 33: 177–178.
— AND M.I. AVERY. 1990. *Forest bird communities.* Forestry Commission Occasional Paper 26. Forestry Commission, Edinburgh, UK.

—, G. SHAW AND D.I.K. ANDERSON. 1994. Value of nestboxes for population studies and conservation of owls in conifer forests in Britain. *J. Raptor Res.* **28**: 134–142.

RATCLIFFE, D.A. 1990. *Bird life of mountain and upland.* Cambridge Univ. Press. Cambridge, UK.

—. 1993. *The Peregrine Falcon* (second edition). Poyser, London, UK.

RATCLIFFE, P.R. AND S.J. PETTY. 1986. The management of commercial forests for wildlife. Pp. 177–187 in D. Jenkins, ed. *Trees and wildlife in the Scottish uplands.* Inst. Terrestr. Ecol., Huntingdon, UK.

SHARROCK, J.T.R. 1976. *The Atlas of Breeding Birds in Britain and Ireland.* British Trust for Ornithology, Tring, UK.

SHAW, G. AND A. DOWELL. 1990. *Barn Owl conservation in forests.* Forestry Comm. Bull. 90. HMSO, London, UK.

SHAWYER, C.R. 1987. *The Barn Owl in the British Isles: its past, present and future.* The Hawk Trust, London UK.

SPENCER, R. AND THE RARE BREEDING BIRDS PANEL. 1993. Rare breeding birds in the United Kingdom in 1990. *Brit. Birds* **86**: 62–90.

TAYLOR, I.R., A. DOWELL, T. IRVING, I.K. LANGFORD AND G. SHAW. 1988. The distribution of the Barn Own *Tyto alba* in south west Scotland. *Scottish Birds* **15**: 40–43.

—, — AND G. SHAW. 1992. The population ecology and conservation of Barn Owls *Tyto alba* in coniferous plantations. Pp. 16–21 in C.A. Galbraith, I.R. Taylor and S. Percival, eds. *The ecology and conservation of European owls.* Joint Nature Conservation Committee, Peterborough, UK.

TAYLOR, K., R. HUDSON AND G. HORNE. 1988. Buzzard breeding distribution and abundance in Britain and Northern Ireland in 1983. *Bird Study* **35**: 109–118.

THOMPSON, D.B.A., D.A. STROUD AND M.W. PIENKOWSKI. 1988. Afforestation and upland birds: consequences for population ecology. Pp. 237–259 in M.B. Usher and D.B.A. Thompson, eds. *Ecological change in the uplands.* Blackwell Sci. Publ., Oxford, UK.

TURNER, R. AND S. HOUSDEN. 1992. The future of forestry. *RSPB Conserv. Rev.* **6**: 3–11.

VILLAGE, A. 1981. The diet and breeding of Long-eared Owls in relation to vole numbers. *Bird Study* **28**: 215–224.

—. 1987. Numbers, territory-size and turnover of Short-eared Owls *Asio flammeus* in relation to vole abundance. *Ornis Scand.* **18**: 198–204.

—. 1990. *The Kestrel.* Poyser, London, UK.

—. 1992. *Asio* owls and Kestrels in recently-planted and thicket plantations. Pp. 11–15 in C.A. Galbraith, I.R. Taylor and S. Percival, eds. *The ecology and conservation of European owls.* Joint Nature Conservation Committee, Peterborough, UK.

WATSON, D. 1977. *The Hen Harrier.* Poyser, Berkhamsted, UK.

WATSON, J. 1992. Golden Eagles *Aquila chrysaetos* breeding success and afforestation in Argyll. *Bird Study* **39**: 203–206.

22

Spotted Owls in Managed Forests of Western Oregon and Washington

Scott P. Horton

Abstract – Successful strategies to conserve northern spotted owls will depend, in part, upon habitat provided in managed forests. Thus it is important to understand their actual and potential relationships with managed forests, defined here as forests from which timber has been harvested and in which future harvests will be conducted. I review and synthesize the knowledge of owls and managed forests in the Western Hemlock Zone of western Oregon and Washington.

Owls selected sites within stands or among similar stands based on their composition, structure and pattern, including: large, decadent trees; snags and canopy diversity; and distance from edges. Owls selected among stand-types such that older unmanaged stands were highly preferred, and younger managed stands were used in proportion to their availability or avoided. Selection became increasingly general from nesting to roosting to foraging to dispersal habitat. Pairs located their ranges in concentrations of old forest, and traversed larger areas where old forests were scarce. Populations of owls were more abundant and productive in concentrations of old, unmanaged forest. But owls appeared to respond favorably to some managed forests in which the structure and function met their life needs.

Forests can be managed to provide favorable conditions for owls. Stands managed to accommodate owls should contain elements of old, unmanaged forests: very large trees; structurally diverse canopies; shade-tolerant conifers; large snags; and fallen trees. Recommended landscape patterns vary with their intended roles in regional conservation plans, but generally call for about half in habitat suitable for dispersing or resident owls. One landowner in the region has implemented an integrated management plan for commercial forests and owls. One research program is evaluating techniques to provide owl habitat in managed forests. Additional management and research programs are in preparation.

Key words: northern spotted owls; nesting; managed forests; Oregon; Washington.

Over the past decade, northern spotted owls have become a powerful symbol of the inability of natural systems to adapt to human alteration. Northern spotted owls (subsequently called "owls" or "spotted owls") are adapted to life in old, unmanaged forests at multiple scales: nesting, foraging, and roosting sites of individuals (Forsman *et al.* 1984, Thomas *et al.* 1990, Carey *et al.* 1992); home ranges of pairs (Carey *et al.* 1992, Lehmkuhl and Raphael 1993); and landscapes inhabited by local populations (Bart and Forsman 1992, Carey et al. 1992).

Populations of spotted owls are declining because management of old, previously unmanaged forests has reduced the quantity and quality of owl habitat (U.S.D.I. 1989). The current distribution and abundance of old, unmanaged forests is believed to be inadequate to maintain viable populations of spotted owls throughout their range (Thomas et al. 1990). Efforts to conserve spotted owls have been difficult because retention of their habitat in old forests conflicts with needs for economic and social benefits derived from harvesting those forests (U.S.D.I. 1992). Successful conservation strategies will be contingent, at least in part, upon owls finding habitat in managed forests (defined here as forests from which timber has been harvested and in which future harvests will be conducted). Thus the question: can spotted owls adapt to life in managed forests? Corollaries to that question are: what evidence exists on the relationships of owls to managed and unmanaged forests?; how might owls adapt to life in managed forests?; what management practices could facilitate the adaptation of owls to managed forests?; and how can we know if owls are adapting to life in managed forests? I will address these questions in this review.

Spotted owls have apparently adapted to at least some managed forests in the coastal redwoods of northwest California (Simpson Timber Co. and RECON 1992), and appear to be living successfully in some managed forests in the eastern Cascades of Washington (Irwin 1993). I will review our knowledge of owls and managed forests in the Western Hemlock Zone (Franklin and Dyrness 1973) of western Oregon and Washington, where the greatest controversy over owls and forest management exists.

FOREST ECOLOGY AND MANAGEMENT

Natural Forests

Before timber harvesting became the dominant influence on the landscape, forest stands in western Washington and Oregon were subject to catastrophic, natural disturbances (fire or wind) at intervals averaging 90–145 yr in drier, fire-prone sites to nearly 750 yr in mesic sites (Agee 1991). While successional pathways are variable, most naturally regenerating stands proceed through a series of structurally distinct stages (see Hall et al. 1985), throughout which substantial numbers of large, standing and fallen dead trees, as well as trees that survived disturbances contribute markedly to the structure and function of the regenerating stands: open stages dominated by shrubs and small saplings (0–30 yr); closed stages in which relatively uniform small ("younger closed") to medium-sized ("older closed") trees have fully occupied the growing space and the little light that reaches the forest floor supports only a depauperate understory (20–80 yr); a mature stage in which differential growth and mortality among the medium to large-sized trees allows the re-initiation and development of an understory (60–200 yr); and an old-growth stage. If undisturbed for >200 yr, stands develop old-growth characteristics (Franklin *et al.* 1981), including: very large live trees (>1 m diameter, >60 m tall); large standing and fallen dead trees; a deep, multi-layered canopy with a midstory of shade-tolerant species; canopy gaps; and a diverse, well-developed understory. The Old-Growth Definition Task Group (1986) provided a definition for old-growth stands. Owl researchers have frequently lumped old-growth stands with other older, unmanaged stands that had many old-growth characteristics but did not fully meet their definition as "old forests". Pre-settlement landscapes were frequently dominated by old forest stands (U.S.D.I. 1989).

Managed Forests

Management of forests for timber production is intended to increase the rate of production of wood fiber per unit time over that in natural stands. The following methods are employed (Scott 1980): stands are completely harvested; regeneration by planting is rapid and careful; competing vegetation and stand density are controlled; stands are harvested at rotation age (usually during the closed stage), 40–80 yr. The intensity of management has varied widely based on the varying objectives of landowners, but in general these techniques cause managed forest stands to be more simple in their structure and composition than natural stands.

Limited timber harvesting began in the region in the late 1700s, but intensive harvesting began in the early 1900s. Lower elevation forests in more gentle terrain, most of which were privately owned, were generally harvested first. Managers of publicly owned (state and federal) forests generally began harvesting in the latter half of the twentieth century, and harvested older stands more slowly in a patchwork pattern. Thus, many private forest landscapes are currently dominated by young-aged managed stands in the open or younger closed stages (regenerated after the second harvest), with lesser amounts of older closed stands at or approaching rotation age (regenerated after the first harvest), while managed, public forest landscapes are generally a patchwork of open and closed managed stands, and old, unmanaged stands.

ON ADAPTATION

Most authors, e.g. Mayr (1970) and Wilson (1975), describe adaptation as an evolutionary phenomenon in which the morphology, physiology and/or behavior of a population changes in response to selective pressures. The positive and negative pressures that human-altered environments present to raptors include alterations to: the composition, structure, and pattern of plant communities and physical features in landscapes; the diversity, abundance, and availability of prey; and ecosystem function, as manifest by new or modified interactions with predators and competitors, and with toxic substances and other hazards.

Forest management in western Oregon and Washington has resulted in profound changes to the environment of spotted owls: altered composition, structure, and pattern of forests, with attendant changes to the prey base, and possibly new interactions with predators and competitors. Forests and prey are elements of owl habitat, and owls have evolved relationships with their habitat that both determine and are determined by their morphology, physiology, and behavior (Cody 1985). And they have evolved mechanisms by which they select the types of environments that are favorable to their phenotype (Hilden 1965). It is unlikely that there has been time for gene frequencies of spotted owl populations to change appreciably in response to the recent alterations of their environment. Thus, whatever adaptation owls are currently exhibiting to managed forests is probably an ecological phenomenon in which the behavioral flexibility of individuals allows them to respond to altered environments.

Habitat Selection

Hilden (1965) reviewed habitat selection in birds and discussed ultimate and proximate factors involved in habitat choices. Ultimate factors are selective pressures, and include: food; water; requirements imposed by the morphological, physiological, and behavioral traits of the species; and shelter from predators and weather. Ultimate factors determining habitat selection by spotted owls are (after Carey 1985) abundant prey, prey that is available based on the owls' morphology and foraging behavior, suitable nesting structures, opportunities for thermoregulation and protection from predation. Proximate factors are environmental stimuli that birds respond to in selecting habitat that provides for their ultimate requirements. Proximate factors can include: general characteristics of the landscape and terrain; unique features such as sites for nesting, roosting, feeding, or singing; the presence of conspecifics or other animals; and food. For example, the abundance and availability of prey are factors ultimately responsible for owls' reproductive success or failure, but owls' immediate selection of habitat for foraging is probably in response to their rates of encountering prey, or aspects of stand structure, or other environmental cues. An internal motivation mediates their responses to these cues, an owl that is more hungry may respond differently to the same proximate cue than an owl that is satiated. Proximate factors might operate independently, hierarchically as an ordered sequence of decisions, or synergistically (Cody 1985) as a 'niche gestalt' (James 1971).

The composition, structure and extent of forests in western Oregon and Washington has resembled that of modern unmanaged forests for the past 6000 yr (Brubaker 1991). Forests are and were the predominant land cover in the region, and it has been estimated that about 70% were older-aged in the mid-1800s (U.S.D.I. 1989). Owls of the genus *Strix* are forest dwellers (American Ornithologist's Union 1983), and spotted owls in the Western Hemlock Zone spent the past 6000 yr of their evolutionary history adapting to life in forests that were predominantly in an old-forest condition. The environmental cues used by owls to select habitat are unknown. Nor is it known whether owls respond to those cues independently, hierarchically, or synergistically. But it is believed that many birds select habitat based on its structural characteristic (Cody 1985). Thus, investigations of habitat selection by spotted owls have focused on forest structure at scales relevant to individual owls, pairs and local populations.

HABITAT RELATIONSHIPS OF SPOTTED OWLS

Spotted owl studies in western Washington and Oregon have been conducted across a range of ecological scales: owls' responses to variations within stands or among stands of similar (fine-grained selection) and different (coarse-grained selection) conditions; characteristics of home ranges of pairs; and characteristics of local populations and the landscapes they inhabit.

Fine-grained Habitat Selection

Spotted owls in the Western Hemlock Zone almost always nest in large cavities in large, old, decadent trees (Forsman *et al.* 1984, Forsman unpubl. data in Thomas *et al.* 1990), even if the nest-tree is in a younger-aged stand. Most nests have been located in stovepipe-like hollows exposed by broken tops of large, live trees sheltered by secondary crowns, or in cavities created when large limbs fell from tree trunks (Forsman *et al.* 1984). Johnson *et al.* (1992) found that 51 nests were farther into old forest stands than randomly selected points.

Carey *et al.* (1992) examined 1007 roost-sites of 62 owls in southwestern Oregon. They found that owls showed a statistically significant tendency to roost in sites that were thermally neutral, based on the owls' positions in the canopy, on the topographic positions of the sites, and the weather at the roost-site. However, only very small amounts of the variation in the characteristics of roost sites could be predicted by regression with weather variables. This suggests that while protection from weather is important in roost-site selection, it is not the most important factor. Their most striking finding was that, although owls roosted preferentially in old, unmanaged forests, the 109 roost-sites in older closed or mature managed and natural stands were similar to roost-sites in old forests in their pronounced vertical diversity of vegetation (which was significantly greater than at random sites in old forests). Thus, owls were able to locate presumably advantageous roost-sites that differed either subtly (in old stands), or more dramatically (in younger stands) from the stands in which they occurred. Johnson *et al.* (1992) found owls selected roost-sites that were at least 100 m interior to edges of forest stands.

North (1993) examined the responses of owls to the subtle variation among stands in a similar condition, mature (60–130 yr old) managed and natural stands. He related the intensity of owl use for foraging and roosting (as determined by radio-telemetry) to stand characteristics, in an attempt to understand the relationships of owls to specific structural attributes of forest stands. He used relocations obtained in separate studies on the Olympic Peninsula (E. D. Forsman unpubl. data) and the North Cascades (Hamer 1988) of Washington to classify 50 stands as receiving high, medium, low, or no use by owls. He measured 65 attributes of the stands, then used a variety of statistical methods to determine those that were most highly correlated with intensity of owl use. Stands that received high use by owls were characterized by greater volumes of standing dead trees (snags), and greater diversity of tree heights.

Canopy diversity, snags, large cavities in trees, and stand interiors i.e. >100 m from edges appear to be important to fine-grained habitat selection by owls. Many of the ultimate factors that determine habitat suitability for owls appear to derive from these characteristics of forest structure and pattern. Canopy diversity may contribute to the availability of prey (Forsman et al. 1984, Carey 1985), and to opportunities for thermoregulation (Forsman et al. 1984, Carey et al. 1992). Northern flying squirrels are the most important prey of owls in the Western Hemlock Zone, and may be limited by the availability of den cavities in snags (Carey 1991). Large cavities are the primary nesting structure for owls in the Western Hemlock Zone. Stand interiors offer thermally neutral conditions (Chen et al. 1993), and may offer protection from predatory great horned owls which are more abundant in forest landscapes with high edge-to-old forest ratios (Johnson 1992) than in continuous-forest landscapes. It can be hypothesized that spotted owls attend to these forest characteristics as the proximate factors for habitat selection.

Coarse-grained Habitat Selection

One of the initial, important issues in owl conservation and management was the degree to which owls depend on old, unmanaged forests for nesting, roosting, and foraging. A partial answer to that question was sought by evaluating differential use of types of forest stands. Selection for nesting habitat has been inferred from the proportion of nests located in different types of stands, and by comparing the context of nest-sites to random locations. Selection of foraging habitat has been evaluated by radio-tracking individual owls, then comparing their proportional use of stand-types (proportion of nocturnal relocations in each type) to the availability of stand-types within their ranges (proportion of each home range) using Bonferroni's inequality (Byers and Steinhorst 1984). Selection among stand-types for roosting has been evaluated both by inspecting the proportion of roosts in different types, and by statistical comparisons similar to studies of foraging habitat selection.

Several researchers have conducted intensive searches for owl nests. Forsman et al. (1984) located 47 nests in Oregon, 90% of which were in old-growth stands. Unpublished data from two studies summarized in Thomas et al. (1990) report 82 of 83 nests in western Oregon and on the Olympic Peninsula of Washington as being in old-growth or other old, unmanaged stands. Ripple et

Table 1. Selection among forest stand types for foraging[a] by radio-tagged spotted owls in western Oregon.

	Numbers of owls[b]					
	Select		Neutral		Avoid	
Stand condition	F	C	F	C	F	C
Old-growth	14/14	39/47	0/14	8/47	0/14	0/47
Other old forest		4/43		31/43		8/43
Mature	1/9	0/20	4/9	12/20	4/9	8/20
Older closed	1/11	1/44	4/11	24/44	7/11	19/43
Younger closed	0/11	0/43	1/11	16/43	10/11	27/43
Older open	0/14	0/44	0/14	13/44	14/14	32/44
Younger open	0/14	0/38	0/14	10/38	14/14	28/38

[a] Simultaneous 95% confidence intervals around proportion of use vs. proportion available, Byers and Steinhorst (1984).
[b] Numerators are numbers of owls in the category, denominators are numbers of owls with the stand type in their ranges; columns labelled F are from Forsman *et al.* 1984, columns labelled C are from Carey *et al.* 1992.

al. (1991) compared circular areas around 30 nests with random locations within the study area, and found that owl nests were located in concentrations of mature and old forests (20–25% greater than the general landscape) at scales ranging from stand-level (260 ha) to greater than home range-level (3588 ha).

In separate studies in western Oregon, Forsman *et al.* (1984) found 95% of 1653 roost-sites in old-growth, and Carey *et al.* (1992) found 88% of 1007 roosts in old forests. Carey *et al.* (1990) reported that nine of nine radio-tagged owls in western Oregon roosted preferentially in old-growth, seven of nine and two of nine roosted respectively in mature and older closed stands in proportion to their availability. Two unpublished studies summarized in Thomas *et al.* (1990) reported that 18 of 18 radio-tagged owls in western Oregon roosted preferentially in old-growth, 13 of 15 and zero of 16 roosted respectively in mature and older closed stands in proportion to their availability.

Two published studies report on habitat selection by foraging owls in western Oregon. Both Forsman *et al.* (1984) and Carey *et al.* (1992) found that forest stands became increasingly attractive as foraging habitat for owls as they proceeded from an open to an old-growth condition (Table 1). Old-growth stands were the only type selected by the majority of owls in either study, but a few owls showed significant selection for mature or older closed stands, and many used those stands in proportion to their availability.

Dispersing juvenile owls apparently move freely throughout forested landscapes (Thomas *et al.* 1990). Little is known about dispersal behavior of subadult and adult owls (Thomas *et al.* 1990), but Carey *et al.* (1992) observed regular use of young and old closed, managed stands by several nomadic subadult owls. Non-territorial owls need habitat that provides protection from predators and the elements, and sufficient prey to maintain themselves.

Old, unmanaged stands are preferred habitat for nesting, roosting and foraging. These stands more frequently contain the abundant snags and diverse canopies (Franklin and Spies 1991) that are probably among the proximate factors determining habitat selection by owls. Old forests also provide a much greater abundance of large snags and large, decadent trees (Spies and Franklin 1991) with the large cavities generally used as nest structures by owls in the Western Hemlock Zone (Forsman et al. 1984). And northern flying squirrels, the primary prey of spotted owls in the Western Hemlock Zone (Forsman et al. 1984, 1991), are often most abundant in old forests (Carey et al. 1992). But it appears that owls' requirements become increasingly general from nesting to roosting to foraging habitat. While few owls have been found nesting outside of old, unmanaged stands, some use younger managed and unmanaged stands for roosting, and many use those stand-types (at least occasionally) for foraging. Although there are few data, the information that does exist suggests that selection among stand-types for dispersal habitat may be especially weak.

Characteristics of Home Ranges

The first, intensive radiotelemetry studies of spotted owls showed that they had large home ranges that encompassed large amounts of old, unmanaged forests (Forsman et al. 1984). These findings were of great concern to those interested in owl conservation and management, and prompted numerous additional investigations.

Researchers usually describe annual home ranges of pairs because spotted owl pairs, the reproductive unit, are resident within their home ranges for life (Forsman et al. 1984), and because range use varies seasonally (Forsman et al. 1984, Carey et al. 1989). Carey et al. (1992) characterize the minimum convex polygon estimate (Hayne 1949), the method adopted as a standard for comparison by owl researchers, as the "area traversed", and consider it the best measure of owl response to the landscape.

Published (Forsman and Meslow 1985, Carey et al. 1990, Carey et al. 1992, Miller et al. 1993) and two unpublished studies summarized in Thomas et al. (1990) describe areas traversed annually by 34 owl pairs (averaged within study areas) in western Oregon of 1200–2500 ha. Pairs traversed much larger areas in Washington, 2500–4000 ha (three unpublished studies of 23 pairs, reported in Thomas et al. 1990).

Mature and old forests, averaged within study areas, comprised 28–61% of areas traversed annually by pairs of owls in western Oregon (Forsman and Meslow 1985, Carey et al. 1990, Carey et al. 1992, and two unpublished studies summarized in Thomas et al. 1990). Carey et al. (1990) found a strong negative relationship between areas traversed and the proportion of those areas in old-growth stands, i.e. owl pairs ranged more widely when their preferred habitat was scarce. Further research (Carey et al. 1992) lent support to that hypothesis. However, Miller et al. (1993) reported preliminary findings of 0–19% old-growth in medium-sized areas (ranges of six pairs in the study averaged 1900 ha) traversed by 14 individual owls.

Areas traversed by owls contain substantial areas that are apparently unused (Carey et al. 1990). Other home range estimators have been used to address this issue. Carey et al. (1990) and Carey et al. (1992) used the modified minimum convex polygon (Harvey and Barbour 1965) and considered it a better estimator of the area used by owls. Owls used 70% of the area they traversed. Owls used 813 ± 133 ha of old forest in western Oregon, and traversed greater areas in order to gain access to that amount of habitat where old forests were scarce. Carey et al. (1992) reported that pairs followed for 2 yr used some areas in the second year such that the cumulative area of old forest used increased by 40%. The few pairs they tracked for 3 yr showed no additional increase in the cumulative area used.

Carey et al. (1992) reported a strong relationship between the areas of old forest used by owl pairs in different forest types and the abundance and diversity of prey in those forests. Flying squirrels and two species of woodrats (*Neotoma* spp.) were abundant in mixed-conifer forests in southwestern Oregon where owls used the least old forest. Flying squirrels were abundant and one species of woodrat was scarce in Douglas-fir dominated forests in the Western Hemlock Zone of western Oregon where owls used moderate amounts of old forest. Flying squirrels were scarce and woodrats were absent in western hemlock dominated forests in western Washington where owls used large amounts of old forest.

Lehmkuhl and Raphael (1993) and Meyer et al. (1993), in a preliminary report, compared the composition and other characteristics of various-sized circles around owl and random sites in Washington and Oregon respectively. Both studies found that owl sites were located in concentrations of old forests at all scales examined. Carey et al. (1990) compared the composition of areas traversed by three pairs of owls with the surrounding landscapes and found that pairs ranged over areas of concentrated old-growth. Carey et al. (1992) studied 23 owl pairs in five landscapes in southwestern Oregon (two landscapes in the Western Hemlock Zone) and found that pairs used areas of concentrated old forest that averaged 1.5 times the level of the surrounding landscape (range = 1.1–2.0). Landscape pattern (the abundance and spatial arrangement to forest stands) was related to the ability of owls to incorporate concentrations of preferred habitat in their ranges.

The home range studies demonstrated a trend for owls to locate their ranges in concentrations of old, unmanaged forests, and to range more widely to gain access to these forests and the resources in them. This suggests that old, unmanaged forests provide very important, if not essential, resources for spotted owls.

Habitat Relationships at the Population Level

Three types of studies bear on the relationships of owl populations to habitat: comparisons of areas used by discrete groups of radio-tagged owl pairs to the

Table 2. The relative importance of the ultimate factors determining habitat selection by spotted owls to functional roles of their habitat.

Habitat functions	Ultimate factors for habitat selection[a]				
	Prey abundance	Prey availability	Predator avoidance	Thermo-regulation	Nest sites
Dispersal	×	×	×	×	
Foraging	××	××	×		
Roosting			××	××	
Nesting	××	××	××	××	××

[a] "××" indicates that the functional role of habitat must be met by resources that are relatively abundant and available within home range-sized portions of the landscape and within and among years; "×" indicates that the functional role of the habitat can probably be met by resources that are less abundant and available both in space and in time.

surrounding landscapes; detection rates of owls from surveys in areas with different forest conditions; and comparisons of density and productivity among owl populations from areas with different forest conditions.

Carey et al. (1992) found that the areas used by discrete groups of four and five owl pairs comprised 1.5 times as much old forest as the general landscapes inhabited by those owls. Surveys in western Oregon (Forsman et al. 1977, Forsman 1988) and Washington (Postovit 1979, unpublished data summarized in Thomas et al. 1990, Irwin et al. 1991) found that owls were rare in areas dominated by closed stands, and that detection rates were nearly 10 times as frequent in areas dominated by old forests. Bart and Forsman (1992) found that owls were about 40 times as abundant in areas of extensive old forest relative to areas devoid of old forest but with extensive older closed (mostly managed) stands. Nearly 50 times as many owls fledged in areas of ≥60% old forest as in areas of ≤20% old forest (Bart and Forsman 1992).

Population-level studies of habitat relationships are consistent with those at finer scales. Subpopulations of owls appear to reside within concentrations of old forest. Areas of old, unmanaged forest not only supported, but produced many more owls than areas of younger, managed forests.

POSSIBILITIES FOR ADAPTATION TO MANAGED FORESTS

The life cycle of spotted owls ideally may be considered to have three stages: a period of dispersal from the natal territory which occurs during the first year of life; a non-territorial period of exploratory wanderings and brief settlings that can last several years; and, for owls that are able to settle in a prime territory, a life-long residential, reproductive period with a single mate. Owls need habitat that provides for their life needs throughout those stages (Table 2). Owls have

Table 3. The importance of forest composition, structure, and pattern to the ultimate factors determining habitat selection by spotted owls.

Ultimate factors for habitat selection	Important forest attributes		
	Composition[a]	Structure[b]	Pattern[c]
Prey abundance	×	×	
Prey availability		×	
Predator avoidance		×	×
Thermoregulation		×	×
Nest sites	×	×	×

[a] Forest composition refers to the abundance and distribution of plant species, standing and fallen dead trees, and other physical and biotic elements of forest stands.
[b] Forest structure refers to the sizes and shapes of vegetation, standing and fallen dead trees, forest openings, and other physical and biotic elements of forest stands.
[c] Forest pattern refers to the abundance, distribution, sizes and shapes of forest stands across the landscape.

evolved behavioral responses to proximate environmental cues that direct their choices among sites, stands, and ranges so that they may survive and reproduce. Humans may be unable to understand the proximate cues that owls respond to in selecting among sites, stands, and areas, but our concepts of the composition and structure of forest stands and the pattern of forested landscapes are probably coincident with at least some of the information owls use. These attributes of forest stands and landscapes are related to factors ultimately important to owls (Table 3).

I propose that spotted owls could be considered to have adapted to managed forests if those forests contribute in a positive manner to the viability of local or regional subpopulations of owls. This contribution could take the form of habitat that provides for survival and movement by non-territorial owls, and/or habitat that provides for survival and reproduction by territorial owls. Habitat that provides for the survival and movement of non-territorial owls can allow for the ready replacement of lost mates at pair sites, re-occupancy of pair sites, establishment of territories in marginal habitat during good years, and demographic connectivity among subpopulations. The apparently lower quality habitat provided by managed forests could support relatively fewer and less productive pairs and still contribute positively to both the fitness of individual owls (Fretwell and Lucas 1969) and to the production of the total population (Brown 1969).

Forest management in western Oregon and Washington generally simplifies the composition and structure of forest stands, and reduces amounts of old forest and forest interior in landscapes. This ongoing process is creating an inhospitable environment for spotted owls. It is likely that some observations of spotted owls in managed forests are of owls that were displaced as habitat was removed (by timber harvest) from their territories in old, unmanaged forests.

Nevertheless, these observations represent the responses of owls to the ranges of conditions encountered in managed-forest landscapes and are a reasonable basis from which to formulate hypotheses that can be tested in forests that are managed for both timber production and owl habitat. It can be hypothesized that owls' use of managed forests is related to the degree that they contain elements characteristic of old, unmanaged forests (Irwin et al. 1993). For example, the use of managed stands for roosting and foraging, particularly those with canopy diversity and snags noted by Carey et al. (1992) and North (1993), may be evidence of owls' ability to adapt to life in managed forests. But owls must use large areas of suitable habitat in order to acquire sufficient prey (Carey et al. 1992), and few managed-forest landscapes currently provide sufficient amounts of forests with diverse canopy structure, snags, decadent large trees with potential nest cavities, abundant prey, and other attributes important to owls. Owls can probably adapt to managed forests at scales from site-level to subpopulation-level if forest composition, structure, and pattern meet their needs. But traditional management regimes are unlikely to provide for these needs. So the question becomes: can forest management adapt to provide for spotted owls in commercial forests?

ADAPTING FOREST MANAGEMENT

Spotted owls respond to physical and biological features of their environment, not to degrees of "naturalness". Management practices can alter the composition, structure, and pattern of forest stands and landscapes to provide either favorable or unfavorable conditions for owls. It is likely that forest managers can manipulate stand composition and structure, and landscape patterns to maintain and restore habitat for dispersal, foraging sites, shelter from the elements and predators, and nest-sites (Tables 2, 3). The general consensus is that manipulations of managed stands should be designed to grow some very large trees, develop or enhance canopy-diversity, encourage shade-tolerant conifer species, and provide large snags and fallen trees (Tappeiner 1992), in other words, to contain at least some elements of old, unmanaged forests. Two proposals recommended that landscape-level management to provide dispersal habitat develop landscapes in which a dominant cover type is young forest stands that are of only marginal value as habitat for territorial owls, but are hypothesized to provide habitat that is adequate for the maintenance and movement of non-territorial owls (Thomas et al. 1990, Beak Consultants, Inc. 1993). Forested landscapes traversed by owl pairs should be managed such that $\geq 40\%$ are in stands that provide suitable foraging and roosting habitat (Carey et al. 1992, Lehmkuhl and Raphael 1993). Larger landscapes that support multiple pairs can be managed under the same guidelines.

There are significant constraints to successful integration of forest and owl management: society's ever-increasing demands for wood fiber; landowners' demands for economic return; workers' demands for employment; other social, economic, and political demands; the uncertainty of forest and wildlife ecologists and managers in predicting outcomes of management plans; and, most important in the short-term, the current inhospitable condition of many forest stands and landscapes. However, the many opportunities to manage forests to provide for owls are enhanced by: substantial areas of old, unmanaged federal forests that are either currently (National Parks or Wilderness Areas) or planned (Forest Ecosystem Management Assessment Team, 1993) to be reserved from timber harvest and will provide, if not a self-sustaining regional population, significant areas where habitat supports good survival and productivity of owls; Western Hemlock Zone forests that are among the most productive in the world (Waring and Franklin 1979); the relatively short history of forest management in the region that has allowed many managed stands to retain some of the "biological legacies" of the original, unmanaged forests that are important to the recovery of ecological function after disturbance (Franklin 1992); theoretical models (Brown 1969, Fretwell and Lucas 1969, Pulliam 1988) that suggest that some local populations in sub-optimal "sink" habitats can serve important and positive roles in overall population function; the substantial information from which to develop a scientific basis for managing commercial forests that provide habitat for owls; and conceptual models that exist with which to initiate management plans that are based on incomplete information (Walters 1986, Thomas et al. 1990).

It is likely that management practices designed to accommodate owls in commercial forests will vary across the Western Hemlock Zone in response to local conditions: current distribution and abundance of owls; land ownership patterns, including the distribution and abundance of reserves on federal land, and ownership patterns of state and private forests; current and potential future conditions of forests; and the intended functions for managed forests in local or regional owl conservation plans. Managed forests could serve a variety of roles in owl conservation: as dispersal areas that allow interchange among local populations; as additions to habitat provided in reserves that will enhance their overall ability to support and produce owls; as habitat areas that in and of themselves support one or more pairs, either for demographic support to local populations or as "stepping-stones" that facilitate interchange among widely-separated local populations that may require demographic support to remain viable; or as combinations of the above. Analyses of the potential contributions of managed forests to spotted owl conservation (U.S.D.I. 1992, Hanson et al. 1993) have suggested that all of these roles are important in the Western Hemlock Zone.

A hierarchical program of investigations can help forest managers understand how to provide owl habitat in commercial forests, and can estimate the contributions of managed forests to the overall viability of spotted owl populations (Thomas et al. 1990, U.S.D.I. 1992). Studies at relatively small spatial scales can evaluate the effects of stand and small drainage-scale management regimes on the ecology of both non-territorial and territorial owls and their prey. Larger-scale studies can evaluate the effects of landscape-level management on the structure and function of local populations of owls and their relationships with potential predators and competitors. Long-term studies of the relationships among local populations of owls in managed and unmanaged forests will be needed to understand the degree to which managed forests are contributing to the viability of owl populations.

One manager of commercial forests in Washington, Murray Pacific Corporation, has implemented a plan to integrate owl conservation on their holdings (Beak Consultants 1993). The state of Oregon is developing such a plan for the Elliott State Forest in the southern Coast Ranges. The state of Washington is developing a plan to integrate owl conservation with management of all state forests (650 000 ha) within the range of the spotted owl. An additional effort by the state of Washington is developing an experimental approach to integrated management of owls, ecosystems in general, and commercial forests on state lands on the western Olympic Peninsula as the Olympic Experimental State Forest. Other large and small owners of managed forests are in various stages of considering or developing their own plans to integrate owl conservation with management of their commercial resources. An ongoing research effort is evaluating management techniques and their outcomes in providing owl habitat in managed stands (Carey and Wunder 1993). The successes and failures of these various efforts will probably become apparent during the lifetimes of today's forest and wildlife ecologists and managers, and will probably determine the ability of spotted owls to adapt to forest management.

ACKNOWLEDGMENTS

I thank Dan Varland, Erran Seaman, and an anonymous reviewer for helpful suggestions on an earlier draft of this manuscript. This work was supported by the Washington State Department of Natural Resources and the Raptor Research Foundation. This is contribution no. 1 of the Olympic Experimental State Forest.

REFERENCES

AGEE, J.K. 1991. Fire history of Douglas-fir forests in the Pacific Northwest. Pp. 25–33 in L.F. Ruggiero, K.B. Aubry, A.B. Carey and M.H. Huff (Tech. coords.). *Wildlife and vegetation of unmanaged Douglas-fir forests*. Gen. Tech. Rep. PNW-GTR-285. U.S. Forest Service, Pacific Northwest Res. Stat., Portland, Oregon, USA.

AMERICAN ORNITHOLOGISTS' UNION. 1983. Check-list of North American birds, 6th edition. American Ornithologists' Union, Allen Press, Lawrence, Kansas USA.

BART, J. AND E.D. FORSMAN. 1992. Dependence of northern spotted owls *Strix occidentalis caurina* on old-growth forests in the western USA. *Biol. Conserv.* 62: 95–100.

BEAK CONSULTANTS, INC. 1993. Habitat conservation plan for the northern spotted owl (*Strix occidentalis caurina*) on timberlands owned by the Murray Pacific Corp. Lewis County, Washington. Murray Pacific Corp., Longview, Washington USA.

BROWN, J.L. 1969. Territorial behavior and population regulation in birds: a review and re-evaluation. *Wilson Bull.* **81**: 293–329.

BRUBAKER, L.B. 1991. Climate change and the origin of old-growth Douglas-fir forests in the Puget Sound lowland. Pp. 17–24 in L.F. Ruggiero, K.B. Aubry, A.B. Carey and M.H. Huff (Tech. coords.). *Wildlife and vegetation of unmanaged Douglas-fir forests.* Gen. Tech. Rep. PNW-GRR-285. U.S. Forest Service, Pacific Northwest Res. Stat., Portland, Oregon USA.

BYERS, C.R. AND R.K. STEINHORST. 1984. Clarification of a technique for analysis of utilization-availability data. *J. Wildl. Manage.* **48**: 1050–1053.

CAREY, A.B. 1985. A summary of the scientific basis for spotted owl management. Pp. 100–114 in R.J. Gutierrez and A.B. Carey (Tech. eds.). *Ecology and management of the spotted owl in the Pacific Northwest.* Gen. Tech. Rep. PNW-185. US Forest Service, Pacific Northwest Res. Stat., Portland, Oregon USA.

—. 1991. *The biology of arboreal rodents in Douglas-fir forests.* Gen. Tech. Rep. PNW-276. U.S. Forest Service, Pacific Northwest Res. Stat., Portland, Oregon USA.

— AND L. WUNDER. 1993. Experimental manipulation of managed stands to provide habitat for spotted owls and to enhance plant and animal diversity. *J. Raptor Res.* **27**: 87.

—, S.P. HORTON AND J.A. REID. 1989. *Optimal sampling for radiotelemetry studies of spotted owl habitat and home range.* Research Paper PNW-416. U.S. Forest Service, Pacific Northwest Res. Stat., Portland, Oregon USA.

—, —, AND B.L. BISWELL. 1992. Northern spotted owls: influence of prey base and landscape character. *Ecol. Monogr.* **62**: 223–250.

—, J.A. REID AND S.P. HORTON. 1990. Spotted owl home range and habitat use in southern Oregon Coast Ranges. *J. Wildl. Manage.* **54**: 11–17.

CHEN, J., J.F. FRANKLIN AND T.A. SPIES. 1993. Contrasting microclimates among clearcut, edge and interior of old-growth forest. *Agric. and For. Meteorol.* **63**: 219–237.

CODY, M.L. 1985. An introduction to habitat selection. Pp. 3–56 in M.L. Cody, ed. *Habitat selection in birds.* Academic Press, New York, USA.

FOREST ECOSYSTEM MANAGEMENT ASSESSMENT TEAM. 1993. *Forest ecosystem management: an ecological, economic, and social assessment.* US Government Printing Office, Salt Lake City, Utah USA.

FORSMAN, E.D. 1988. A survey of spotted owls in young forests in the northern Coast Range of Oregon. *Murrelet* **69**: 65–68.

— AND E.C. MESLOW. 1985. Old-growth retention for spotted owls – how much do they need? Pp. 58–59 in R.J. Gutierrez and A.B. Carey (Tech. eds.). *Ecology and management of the spotted owl in the Pacific Northwest.* Gen. Tech. Rep. PNW-185. US Forest Service, Pacific Northwest Res. Stat., Portland, Oregon USA.

—, —, AND M.J. STRUB. 1977. Spotted owl abundance in young versus old-growth forests, Oregon. *Wildl. Soc. Bull.* **5**: 43–47.

—, E.C. MESLOW AND H.M. WIGHT. 1984. Distribution and biology of the spotted owl in Oregon. *Wildl. Monogr.* **87**: 1–64.

—, I. OTTO. AND A.B. CAREY. 1991. Diet of spotted owls on the Olympic Peninsula, Washington and the Roseburg District of the Bureau of Land Management. Pp. 527 in L.F. Ruggiero, K.B. Aubry, A.B. Carey and M.H. Huff (Tech. coords.). *Wildlife and vegetation of unmanaged Douglas-fir forests.* Gen. Tech. Rep. PNW-GTR-285. US Forest Service, Pacific Northwest Res. Stat. Portland, Oregon USA.

FRANKLIN, J.F. 1992. Scientific basis for new perspectives in forests and streams. Pp. 25–72 in R.J. Naiman, ed. *Watershed management: balancing sustainability and environmental change.* Springer-Verlag, New York, NY, USA.

— AND C.T. DYRNESS. 1973. *Natural vegetation of Oregon and Washington.* Gen.

Tech. Rep. PNW-8. US Forest Service, Pacific Northwest Res. Stat., Portland, Oregon USA.

—, K. CROMACK, JR., W. DENISON, A. MCKEE, C. MASER, J. SEDELL, F. SWANSON AND G. JUDAY. 1981. *Ecological characteristics of old-growth Douglas-fir forests.* Gen. Tech. Rep. PNW-118. US Forest Service, Pacific Northwest Res. Stat., Portland, Oregon USA.

— AND T.A. SPIES. 1991. Composition, function, and structure of old-growth Douglas-fir forests. Pp. 71–77 in L.F. Ruggiero, K.B. Aubry, A.B. Carey, and M.H. Huff (Tech. coords.), *Wildlife and vegetation of unmanaged Douglas-fir forests.* Gen. Tech. Rep. PNW-GTR-285. US Forest Service, Pacific Northwest Res. Stat., Portland, Oregon USA.

FRETWELL, S.D. AND H.L. LUCAS, JR. 1969. On territorial behavior and other factors influencing habitat distribution in birds. I. Theoretical development. *Acta Biotheor.* **19**: 16–36.

HALL, F.C., L.W. BREWER, J.F. FRANKLIN AND R.L. WERNER. 1985. Plant communities and stand conditions. Pp. 17–31 in E.R. Brown, ed., *Management of wildlife and fish habitats in forests of western Oregon and Washington.* R6-F&WL-192-1985. US Forest Service, Pacific Northwest Region, Portland, Oregon USA.

HAMER, T.E. 1988. Home range size of the northern barred owl and northern spotted owl in western Washington. M.Sc. thesis, Western Washington Univ., Bellingham, Washington, USA.

HANSON, E., D. HAYS, L. HICKS, L. YOUNG AND J. BUCHANAN. 1993. Spotted owl habitat in Washington: a report to the Forest Practices Board. Washington Forest Practices Board, Olympia, Washington, USA.

HARVEY, M.J. AND R.W. BARBOUR. 1965. Home range of *Microtus ochrogaster* as described by a modified minimum area method. *J. Mammal.* **46**: 398–402.

HAYNE, D.W. 1949. Calculation of size of home range. *J. Mammal.* **30**: 1–18.

HILDEN, O. 1965. Habitat selection in birds. *Ann. Zool. Fenn.* **2**: 53–75.

IRWIN, L.L. 1993. *Habitat conditions, wildfire risk, and demography of northern spotted owls in the eastern Cascade Mountains, Washington.* NCASI Special Report 93-04:20-23, Natl. Counc. Paper Industry for Air and Stream Improvement, Inc., New York, New York, USA.

—, D.F. ROCK, T.D. WALLACE AND G.P MILLER. 1993. *Habitat structure of stands used by spotted owls for foraging in managed and fire-regenerated forests, western Oregon.* NCASI Special Report 93-04:18-20, Natl. Counc. Paper Industry for Air and Stream Improvement, Inc., New York, New York, USA.

—, T.L. FLEMING, S.M. SPEICH AND J.B. BUCHANAN. 1991. *Spotted owl presence in managed forests of southwestern Washington.* Tech. Bull. No. 601. Natl. Counc. Paper Industry for Air and Stream Improvement, Inc., Corvallis, Oregon USA.

JAMES, F.C. 1971. Ordinations of habitat relationships among breeding birds. *Wilson Bull.* **83**: 215–236.

JOHNSON, D.H. 1992. Spotted owls, great horned owls, and forest fragmentation in the central Oregon Cascades. M.Sc. thesis, Oregon State Univ., Corvallis, Oregon USA.

—, G.S. MILLER AND E.C. MESLOW. 1992. Edge effects and the northern spotted owl. Appendix C in D.H. Johnson. Spotted owls, great horned, owls and forest fragmentation in the central Oregon Cascades. M.Sc. thesis, Oregon State Univ., Corvallis, Oregon USA.

LEHMKUHL, J.F. AND M.G. RAPHAEL. 1993. Habitat pattern around northern spotted owl locations on the Olympic Peninsula, Washington. *J. Wildl. Manage.* **57**: 302–315.

MAYR, E. 1970. Populations, species, and evolution. Harvard Univ. Press, Cambridge, Massachusetts USA.

MEYER, J.S., L.L. IRWIN AND M.S. BOYCE. 1993. *Influence of habitat fragmentation on spotted owl site location, site occupancy, and reproductive status in western*

Oregon: 1991 progress report. NCASI Special Report 93-04:1–6, Natl. Counc. Paper Industry for Air and Stream Improvement, Inc., New York, NY USA.

MILLER, G.P., J. CHATT AND D.F. ROCK. 1993. *Spotted Owl occupancy and habitat use in intermediate-aged forests, western Oregon: 1991 progress report.* NCASI Special Report 93-04:12–15, Natl. Counc. Paper Industry for Air and Stream Improvement, Inc., New York, NY USA.

NORTH, M. 1993. Stand structure and truffle abundance associated with northern spotted owl habitat. Ph.D diss., Univ. Washington, Seattle, Washington, USA.

OLD-GROWTH DEFINITION TASK GROUP. 1986. *Interim definitions for old-growth Douglas-fir and mixed-conifer forests in the Pacific Northwest and California.* Res. Note PNW-447. US Forest Service, Pacific Northwest Res. Stat., Portland, Oregon USA.

POSTOVIT, H.R. 1979. *A survey of the spotted owl in northwestern Washington.* Natl. For. Products Assoc., Washington, D.C. USA.

PULLIAM, H.R. 1988. Sources, sinks, and population regulation. *Am. Nat.* **132**: 652–661.

RIPPLE, W.J., D.H. JOHNSON, K.T. HERSHEY AND E.C. MESLOW. 1991. Old-growth and mature forests near spotted owl nests in western Oregon. *J. Wildl. Manage.* **55**: 316–318.

SCOTT, D.R. 1980. The Pacific Northwest region. Pp. 447–493 in J.W. Barrett, ed., *Regional silviculture of the United States.* John Wiley & Sons, New York, NY USA.

SIMPSON TIMBER COMPANY AND RECON (REGIONAL ENVIRONMENTAL CONSULTANTS, INC.). 1992. *Habitat conservation plan for the northern spotted owl on the California timberlands of Simpson Timber Company.* Simpson Timber Co., Arcata, California USA.

SPIES, T.A. AND J.F. FRANKLIN. 1991. The structure of natural young, mature and old-growth Douglas-fir forests in Oregon and Washington. Pp. 91–109 in L.F. Ruggiero, K.B. Aubry, A.B. Carey and M.H. Huff (Tech. coords.), *Wildlife and vegetation of unmanaged Douglas-fir forests.* Gen. Tech. Rep. PNW-GTR-285. US Forest Service, Pacific Northwest Res. Stat., Portland, Oregon USA.

TAPPEINER, J. 1992. Managing stands for northern spotted owl habitat. Appendix G in U.S.D.I. 1992. Recovery plan for the northern spotted owl – draft. US Dep. Interior, Portland, Oregon USA.

THOMAS, J.W., E.D. FORSMAN, J.B. LINT, E.C. MESLOW, B.R. NOON AND J.R. VERNER. 1990. *A conservation strategy for the northern spotted owl.* Report of the interagency committee to address the conservation strategy of the northern spotted owl. US Forest Service, Portland, Oregon USA.

UNITED STATES DEPARTMENT OF INTERIOR. 1989. Endangered and threatened wildlife and plants: proposed threatened status for the northern spotted owl; proposed rule. *Federal Register* **54**: 26666–26677.

—. 1992. Recovery plan for the northern spotted owl – draft. US Dept. Interior, Portland, Oregon USA.

WALTERS, C.J. 1986. *Adaptive management of renewable resources.* MacMillan Publishing Co., New York, NY USA.

WARING, R.H. AND J.F. FRANKLIN. 1979. Evergreen coniferous forests of the Pacific Northwest. *Science* **204**: 1380–1386.

WILSON, E.O. 1975. *Sociobiology: the new synthesis.* Harvard Univ. Press, Cambridge, Massachusetts USA.

23

Goshawk Adaptation to Deforestation: Does Europe Differ From North America?

Robert E. Kenward

Abstract – In North America, northern goshawks nest mainly in areas of continuous woodland. Protection effort concentrates on preserving this habitat. In Europe, however, goshawks nests are at a lower density, wintering juveniles less prevalent and ranges of radio-tagged hawks larger in boreal forest than in sub-boreal areas of woodland/farmland mosaic. Radio-tracking also shows that goshawks prefer to hunt, have the smallest foraging ranges and accumulate as juveniles in areas where food is most abundant. It seems that woodland/farmland mosaics are optimal goshawk habitat in Europe, so why not in North America? Food availability for breeding goshawks is probably at least as good in North America as in Europe, but there may be less winter food in sub-boreal regions, especially for male hawks. Nesting goshawks in North America may also face more problems than in Europe from competition by *Bubo* and *Buteo* species. Similar difficulties may affect goshawks when felling creates clearings in forests. Conservation management in North American woodland requires more study of interspecific nest relationships, winter diet of goshawks and their performance between fledging and breeding.

Key words: northern goshawk; deforestation; radio-tracking; breeding; prey; competition.

In North America, there is concern about the impact of forestry and management on the northern goshawk. In the United States, attempts have even been made to list the species as endangered, e.g. USFWS 1992, Greater Gila Biodiversity Project 1994, on the basis that the Endangered Species Act would then preserve mature woodland as critical goshawk habitat. The goshawk was formerly perceived as a species dependent on the extensive northern forests in Europe (Brown 1976), as in North America (Bent 1937). However, recent studies of goshawks in Europe have shown that the species reaches its highest densities in sub-boreal (temperate and more southerly) regions where continuous forest is fragmented by creation of farmland. In this paper I will:

(1) review ways in which goshawk habitat use, juvenile movements and area requirements relate to food supplies in Europe;
(2) indicate how these relationships may explain the success of goshawks in human-altered habitats in Europe;

(3) speculate why goshawks may not adapt so readily to such habitats in North America, and
(4) provide some suggestions for further research.

GOSHAWKS IN EUROPEAN BOREAL AND SUB-BOREAL REGIONS

In Europe, goshawks appear to favour hunting in those habitats where suitable prey are most abundant (Kenward and Widén 1989). Thus, goshawks studied in Swedish boreal forest, where red squirrels were the main prey, hunted mainly in the mature woodland where squirrel densities were highest (Widén 1989). In contrast, in Swedish landscapes of woodland/farmland mosaic, most goshawks were recorded in woodland edge zones, where their hunting success was greatest (Kenward 1982). Here most prey biomass was provided by edge-dwelling prey, including hares taken by female hawks, and ring-necked pheasants taken by males, which did not kill full-grown hares (Kenward et al. 1981).

In the woodland/farmland areas, although the area of goshawk foraging ranges varied up to five-fold in each of three study areas, the length of woodland edge within each range remained relatively constant (Kenward 1982). The mean length of woodland edge per range was greatest in a region with relatively few pheasants, all of which were wild, and lowest in an area where many pheasants were released each year. The size of foraging ranges followed the same pattern, averaging 20 km^2 in a pheasant-enriched area compared with 44 km^2 in a nearby area without released pheasants. Moreover, range sizes were smaller still, at 16 km^2, on the Swedish island of Gotland, where snowcover was short enough for rabbits to become another abundant edge-living prey and the main food for female goshawks in winter (Marcström et al. 1990). Even smaller ranges were recorded for goshawks released in lowland Britain, where woodpigeons were another important goshawk prey (Kenward 1979), taken mainly in edge zones.

On the 3100 km^2 island of Gotland, more than 350 goshawks were radio-tagged during seven year to study behavior and population dynamics. Young hawks from inland nests dispersed at a younger age and farther than those from nests in a coastal area where rabbits were abundant. Feeding experiments showed that early dispersal resulted from food shortage, and as young rabbits reached full size the male hawks (but not females) from the rabbit-rich area dispersed as far as those from areas with few rabbits (Kenward et al. 1993). Nevertheless, there could be as many as one hawk per 2 km^2 during winter in the rabbit-rich area, close to the density of the one hawk per 1.5 km^2 around released pheasants on the mainland, compared with one hawk per 5–10 km^2 elsewhere in Central Sweden (Kenward et al. 1981, Kenward, 1985). Foraging ranges overlapped extensively in all areas.

Table 1 combines data from winter foraging ranges of radio-tagged hawks,

Table 1. Breeding densities, winter range sizes from radio-tracking, and juvenile persistence of northern goshawks in Europe

Habitat type	Km² for each		% Juveniles among hawks trapped in winter
	Breeding pair	Hawk in winter (mean min–max)	
Boreal forest	40–100	57 (18–92)	20
Sub-boreal region with snowcover	20–30	44 (13–101)	53
Sub-boreal region with snowcover and pheasants	NA	20 (7–46)	83
Sub-boreal maritime region (Gotland, Sweden)	8–25	16 (2–129)	50

Sources of data: Höglund (1964), Bednarek (1975), Kenward et al. (1981, 1993, unpublished), Marcström and Kenward (1981b), Wikman and Lindén (1981), Kenward (1982), Widén (1985, 1987), Link (1986).

from winter trapping and from some studies of goshawk nesting density in Sweden. In regions with woodland and farmland, the average area per active nest and winter ranges were smallest where prey was most abundant. Thus, there was one nest per 8 km² in the rabbit-rich area on Gotland, compared with one nest per 25 km² farther inland, which was in turn comparable with other woodland/farmland areas in Scandinavia (Sulkava 1964, Lindén and Wikman 1983). Areas with close to 10 km² per nest have also been recorded in German regions with only 12–15% woodland but with abundant prey (Bednarek 1975, Link 1986). In particular, there are high densities of feral pigeons near German cities; feral pigeons and woodpigeons are frequent prey of goshawks in Central Europe (Brüll 1964, Opdam 1975).

Just as radio-tagging showed that juvenile goshawks tended to move to rabbit-rich areas on Gotland, so the proportion of juvenile hawks was highest at sites with most prey on mainland Sweden. Juveniles comprised 83% of the hawks trapped on estates with released pheasants, compared with 53% at other sites in central and southern Sweden (Marcström and Kenward 1981a). In contrast, juveniles were only 20% of the hawks caught in boreal forest during winter (Widén 1985). Banding shows that juvenile goshawks from boreal regions in northern Scandinavia tend to move south to the sub-boreal in winter, with much less movement by adults, and some subsequent return movement north by first winter hawks (Marcström and Kenward 1981a). Thus the trapping data (Table 1), combined with banding recoveries, suggest that boreal forest is not a very favourable habitat for juvenile goshawks in winter. The relatively large ranges and low nest densities in boreal forest are further evidence that European goshawks fare best in human-altered habitats.

GOSHAWKS IN NORTH AMERICAN BOREAL AND SUB-BOREAL REGIONS

It would be convenient for forestry in the United States if North American goshawks adapted so well to forest fragmentation. However, although goshawks are by no means limited to conifer woodland in North America, and a small population has recently been found breeding (but not wintering) in scattered steppland aspen woods (J. Yonk and M. Bechard pers. comm.), North American goshawks seem scarce away from areas of continuous forest. Why should this be? Let us consider the following three hypotheses.

Hypothesis I: Other North American Raptors Displace Goshawks from Fragmented Woods

There are marked guild differences between sub-boreal Europe and America in the raptors which might compete with nesting goshawks, and thus perhaps displace nesting pairs (Table 2). The red-tailed hawk is a common potential

Table 2. The mass, percentage of birds in the nesting diet, and density of bird-eating raptors which weigh 300–3000 g and are common in woodland in sub-boreal Europe and North America

Species	Sex	Europe			North America		
		Mass (g)	Birds in diet (%)	Pairs 100 km^{-2}	Mass (g)	Birds in diet (%)	Pairs 100 km^{-2}
Goshawk	(m)	850	73–94	3–12	850	4–94	1–4
	(f)	1250	—	—	1100	—	—
Cooper's hawk	(m)	—	—	—	350	53–88	4–8
	(f)	—	—	—	550	—	—
Buzzard	(m)	800	5–30	10–30	—	—	—
	(f)	900	—	—	—	—	—
Red-tailed hawk	(m)	—	—	—	1000	17–33	5–12
	(f)	—	—	—	1200	—	—
Red-shouldered hawk	(m)	—	—	—	550	21–29	15–20
	(f)	—	—	—	700	—	—
Bubo spp.	(m)	2300	8–20	2–7	1300	30–60	5–10
	(f)	2800	—	—	1700	—	—

Source of data: Craighead and Craighead (1956), Brüll (1964), Tubbs (1974), Bednarek (1975), Opdam (1975), Reynolds (1975), Bartelt (1977), Cramp and Simmons (1980), Marcström and Kenward (1981b), Wikman and Lindén (1981), Marquiss and Newton (1982), Reynolds and Meslow (1984), Widén (1985), Goszczynski and Pilatowski (1986), Link (1986), J. Yonk and M. Bechard (pers. comm.)

competitor of the northern goshawk in sub-boreal woodlots and forest across North America. Red-tails are larger than their closest European equivalent, the common buzzard, commonly take larger prey, and tend to nest before goshawks (Craighead and Craighead 1956). In Europe, common buzzards nest later than goshawks (Cramp and Simmons 1980) and tend to avoid goshawk nests (Kostrzewa 1987). There is also a difference between Europe and North America in the *Bubo* spp. owls, which regularly kill goshawks (Mikkola 1983). The great horned owl is a common tree-nesting species in woodlands and fragmented forests across North America. The European eagle owl is twice the weight of the great horned owl and very seldom nests in trees. Eagle owls usually nest in rock locations, typically on or near cliffs (Mikkola 1983), so lack of suitable nest sites probably explains why eagle owls are uncommon in the woodland/farmland areas where goshawks are most abundant. Another guild difference between the two continents is the lack of a medium-sized accipiter in Europe. However, it seems unlikely that goshawks would be displaced from North American woodlots by Cooper's hawks, which are smaller and nest later than goshawks. It is more likely that Cooper's hawks would avoid goshawks (Reynolds 1975). North America also lacks two kites that nest in European woodland used by goshawks, but it seems unlikely that their absence would adversely affect goshawks. In summary, the impact on goshawks of aggression and predation by red-tailed hawks and great horned owls in North American woodlots and fragmented forest may well be more substantial than the effect of competition with common buzzards and eagle owls in European sub-boreal habitats.

Hypothesis II. Goshawks Lack Food for Breeding in North American Sub-boreal Habitats

Table 2 suggests that North American goshawks might face relatively more competition than their European counterparts for avian prey during breeding. Competition from *Buteo* species might not differ markedly between the continents because the proportion of birds in the diets of all species are fairly similar and the combined density of red-tailed and red-shouldered hawks is similar to the maximum density of common buzzards. Although great horned owls take more avian prey than their European counterparts, young thrushes and corvids which form the bulk of summer diet for sub-boreal goshawks in Europe seem abundant in American woods too, with a much greater diversity of sciurids, including chipmunks *Citellus Tamius* spp. and ground squirrels (*Spermophilus* spp.), than in Europe. Corvids, thrushes, woodpeckers and sciurids dominate the nesting diet of American goshawks (Craighead and Craighead 1956, Schnell 1958, Meng 1959, Storer 1966, Bartelt 1977, Allen 1978, Reynolds and Meslow 1984, Bloom et al. 1985, Kennedy 1991). Therefore, it seems unlikely that shortage of food during breeding could explain the absence of goshawks from sub-boreal woodland mosaics in North America. In winter however, the picture is very different, leading to a third hypothesis.

Table 3. The prey groups commonly present as >10% of items taken by northern goshawks in three habitat types in Europe and North America

Habitat type	Breeding (nest survey)	Winter (radio-tracking)
Boreal forest	Tetraonid Corvid Thrush	Squirrel Tetraonid Hare
Sub-boreal with snowcover	Thrush Corvid Columbid	Squirrel Phasanid Hare
Sub-boreal maritime	Thrush Corvid Columbid Rabbit	Squirrel Columbid Rabbit Phasanid Hare

Sources of data: Craighead and Craighead (1956), Schnell (1958), Meng (1959), Eng and Gullion (1962), Brüll (1964), Höglund (1964), Sulkava (1964), Storer (1966), McGowan (1975), Opdam (1975), Bartelt (1977), Allen (1978), Kenward et al. (1981, 1993), Marquiss and Newton (1981), Wikman and Lindén (1981), Ziesemer (1983), Reynolds and Meslow (1984), Bloom et al. (1985), Goszczynski and Pilatowski (1986), Widén (1987), Kennedy (1991), F.I.B. Doyle (pers. comm.).

Hypothesis III. Goshawks Lack Winter Food in North American Sub-boreal Habitats

Table 3 shows prey groups which regularly formed more than 10% of the goshawk diet in European studies. There are marked differences between diet in summer and winter. For example, young thrushes and corvids form an important part of the prey at goshawk nests throughout Northern and Central Europe (Brüll 1964, Höglund 1964, Sulkava 1964, van Beusekom 1972, Opdam 1975, Marquiss and Newton 1982, Lindén and Wikman 1983, Goszczynski and Pilatowski 1986, Mañosa 1991, Tornberg and Sulkava 1991), but these species are much less frequently killed in winter (Kenward 1979, Kenward et al. 1981, Ziesemer 1983, Widén 1987).

In Europe, squirrels, tetraonids and hares dominate winter prey of goshawks in boreal areas (Widén 1985, R. Tornberg pers. comm.). In North America, boreal forests provide similar prey, including an even greater abundance of lagomorphs. Whereas adult varying hares average 3500 g in Europe and are too large a prey for male goshawks, snowshoe hares are small enough (900–1600 g) to be common prey at goshawks' nests (McGowan 1975) and for both goshawk sexes in North America (F.I.B. Doyle pers. comm.). The relatively small size of this important lagomorph, whose population cycles are associated with irruptions of boreal goshawks (Mueller et al. 1977, Keith and Rusch 1988), may also explain why female goshawks are smaller in North America than in northern Europe and why the species is thus less sexually dimorphic (Table 2, see also McGowan 1975). The American pine squirrel is also smaller, at 140–250 g (Whitaker 1980), than its Eurasian equivalent red squirrel, which weighs 250–400 g (Van den Brink 1967).

In sub-boreal areas that lack snowshoe hares or red squirrels, *Tamiasciurus* spp. there appears to be less food for North American goshawks than for their European counterparts. Red squirrels occur in all woodland types throughout sub-boreal Europe, which did not evolve an equivalent for the North American grey squirrel, and are important winter prey for both goshawk sexes (Kenward *et al*. 1981). Grey squirrels replace *Tamiasciurus* in North American sub-boreal deciduous woods, and were introduced in Britain where they are killed readily by female goshawks. However, the relatively large grey squirrels (450–700 g) prove a difficult prey for male goshawks in Britain (Kenward *et al*. 1991), and presumably in North America too, where *Citellus* and *Spermophilus* species are below ground in winter. North America has cottontail rabbits (*Sylvilagus* spp.), but comparison with Europe suggests that these and other large sub-boreal lagomorphs would also be difficult prey for male goshawks in winter. Native phasanids may now be less available for sub-boreal goshawks in North America than the grey and red-legged partridges in Europe, but perhaps phasanids were equally available in the past; the introduced wild pheasants are now declining on both continents. The other major winter food of sub-boreal goshawks in Europe is the migratory, flocking woodpigeon. North America has lost its vast flocks of passenger pigeons, although introduced feral pigeons are common in some woodland/farmland regions.

CONCLUSIONS AND FURTHER QUESTIONS

In summary, radio-tracking has shown that European goshawks prefer to hunt in habitats where success is highest, and that their ranges are smallest where there is most food. Juvenile goshawks disperse furthest from areas with little food, and tend to accumulate in areas with much food. Goshawk ranges are large, nests are sparse and wintering juveniles few in European boreal forests. However, goshawks thrive in cultivated sub-boreal areas, especially where human activities have resulted in abundant columbid and phasanid prey.

Food availability for breeding goshawks is probably at least as good in North America as in Europe, especially among mammal prey, because lagomorphs (snowshoe hares) can be highly abundant in the boreal forest, and sciurids farther south. However, there may be less winter food for sub-boreal goshawks in North America than in Europe, especially the male hawks. The males must use winter prey to attract mates for breeding, during which they may also face problems from aggression by great horned owls and red-tailed hawks.

The reasons for goshawks adapting so well in Europe, but not North America, to the woodland/farmland mosaics created by humans, may apply to goshawks in managed forest too. Changing the structure of forests by clear-cutting is likely to change prey availability and may also increase the opportunities for *Buteo* species to hunt in cleared areas. Does this increase abundance of aggressive species which may compete with goshawks for remaining stands of mature trees? Or could habitats be managed so that edge-dwelling goshawk prey increase to compensate for effects of increased competition?

Questions of interspecific aggression or competition can be addressed by comparing spacing and nesting success in areas with varying degrees of management, extending the pioneering work of Reynolds (1975). However, as noted by Wikman and Lindén (1981), when faced with an inexplicably declining goshawk population, for which the nesting biology was well understood, there is a risk of "beating around the bush studying hawks in summer, while the most important things are happening in winter." Based on findings in Europe, several questions spring to mind. What are the main winter foods of North American goshawks in sub-boreal habitats? What do goshawk movements tell us about winter habitat requirements? In particular, what do movements and survival before first breeding indicate about habitats, prey availability, and population developments? An emphasis on pre-breeders is both appropriate and practical. It is appropriate, because problems with prey or habitats may be more obvious in potential recruits to a population than among settled adults, which have already overcome many problems. It is practical, because earlier radio-tagged studies, such as by Kennedy (1991), can now be extended by marking nestling hawks with improved tags to monitor predation, movements and survival during their first 3–4 yr of life.

The need for more knowledge of goshawk demography and winter foraging is recognised (Reynolds *et al*. 1992). Although research has yet to explain how modern silvicultural practices in North America influence goshawk occupancy and productivity, it is likely that food supplies, competition or both factors limit densities. It would be wise to manage forests to produce landscapes capable of supporting goshawks and other wildlife.

ACKNOWLEDGMENTS

I am greatly indebted to V. Marcström and M. Karlbom for their collaboration and friendship during many years of studying goshawks in Sweden, and to C.M. Perrins, I. Newton, D. Jenkins and J.D. Goss-Custard for support and encouragement at other times. For improvements to the manuscript, I thank J. Buchanan, A.J. Gray, S. Petty, D. Varland and S.S. Walls.

REFERENCES

ALLEN, B.A. 1978. Nesting ecology of the Goshawk in the Adirondacks. M.Sc. thesis, State Univ. New York, Syracuse, New York USA.

BARTELT, P.E. 1977. Management of the American goshawk in the Black Hills National Forest. M.Sc. thesis, Univ. Iowa City, Iowa USA.
BEDNAREK, W. 1975. Vergleichende Untersuchungen zur Populationsökologie des Habichts (*Accipiter gentilis*): Habitatbesetzung und Bestandsregulation. *Deutscher Falkenorden jb* (1975): 47–53.
BENT, A.C. 1937. Life histories of North American birds of prey, part 1. *U.S. Nat. Mus. Bull.* 167, Washington, D.C. USA.
BLOOM, P.H., G.R. STEWARD AND B.J. WALTON. 1985. *The status of the Northern Goshawk in California 1981–83*. Wildl. Mgt. Branch Admin. Rep. 85–1. Dep. Fish and Game, Sacramento, California USA.
BROWN, L. 1976. *British birds of prey*. Collins, London, UK.
BRÜLL, H. 1964. *Das Leben deutscher Greifvögel*. Fischer, Stuttgart, Germany.
CRAIGHEAD, J.J. AND F.C. CRAIGHEAD. 1956. *Hawks, owls and wildlife*. Stackpole Co., Harrisburg, Pennsylvania USA.
CRAMP, S. AND K.E.L. SIMMONS. 1980. *Handbook of the birds of Europe, the Middle East and North Africa*. Vol. 2. Oxford Univ. Press, Oxford, UK.
ENG, R.L. AND G.W. GULLION. 1962. The predation of goshawks upon ruffed grouse on the Cloquet Forest Research Center, Minnesota. *Wilson Bull.* 74: 227–242.
GOSZCZYNSKI, J. AND T. PILATOWSKI. 1986. Diet of common buzzards (*Buteo buteo*) and goshawks (*Accipiter gentilis* L.) in the nesting period. *Ekol. Polska* 34: 655–667.
GREATER GILA BIODIVERSITY PROJECT. 1994. Petition to list the Queen Charlotte goshawk (*Accipiter gentilis laingi*) as a federally endangered species. Endangered Species Series No. 9, Silver City, New Mexico USA.
HÖGLUND, N. 1964. Über die Ernährung des Habichts (*Accipiter gentilis* L.) in Schweden. *Viltrevy* 2: 271–328.
KEITH, L.B. AND RUSCH, 1988. Predation's role in the cyclic fluctuations of ruffed grouse. Pp. 699–732 in H. Ouellet, ed. *Acta XIX Congressus Internationalis Ornithologicus*. Univ. Ottawa Press, Ottawa, Canada.
KENNEDY, P.L. 1991. Reproductive strategies of northern goshawk and Cooper's hawks in north-central New Mexico. Ph.D. thesis, Utah State Univ., Logan, Utah USA.
KENWARD, R.E. 1979. Winter predation by goshawks in lowland Britain. *Brit. Birds* 72: 64–73.
—. 1982. Goshawk hunting behaviour, and range size as a function of habitat availability. *J. Anim. Ecol.* 51: 69–80.
—. 1985. Problems of goshawk predation on pigeons and some other game. Pp. 666–678 in V.D. Ilyichev and V.M. Gavrilov, eds. *Acta XVIII Congressus Internationalis Ornithologicus*. Acad. Sci. USSR, Moscow, Russia.
—, V. MARCSTRÖM AND M. KARLBOM. 1981. Goshawk winter ecology in Swedish pheasant habitats. *J. Wildl. Manage.* 45: 397–408.
—, — AND — 1993. Post-nestling behaviour in Goshawks, (*Accipiter gentilis*): I. The causes of dispersal. *Anim. Behav.* 46: 365–70.
—, T.P. PARISH AND P.A. ROBERTSON. 1991. Are tree species mixtures too good for grey squirrels? Pp. 243–253 in M.G.R. Cannell, P.A. Robertson and D.C. Malcolm, eds. *Ecology of mixed species stands of tress*. Blackwell Scientific Publ., Oxford, UK.
— AND P. WIDÉN. 1989. Do goshawks need forests? Some conservation lessons from radio tracking. Pp. 561–567 in B.-U. Meyburg and R.D. Chancellor, eds. *Raptors in the modern world*. World Working Group on Birds of Prey, Lentz Druck, Berlin, Germany.
KOSTRZEWA, A. 1987. Quantitative analysis of nest-habitat separation in common buzzard (*Buteo buteo*), goshawk (*Accipiter gentilis*) and honey buzzard (*Pernis apivorus*). *J. Ornithol.* 128: 209–229.
LINDÉN, H. AND M. WIKMAN. 1983. Goshawk predation on tetraonids: availability of prey and diet of the predator in the breeding season. *J. Anim. Ecol.* 52 953–968.

LINK, H. 1986. Untersuchungen am Habicht (*Accipiter gentilis*) Ph.D. thesis, Friedrich-Alexander Univ., Erlangen-Nürnberg, Germany.
MAÑOSA, S. 1991. Biologia trofica, us de l'habitat de la reproduccio de l'Astor *Accipiter gentilis* (Linnaeus, 1758) a la Segara. Ph.D. thesis, Univ. Barcelona, Barcelona, Spain.
MARCSTRÖM, V. AND R.E. KENWARD. 1981a. Movements of wintering goshawks in Sweden. *Swedish Game Res.* **12**: 1–35.
— AND —. 1981b. Sexual and seasonal variation in condition and survival of Swedish goshawks (*Accipiter gentilis*). *Ibis* **123**: 311–327.
—, — AND M. KARLBOM. 1990. *Düvhöken och dess plats i naturen.* Trycksaker, Norrköping, sweden.
MARQUISS, M. AND I. NEWTON. 1982. The goshawk in Britain. *Brit. Birds* **75**: 243–260.
McGOWAN, J.D. 1975. *Distribution, density and productivity of goshawks in interior Alaska.* Final Rep. Fed. Aid in Wildl. Restor. Proj. W-17-3,4,5,6. Alaska Dept. Fish and Game, Juneau, Alaska USA.
MENG, H. 1959. Food habits of nesting Cooper's Hawks and goshawks in New York and Pennsylvania. *Wilson Bull.* **71**: 169–174.
MIKKOLA, H. 1983. *Owls of Europe.* T. & A. D. Poyser, Calton, UK.
MUELLER, H.C., D.D. BERGER AND G. ALLEZ. 1977. The periodic invasions of goshawks. *Auk* **94**: 652–663.
OPDAM, P. 1975. Inter- and intraspecific differentiations with respect to feeding ecology in two sympatric species of the genus *Accipiter. Ardea* **63**: 30–54.
REYNOLDS, R.T. 1975. Distribution, density and productivity of three species of *Accipiter* hawks in Oregon. M.Sc. thesis, Oregon State Univ., Oregon USA.
— AND E.C. MESLOW. 1984. Partitioning of food and niche characteristics of coexisting *Accipiter* during breeding. *Auk* **101**: 761–779.
—, R.T. GRAHAM, M.H. REISER, R.L. BASSETT, P.L. KENNEDY, D.A. BOYCE, G. GOODWIN, R. SMITH AND E.L. FISHER. 1992. *Management recommendations for the northern goshawk in the Southwestern United States.* Gen. Tech Rep. RM-217, U.S. For. Serv., Ft. Collins, Colorado USA.
SCHNELL, J.H. 1958. Nesting behaviour and food habits of goshawks in the Sierra Nevada of California. *Condor* **60**: 377–403.
STORER, R.W. 1966. Sexual dimorphism and food habits in three North American accipiters. *Auk* **83**: 423–436.
SULKAVA, S. 1964. Zur Nahrungsbiologie des Habichts, *Accipiter gentilis* (L). *Aquilo Ser. Zool.* **3**: 1–103.
TORNBERG, R. AND S. SULKAVA. 1991. The effect of changing tetraonid populations on the nutrition and breeding success of the goshawk (*Accipiter gentilis* L.) in Northern Finland. *Aquilo Ser. Zool.* **28**: 23–33.
TUBBS, C.R. 1974. *The buzzard.* David and Charles, London, UK.
USFWS. 1992. *90-day finding on petition to list the Northern Goshawk as an endangered species in the southwestern United States.* U.S. Fish Wildl. Serv., Albuquerque, New Mexico USA.
VAN BEUSEKOM, C.F. 1972. Ecological isolation with respect to food between sparrowhawk and goshawk. *Ardea* **60**: 72–96.
VAN DEN BRINK, F.H. 1967. *A field guide to the mammals of Britain and Europe.* Collins, London, UK.
WHITAKER, J.O. 1980. *The Audubon Society field guide to North American mammals.* Knopf, New York, New York USA.
WIDÉN, P. 1985. Population ecology of the goshawk (*Accipiter gentilis* L.) in the boreal forest. Ph.D. thesis, Univ. Uppsala, Sweden.
—. 1987. Goshawk predation during winter, spring and summer in a boreal forest area of Central Sweden. *Holarct. Ecol.* **10**: 104–109.
—. 1989. The hunting habits of goshawks *Accipiter gentilis* in boreal forests of Central Sweden. *Ibis* **131**: 205–213.

WIKMAN, M. AND H. LINDÉN. 1981. The influence of food supply on goshawk population size. Pp. 105–113 in R.E. Kenward and I.M. Lindsay, eds. *Understanding the goshawk*. Int. Assoc. of Falconry and Conservat. of Birds of Prey, Oxford, UK.

ZIESEMER, F. 1983. *Untersuchungen zum Einfluß des Habichts* (Accipiter gentilis) *auf Population seiner Beutetiere*. Beitr. Wildlbiologie 2. Hartmann, Kronshagen, Germany.

24

Rain Forest Raptor Communities in Sumatra: The Conservation Value of Traditional Agroforests

Jean-Marc Thiollay

Abstract – Managed agroforests increasingly replace natural forests in western Indonesia. The raptor community of three of the richest types of agroforests was compared with that of the primary forest and to the open cultivated areas, using 1-km^2 sample plots. Both species richness and density in agroforests were more than twice higher than in cultivated areas, but they were only half as high in primary forests. The 12 raptor species recorded were divided into 4 groups according to their increasing tolerance to forest degradation or management. Six species had no viable populations outside mature natural forest and three species were more abundant in primary forest than elsewhere. The last three species were more frequent in agroforests, but only one of them was absent from the primary forest. It is concluded that agroforests conserve no more than a quarter of the original forest raptor community and provide an adequate habitat for only one additional open woodland species. An even smaller subset of species was found in the sparsely wooded cultivated areas.

Key words: rain forest; raptor; community; Sumatra; agroforest.

The remaining tropical forests are now increasingly disturbed and only a minor proportion of the still primary tracts are actually protected. The pressure of human population is especially serious in Southeast Asia and consequently so is the rate of deforestation or forest conversion. Moreover, the majority of resident raptor species in tropical Asia are forest species because the relative area of open grasslands and wetlands was historically low in this region. Primary forests may be converted into a wide variety of forest types from regenerating stands after selective logging to permanently and intensively disturbed secondary forests or artificial plantations of exotic trees for industrial purposes. Therefore, it is important to know:

(1) how exploited, disturbed, managed or planted forests meet the requirements of forest raptor species, i.e. how vulnerable are these species to habitat degradation;
(2) how modified woodlands or forests may be acceptable substitutes for, or play a role in, the survival of primary forest species.

The traditional agroforests of Indonesia, which occupy over ten million hectares in Java, Sumatra and Kalimantan, are man-made reconstructions of forest ecosystems that often look like natural secondary forests. They combine relatively high levels of economic productivity and floristic diversity. That is why they are cited as one of the best compromises between the maintenance of high human population and the conservation of tropical forests (Michon and Bompard 1987). However, they are devised mostly for cash income (resin, latex, fruits) and year-long availability of various other minor products, e.g. spices, medicines, timber, flowers. Their biological richness is only an unintended byproduct of their economic uses.

In a cooperative research project, three representative agroforest types were compared with the nearby primary forest to assess the relative contribution of agroforests to the conservation of the initial biodiversity. The whole bird community was investigated (Thiollay 1995) and the Falconiforms are here singled out because of their sensitivity to habitat changes, the different sampling methods used, and their necessarily larger study area. The problem of raptor conservation in Sumatra is now pressing since >90% of the lowlands have been deforested and 60% of the cultivated areas are occupied by agroforests. It is thus important to assess the value of this habitat.

STUDY AREA

Three different areas of western Sumatra (Fig. 1) were surveyed for both agroforests (58 plots), openfields (17 plots) and lowland (4 plots) or hill primary forests (18 plots). They were centred on the small towns of Krui (Damar agroforests and nearby Barisan Selatan National Park) Rantaupandan (*Hevea* agroforests and eastern Mount Kerinci National Park) and Maninjau (Durian agroforests and surrounding slopes up to Bukittingi and to the Rimbo Panti forest reserve). In a fourth area (Gunung Leuser National Park) only primary forest was surveyed: two lowland plots (<500 m), two hill plots (500–1000 m) and two submontane plots (1000–1500 m).

All agroforest and open cultivated areas were below 500 m (mostly 150–400 m) in elevation and were classified as lowlands (Van Marle and Voous 1988, Laumonier 1991). Conversely, most primary forest plots (22) were above 500 m because very few natural forest patches remained below this elevation.

Habitat types

(1) **Cultivated areas.** The densely populated and irrigated farmlands included mostly rice fields, dotted with palms or fruit trees, well-wooded home gardens and some patches of grasslands, coffee plantations or secondary growth.

(2) **Agroforests.** All come from clear-felled natural forest, first burned and then cultivated with dry rice for 1–3 yr. Then seedlings of valuable trees are planted

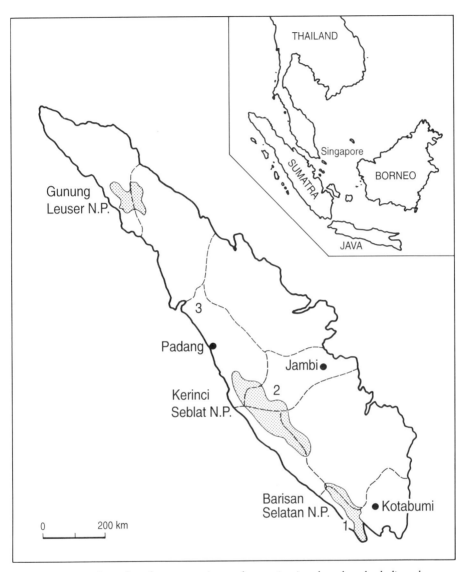

Figure 1. Location of study areas, primary forests (national parks, shaded) and agroforests (1 = Krui; 2 = Rantaupandan; 3 = Maninjau).

together with vegetables, bananas and coffee which are harvested until yr 7–8. Natural vegetation is then left growing for 10 (rubber) to 20 yr (damar) before the successional vegetation is cut, except for useful trees, and exploitation begins in a regularly managed agroforest. Aging trees, i.e. from 30 (rubber) to 70 yr (damar) old, are felled for timber and replaced. These agroforests are multipurpose, highly diversified tree plantations providing mainly cash crops, e.g. resin, latex, fruits, spices, timber. They are structurally close to natural secondary

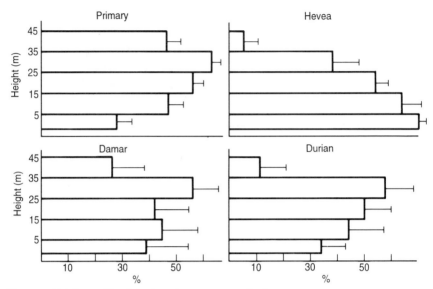

Figure 2. Height profile of the four forest types studied in western Sumatra. Mean (±1 SE) percent foliage cover visually estimated within the five main strata (between 0.5–45 m) at 112 points selected at random within each forest type (see Thiollay 1995).

forests and have very little resemblance with industrial plantations. Most of them however, are spatially heterogeneous, being mosaics of different ages, dominated by different proportions of the various species planted. Many of them have surprisingly high densities of wild pigs, monkeys (*Presbytis, Macaca* spp.), gibbons (*Hylobates, Symphalangus* spp.) and squirrels (*Callosciurus, Ratufa, Petaurista* spp.), but birds are often heavily hunted. Detailed descriptions are included in Torquebiau (1984), Whitten et al. (1984), Michon (1985 and 1991), Michon et al. (1986), Michon and Bompard (1987), Laumonier (1991) and Thiollay 1995.

(a) Damar agroforests. The Damar tree agroforests, tapped for resin, are the tallest (35–50 m) and densest of the agroforests studied (245–500 trees/ha at canopy level). Their height profile is the most similar to that of the primary forest (Fig. 2). Though 56–80% of trees >20 cm dbh are damars, at least 39 other species of trees occur. These are mostly fruit trees with some patches of clove trees. The usually open understory is rich in palms and bamboos and the ground cover includes many ferns, Melastomaceae, Marantaceae, Acanthaceae and *Piper* spp.

(b) Hevea agroforests. The rubber tree agroforests are lower (closed canopy at 20–30 m) with a denser mid and low understory than the other two agroforests. Two-thirds of the trees are the exotic rubber tree mixed with up to 91 other tree species, including patches of bamboos, coffee, cocoa, clove, cinnamon, nutmeg and palms. They cover several million hectares in Sumatra and Kalimantan, i.e. more than any other agroforest type.

(c) Durian agroforests. They are dominated by durian and secondarily, as in

other agroforests, by other fruit trees (*Mangifera, Parkia, Dinocarpus, Baccaurea, Nephelium*), timber tree species (*Actinodaphne, Alangium, Litrea, Toona, Pterospermum*) and self-established useful trees (*Terminalia, Ficus, Artocarpus, Octomeles*). The canopy is intermediate in height (30–40 m) between the other two agroforests and usually is more open (350 trees ha^{-1}). Dense patches of clove, cinnamon, nutmeg, coffee, bamboos or *Pandanus* are found in the understory.

(3) **Primary forest.** This wet or moist lowland or hill mixed dipterocarp rain forest is the richest vegetation in Sumatra in terms of plant as well as bird species (Whitten *et al.* 1984, Whitmore 1985, Collins *et al.* 1991). Up to 382 plant species, including 71 tree species, have been counted along a 100-m transect in the primary forest adjoining the Rantaupandan area. The tree species richness alone may be two to four times greater in the primary forest than in any of the agroforests. The almost continuous canopy, which culminates at 45–55 m, includes many *Dipterocarpaceae*, as well as *Cesalpiniaceae, Myristicaceae, Sapotaceae* and *Moraceae*. The understory is darker than in most agroforests. Gaps from tree falls or landslides are frequent because most of the remaining forests are on steep slopes and in high rainfall areas. Almost all sample plots were in protected areas, but they sometimes included also logged and high secondary forests.

METHODS

The sampling technique was a relatively quick and simple method designed for large-scale surveys in tropical countries. It was an adaptation of the classical point count method to the average density and detectability of diurnal raptors. All individuals, present within as many random plots as possible, and during a limited period of time, were counted and then pooled to assess the overall raptor community in the habitat involved. The study was conducted during 9 wk from May to July 1991 and 1992, in a relatively dry or low rainfall season where most raptors were breeding (based on observations of occupied nests, fledged young or adults carrying prey).

A sample plot was a 1-km^2 area. Whenever possible, it was a 1 × 1-km square. Depending on field accessibility, it was sometimes (five plots) a 120° angle wedge-shape area with a 1-km radius, or (three plots) a 2 × 0.5-km rectangle. The general locations of plots were taken at random, but their exact limits were constrained by three prerequisites. A plot had to be:

(1) wholly visible from a dominant lookout or a viewpoint outside or above the forest;
(2) crossed by at least one track or foot path; and
(3) covered by a dominant habitat type over at least 60% of its area.

The boundaries of sample plots were recognized using maps, prominent field marks and visual estimates of distances based on repeated training with known distances. Planimetric measures, independent from the relief, were used. A total of 103 plots was censused.

Two complementary searching methods were used. First, most data were obtained in a 2-hr watching session between 0800 h and 1200 h from a vantage point (clearing, gap, or ridge) overlooking the entire plot. Then at least 2 hr more were spent walking through the area or searching from convenient openings in early morning and in the afternoon, between sunrise and sunset. During the crucial 0800–1200 h morning period, at least one adult of every resident pair soared, calling or displaying, over the forest at least once a day for 5–10 min. In Southeast Asia, every forest raptor species, except the falconets, exhibit this territorial behavior unlike in other tropical rain forests where some species never soar. The 2-hr watching session often proved to be long enough to record most, if not all, pairs whose home range was centered on the plot. Increasing this duration tended to inflate the number of birds by allowing more marginal individuals to enter the limits of the plot. At other times of the day, some species still readily flew over the forest (*Pernis, Ictinaetus*) while *Accipiter* spp. were occasionally glimpsed in the understory and more vocal species (*Spilornis, Spizaetus*) could often be heard. The abundance of the black-thighed falconet was probably underestimated in forest because this species is very small, does not fly often and is usually perched in the tree tops.

Only the birds in adult plumage perching, displaying or foraging well within the plot were counted. Young birds or individuals just flying high above or crossing the quadrat or entering the plot for a short time or chased out by a resident pair were not considered. This minimum number of different individuals recorded on the plot on only one occasion during 4 hr during a single day was used as an abundance index. Because of different specific conspicuousness, stages of breeding cycles, home range sizes, activity, movements and social behavior, this index was not comparable between species and could not be used to calculate a true density. It was only assumed to reflect the changes of abundance of a given species between habitats and plots.

As far as possible, plots were selected randomly but under the constraint of accessibility. For logistical reasons and to gain time, two or even three sample plots were often contiguous or close to each other. This violates the assumption of independence between plots required for usual statistical treatments. In a small number of such cases, the same individual may have been recorded twice on adjacent plots.

Habitat Categories and Vegetation Structure

Six main habitat categories were defined:

CULT = Open, inhabited and cultivated areas with fields, pastures and few scattered bushes and trees (mostly coconut palms and mango trees).
TREE = Wooded grasslands, small fields, gardens and road verges with large

trees, hedgerows, woodlots, forest edges, open tree plantations (coconut, mango, coffee, clove, ...) and clearings or large forest gaps.

FOR 1 = low dense shrubby secondary growth (*Lantana, Eupatorium, Bishoffia, Dillenia*, ...) dominated by sparse trees, dense plantations of young trees (palms, cinnamon, *Casuarina*, ...) and small, fragmented patches of low or highly degraded forest.

FOR 2 = Heavily logged or naturally disturbed, open canopy secondary forest, well developed timber tree plantations and agroforests.

FOR 3 = Little broken canopy, evergreen natural forest but selectively logged, or fragmented, or mostly lower than 25 m or largely moist deciduous forest type.

FOR 4 = Large tracts of closed canopy, high primary evergreen wet forest.

This classification is based on an increasing cover of high canopy trees (above 25 m). On every plot, the percent cover of each habitat category, rounded to the nearest 10%, was estimated using a grid-cell map, topographical features and field-drawn contours. To facilitate analysis, plots were all homogeneous enough to be classified into a single dominant habitat category including ⩾60% of their total area (CULT + TREE, or FOR 1 + FOR 2, or FOR 3 + FOR 4).

To describe each plot by a single index of forest cover, including both area and quality of woodlands, I attributed to each habitat category a fixed coefficient roughly proportional to its mean estimated upper canopy cover (⩾25 m), i.e. from (CULT) and 1 (TREE) to 5 (FOR 4). Thus, Forest Cover Index = Σ (HAB$_i$ × CAN$_i$), where HAB$_i$ was the percent cover of habitat category i (1–10) and CAN$_i$ was its canopy cover (1–5). The maximum index would be 50, i.e. the entire plot (10) covered by primary forest (5).

Statistical Analysis

To assess first whether the differences of species density and frequency of occurrence were non-random, I used a general linear model for analysis of variance, and repeated measures ANOVA, with forest types as the main units and plots as the subunits. This analysis was performed on Log_{10} $(x + 1)$ transformed data to meet variance assumptions of parametric analysis. Then between-species density and species richness were compared by a non-parametric multiple comparison procedure based on two-tailed Mann-Whitney U tests performed on all possible pair-wise comparisons (Sokal and Rohlf 1981). Frequencies were compared by χ^2 tests on transformed data. To single out the most important habitat for each species, I tested the effect of the percent cover of all six habitat categories, separately on the species density index in least square regressions with transformed data. This procedure was preferred to a multiple regression because the habitat covers in each plot were interdependent. A product moment correlation coefficient was used to assess the relationship between forest cover index and the abundance of each species.

The habitat niche breadth was calculated by the Levins formula:

$$\beta = \sum_{1}^{n} (p_i^2)^{-1},$$

where p_i is the proportion of records in the ith habitat. The inter-habitat turnover of community structure was given by the similarity index:

$$C_{ij} = 1 - 0.5 \sum_{i}^{s} |p_{ik} - p_{jk}|,$$

where p_{ik} and p_{jk} are the numerical proportions of each community that comprised the kth species (Hanski 1978).

RESULTS

Community Composition and Structure

Sixteen species of diurnal raptors are breeding and resident in Sumatra (Van Marle and Voous 1988, Andrew 1992). All of them were recorded in at least one study area. They can be divided into three groups:

(1) Eight species (Table 1) were common or regular in their preferred habitat type in all four study areas. They made up the main part of the analysis.
(2) Four additional species were seen only in the primary forest plots. Two of them, the Jerdon's baza and the bat hawk were rare or local in lowland and hill forests. The other two species were probably frequent within their restricted habitat and altitudinal ranges since they were recorded respectively in all four lowland primary forest plots (the Wallace's hawk-eagle), and in the only submontane forest plots surveyed (the besra).
(3) The last four species were excluded from the analysis because they were only seen flying over or outside the sample plots. They included the black-shouldered kite, an open field specialist, and three species associated with rivers and inundated ricefields (the lesser fish-eagle) or lakes and sea coast (the brahminy kite and the white-bellied sea-eagle).

The species distribution and abundance were derived from the mean number of individuals recorded in sample plots and from the frequency of occurrence of the species in each habitat (Table 1). Both frequency and density indices of every species changed significantly from the cultivated areas to the primary forest and between the three agroforests types (Table 1, two-way ANOVAs, $P < 0.05$). The number of coexisting species was minimal in the fields where over half of the plots had no raptor at all and none had more than two species. Conversely, the primary forest included 11 of the 12 species, with at least one species in every plot and up to 3–4 on 65% of them. All agroforests were intermediate, from damar (0–2 species/plot) to durian-dominated type (1–4 species/plot). No

Table 1. Distribution of wide-ranging species among habitat types (Open cultivated areas, *Hevea*, Damar or Durian dominated agroforests, Primary forest).

	Fields		*Hevea*		Damar		Durian		Primary	
	D	F	D	F	D	F	D	F	D	F
Changeable hawk-eagle	18	12	33	28	*82	*64	67	44		
Crested serpent eagle	24	18	50	33	36	32	*133	*78	38	27
Black-thighed falconet	35	18	28	17	9	5	*44	*22	15	8
Oriental honey buzzard			22	17	5	5	6	6	*23	*23
Crested goshawk			17	17	5	5			*54	42
Indian black eagle					5	5			*46	*43
Rufous-bellied eagle			6	6			6	6	*27	*23
Blyth's hawk eagle							6	6	*138	*92

D (density index) = mean number of individuals per 100 1-km^{-2} plots
F (frequency) = percentage of plots where the species was recorded. Values rounded to the nearest unit. Significantly highest values in bold types ($P < 0.05$, see methods)
* = significantly highest values for the species ($P < 0.05$, see methods)

Table 2. Raptor community structure of main habitat types in Western Sumatra.

	Fields/ pastures	*Hevea* agroforests	Damar agroforests	Durian agroforests	Primary forests
Sample size[a]	17	18	22	18	28
Total species richness[b]	3	6	6	6	9–11
Mean richness/plot[c]	0.47	1.17	1.18	1.61	2.68
(SD)	(0.62)	(0.85)	(1.17)	(1.08)	(0.77)
Mean density/plot[d]	0.76	1.56	1.45	2.61	3.46
(SD)	(1.09)	(1.24)	(0.74)	(1.22)	(1.07)
Diversity H'[e]	1.531	2.373	1.722	1.806	2.642

[a] Number of 1-km^2 plots
[b] Number of species recorded over all plots
[c] Mean number of species per plot
[d] Mean number of individual raptors per plot
[e] Shannon's diversity index

raptor was found on 10% of agroforest plots and only 1–2 species were found on 85% of them. Overall, both species richness and density index in agroforests were at least twice as high as in cultivated areas, but they remained about half that of primary forests (Table 2, one-way ANOVAs, $P < 0.01$).

All three agroforests had the same three dominant species (72–93% of all individuals recorded), with a particular abundance of the crested serpent-eagle and the black-thighed falconet in the durian forest and of the changeable hawk-eagle in the damar forest. The same three species alone survived in agricultural areas but at much lower densities than in agroforests.

The five dominant species of the primary forest (see Table 1) comprised 85%

of individual raptors recorded between 500–1000 m. They were all occasionally sighted in at least one agroforest, but only the oriental honey buzzard and the crested goshawk occurred there in substantial numbers, and mostly in *Hevea* (rubber) forests (25% of all raptors). Even in *Hevea* agroforests they seemed to do hunting incursions from nearby natural forests where they were most likely to breed (based on observations of prey items being carried out of the agroforests).

The proportion of the two most numerous species in the total population is a simple index of dominance and the first indicator of community diversification. It was 77–83% in fields and two agroforests, against 53% in the primary forest, as well as in the *Hevea* agroforest which held the richest plant and bird communities of the three managed woodlands (Thiollay 1995). The Shannon's diversity index was 35% higher in the primary forest community than the mean of all agroforests because of both a higher species richness and a higher equitability of species abundances.

Though mostly derived from the primary forest, the raptor community of any agroforest was more similar to that of the cultivated areas than to the primary forest (similarity index C_{ij} = 0.544–0.709 vs. 0.207–0.347). This suggests how significantly the raptor population is affected by intensive forest management.

Species Habitat Selection

Each species has its own habitat requirements and its own degree of tolerance or adaptability to human-altered woodlands, i.e. here to forest fragmentation, management degradation or opening (see Appendix). This can be illustrated for each species by the mean forest cover index of all sample plots where they occurred (Fig. 3) or by the distribution of specific records among habitat categories (Fig. 4). The succession of species among habitats is continuous with substantial overlaps. For convenience, the raptor community may be divided into four species groups according to their pattern of habitat distribution in the man-made gradient, from the dense primary forest to the little-wooded cultivated areas.

(1) Species typically associate with dense primary forest where they reach, by far, their highest density. They may tolerate some form of natural or artificial disturbance and venture occasionally in surrounding agroforests, but they probably cannot survive entirely without any patch of mature forest. The rufous-bellied eagle and the Blyth's hawk-eagle are most representative of this category. They have also the lowest niche habitat breadth (Fig. 4) and they are large predators of vertebrates which presumably require large hunting ranges. Three other species belong to this category: the Jerdon's baza, unaccountably rare, the bat hawk, mostly found where numerous bat caves provide an abundant food supply and the Wallace's hawk-eagle, restricted to the lowland rain forest where it replaces the Blyth's hawk-eagle, which inhabits the hill and montane forests.

(2) Other dense forest species definitely are more tolerant of forest disturb-

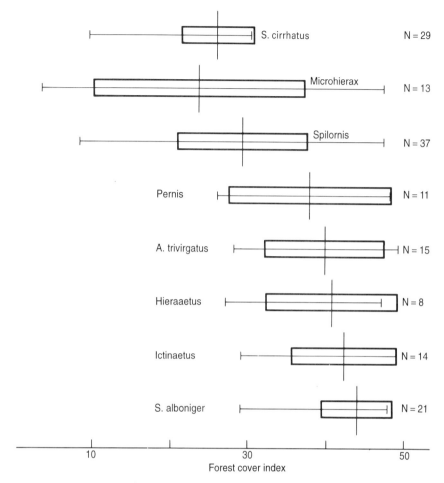

Figure 3. Forest cover index (see Methods) of all the sample plots (N) where the main species were recorded (mean ± 2SD and range).

ance and fragmentation. They were even found in secondary forests far from any primary tract and they ventured regularly into some agroforests where their frequency remained. Their abundance however, was much lower than in primary forests. In this category, the crested goshawk was a secretive forest interior species whereas the Indian black eagle foraged above the canopy which induced a wider adaptability to open habitats than that exhibited by forest understory species. The endemic subspecies of the besra was found only above 1000 m, not overlapping with the range of the crested goshawk, although outside Sumatra the two species occur together (Mees 1980, Thiollay 1993).

(3) Two species were even more tolerant of openings: the oriental honey buzzard (sighted in all agroforests) and the crested serpent eagle (the most abundant raptor in two of the three agroforest types). Both were found in

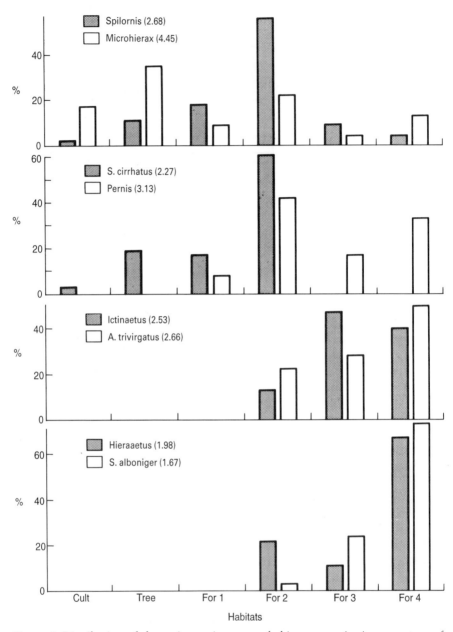

Figure 4. Distribution of the main species among habitat categories in percentage of individuals recorded. Habitat niche breadth in parentheses.

primary forest mostly around natural gaps, disturbed patches or edges. They have specialized diets (bees nests and snakes, respectively).

(4) The most regular species in the open cultivated woodlands were the changeable hawk-eagle, the only species never seen in pristine natural forests,

Table 3. Highest determinants of species abundance in regressions of habitat cover in all sample plots against density index (R^2) and correlation coefficient between forest cover and density index of species (r).

	Regression coefficient R^2 of habitat cover/species density						r
	CULT	TREE	FOR1	FOR2	FOR3	FOR4	Forest cover
Changeable hawk-eagle			0.17	0.67			−0.42**
Crested serpent eagle			0.18	0.32			−0.65**
Black-thighed falconet	0.11	0.25					−0.48**
Oriental honey buzzard					0.36*	0.35*	+0.75**
Crested goshawk					0.76**	0.87**	+0.80**
Indian black eagle					0.89**	0.98**	+0.77**
Rufous-bellied eagle					0.95**	0.87**	+0.82**
Blyth's hawk-eagle					0.84**	1.00**	+0.86**

* $P < 0.05$
** $P < 0.01$

and the black-thighed falconet, which had the widest habitat niche. The latter feeds on flying insects at tree-top level. It was found in primary forest on dead emergent trees of the upper canopy or in gaps, as well as in all other habitats studied, even on single isolated dead palms in fields or large clearings. It is noteworthy that this group involves both one of the largest and most diet generalist species (hawk-eagle) and the smallest and rather specialized species (falconet).

The two habitat categories most significantly affecting the presence and abundance of each species are given in Table 3, as well as the overall response of species to the forest index cover. Species of group 1 (Blyth's hawk-eagle, rufous-bellied eagle) are highly associated with the most mature forest type. This trend decreases slightly in group 2 (black eagle, crested goshawk) and still further in group 3 (oriental honey buzzard) where crested serpent eagle is more associated with agroforests than with natural forest. As expected, the last two species (black-thighed falconet, changeable hawk-eagle) are negatively associated with dense forest and primarily dependent upon the most open woodlands.

DISCUSSION

Although the sample size in each habitat was limited to 17–28 km², all the breeding raptor species known from Sumatra were recorded in at least one of the study areas, if not in a sample plot. A shortcoming of this work was the lack of a sufficient area of lowland primary forest to compare with each agroforest type. Instead I had to pool separate primary forest patches, sampled over a wider elevation range and spread over four areas, one of which was not associated with an agroforest. One species (the besra) never occurred in the lowlands where the Wallace's hawk-eagle replaced the Blyth's hawk-eagle, which was found almost exclusively at higher elevation. Pooling such widespread data may overestimate the species richness of the natural forest. However, when the subset of 20 hill forest plots was compared either with the similar sample size of each agroforest (18–22 plots) or with agroforests pooled (48 plots), the natural forest always remained richer than agroforests. The overall bird communities of lowland areas, where all the agroforests are found, are known to be richer than upper level avifaunas (Van Marle and Voous 1988, Thiollay 1995). Considering the larger sample size in agroforests, the difference between the two groups should be presumably greater if about 20 lowland primary forest plots could have been compared with each agroforest sample, i.e. the relative impoverishment of agroforests could only be accentuated.

The individual species, or groups of species, defined according to their tolerance to forest degradation, may be used as indicators of forest disturbance and fragmentation in assessments of the degree of alteration of a forested area and its potential conservation value. The exact environmental factors to which the species are responding may be determined only with much larger sets of records and more detailed measurements of vegetation structure, floristic diversity, degree of fragmentation, prey availability and human disturbance. Indeed this study is a preliminary assessment of the relative richness of intensively managed vs. natural mature forest, using a group of top predators to illustrate trends that were already found to be similar among the whole bird community (Thiollay 1995).

All but one of the 12 species of the gradient occurred in the primary forest. Seven of these species were mostly, if not only, restricted to primary or little-disturbed natural forests. Two additional species were found in the primary forest at higher density than in any of the modified habitat. The raptor community of agroforests was only a subset of that of the primary forest, except for the addition of the changeable hawk-eagle. From their mean densities and frequency distributions, only three species have viable populations in agroforests and were truly independent of the proximity of any primary forest.

If I arbitrarily set from my samples (Table 1) a level of population viability at >25 adults per 100 km^2 with an occurrence of >20% of 1-km^2 plots, then six species reached this level in the primary forest, two in every agroforest (and a third species in a single agroforest) and almost none in the cultivated areas. Five other forest species extended their foraging range occasionally into some agroforests, but they were unlikely to live in them permanently.

The only species found in open cultivated areas were the dominant ones of the agroforests. They were indeed woodland species which maintained patchily distributed and marginal populations in farmlands. Only the black-shouldered kite was primarily associated with open fields and grasslands. It was, however, uncommon or local in Western Sumatra and was absent from all the sample plots. This was perhaps because of the recent deforestation of many areas, and hence their late colonization by this kite.

Historically, Sumatra was mostly covered by dense rain forest, and few of the original species were pre-adapted to colonize the open woodlands. The species which maintained significant populations in human-altered habitats were those which inhabited marginal parts of the primary forests, i.e. edges, gaps and upper canopy. A similar evolution was found in many other tropical countries, including nearby southern India, Malaysia, Borneo and Java (Thiollay 1983, 1993, Thiollay and Meyburg 1988).

Conservation Implications

Many traditional agroforests are among the richest man-made, sustainably used and most forest-like tropical habitats. They provide however, an adequate habitat for viable populations of little more than a quarter of the original primary forest species. Clearly they are a poor substitute for primary forests. They cannot replace the latter in a conservation strategy at the regional scale and they must be viewed instead as a complement to natural forest reserves. When they surround some tracts of primary forest, as was the case in most study areas, they may allow forest species to extend their hunting range and help maintain their population in otherwise too small natural forest patches. Agroforests are not permanently inhabited, well-wooded buffer zones between the deforested areas around villages and the remaining, mostly protected, primary forests. Nevertheless, raptors may be more affected by transformation of primary forests into agroforests than by selective logging alone (Johns 1992).

These results question the myth of the compatibility of the conservation of many forest species with the sustainable use of forest resources, especially when this exploitation must be economically significant for an ever-growing human population, with increasing needs and standards of living. Trends toward an intensification of the management of agroforests are already apparent and are almost unavoidable in the long term.

Another cause for concern is that these agroforests have superseded the lowland forests, which are known as the richest forest type in Sumatra (Collins *et al.* 1991, Laumonier 1991) as in many tropical countries. Almost all the existing protected areas include either mangrove and peat swamp forest or upper hill and montane forests, but not lowland, well-drained forest. The Wallace's hawk-eagle, the most narrowly associated with the lowland rain forest, is likely to be the first local raptor species to be extirpated.

ACKNOWLEDGMENTS

This study was funded by the French Ministry of Environment (DRAEI) and carried out with a French (ORSTOM) and Indonesian team at the BIOTROP Institute of Bogor (Java). I thank very much H. de Foresta, G. Michon and my wife Françoise for their invaluable help at various stages of this work.

REFERENCES

ANDREW, P. 1992.The birds of Indonesia. A checklist. Kukila checklist no. 1, Indonesian Ornithological Society, Jakarta, Indonesia.
COLLINS, N.M., J.A. SAYER, AND T.C. WHITMORE. 1991. *The conservation atlas of tropical forests*. Asia and the Pacific. IUCN and MacMillan Press, London, UK.
HANSKI, I. 1978. Some comments on the measurement of niche metrics. *Ecology* 59: 168–174.
JOHNS, A.D. 1992. Species conservation in managed tropical forests. Pp. 15–53 in T.C. Whitmore and J.A. Sayer, eds. *Tropical deforestation and species conservation*. Chapman and Hall and IUCN, London, UK.
LAUMONIER, Y. 1990. Search for phytogeographic provinces in Sumatra. Pp. 193–211 in P. Baas, V. Kalkman and R. Geesink. eds. *The plant diversity in Malaysia*. Kluwer Academic Publishers, Dordrecht, Netherlands.
—. 1991. Vegetation de Sumatra: écologie, flore, phytogéographie. Doct. Thesis. Univ. of Toulouse, Toulouse, France.
MEES, G.F. 1980. Sparrowhawks of the Andaman islands. *J. Bombay Nat. Hist. Soc.* 77: 371–412.
MICHON, G. 1985. De l'homme de la forêt au paysan de l'arbre: agroforesteries indonésiennes. Doct. Thesis, Univ. des Sciences Techniques, Montpellier, France.
—. 1991. The Damar gardens: existing buffer zones at Pesisir area of Sumatra Selatan National Park; Lampung. *Proceedings of the Symposium on Rain Forest Protection and National Park Buffer Zones*. Directory General of Nature Conservation and Forest Protection Nat. Park. Devel. Proj. Buffer Zone Research and Management. Biotrop, Jakarta, Indonesia.
—, F. MARY AND J.M. BOMPARD. 1986. Multistoried agroforestry garden system in West Sumatra, Indonesia. *Agroforestry Systems* 4: 315–338.
— AND J.M. BOMPARD. 1987. Agroforesteries indonésiennes: contributions paysannes à la conservation des forêts naturelles et de leurs ressources. *Revue d'Ecologie (Terre et Vie)* 42: 3–37.
SOKAL, R.R. AND F.J. ROHLF. 1981. *Biometry*. Freeman, San Francisco, USA.
THIOLLAY, J.M. 1983. Evolution actuelle du peuplement de rapaces diurnes dans le nord de Bornéo. *Alauda* 51: 109–123.
—. 1989. Censusing of diurnal raptors in a primary rain forest: comparative methods and species detectability. *J. Raptor Res.* 23: 72–84.
—. 1993. Response of a raptor community to shrinking area and degradation of tropical rain forest in the south western Ghâts (India). *Ecography* 16: 97–110.
—. 1995. The role of traditional agroforests in the conservation of rain forest bird diversity in Sumatra. *Conserv. Biol.* 9: 335–353.
— AND B.U. MEYBURG. 1988. Forest fragmentation and the conservation of raptors: a survey on the island of Java. *Biol. Conserv.* 44: 229–250.
TORQUEBIAU, E. 1984. Man-made Dipterocarp forest in Sumatra. *Agroforestry Systems* 2: 103–128.
VAN MARLE, J.G. AND K.H. VOOUS. 1988. The birds of Sumatra. B.O.U. Check-list no. 10, British Ornithologists' Union, Tring, UK.
WHITMORE, T.C. 1985. *Tropical rain forests of the Far East*. Clarendon Press, Oxford, UK.
WHITTEN, A.J., S.J. DAMANIK, J. ANWAR AND N. HISYAM. 1984. *The ecology of Sumatra*. Gadjah Mada Univ. Press, Yogyakarta, Indonesia.

APPENDIX

Habitat requirements of raptors in western Sumatra

Species	Population abundance	Altitudinal distribution	Natural habitat	Optimal habitat in study areas	Tolerance to forest degradation
Jerdon's baza	rare/local	lowlands and hills	primary forest	primary forest	low
Oriental honey buzzard	frequent	lowlands and hills	most forest types	primary and secondary forests	moderate
Bat hawk	local	lowlands and hills	natural forests with gaps	broken forest near limestone caves	moderate
Crested serpent eagle	common	lowlands to montane	forest edges and gaps	agroforests, secondary forests	high
Besra	local	montane	primary forest	forest, including edges	?
Crested goshawk	frequent	lowlands and hills	dense forests	primary and closed secondary forests	rather low
Black eagle	frequent	lowlands to montane	most forest types	primary and disturbed forests	moderate
Rufous-bellied eagle	rather frequent	lowlands to submontane	primary forest	little disturbed forests	low
Changeable hawk-eagle	rather common	lowlands to submontane	open woodlands	agroforests	high
Blyth's hawk-eagle	common	hill and montane	primary forest	primary and closed secondary forests	low
Wallace's hawk-eagle	local	lowlands only	primary forest	primary forest	very low
Black-thighed falconet	widespread	lowlands and hills	forest upper canopy	agroforests, fields	high

25
Diurnal Raptors in the Fragmented Rain Forest of the Sierra Imataca, Venezuela

Eduardo Alvarez, David H. Ellis, Dwight G. Smith and Charles T. Larue

Abstract – The rain forest of the Sierra Imataca in eastern Venezuela has been subjected to extensive deforestation for pastures and agricultural settlements. In the last decade the opening of access roads combined with intensified logging and mining activities have fragmented a significant portion of the remaining forest. We noted local distribution and habitat use for 42 species of diurnal raptors observed in affected areas in this region. We observed some raptors considered as forest interior species and other open country species foraging and roosting in man-made openings inside the forest.

Key words: Neotropical raptors; rain forest fragmentation; road survey; Sierra Imataca: Venezuela; logging.

In the late 1980s the Venezuelan government created a series of extensive Forest Management Units (FMUs), each with >100 000 ha and some exceeding twice that area, encompassing much of the remaining lowland forests north of Canaima National Park. These large logging concessions extend from the eastern shore of the Guri Lake (lower Caroní River, Bolívar) to the Orinoco River delta and the southeastern portions of Amacuro and Bolívar state which border with Guyana (Gorzula and Medina-Cuervo 1986).

The Imataca Forest Reserve, with >3 million ha allocated to logging interests (Luy 1992), is a scarcely explored range of low mountains and foothills, including the Sierra Imataca mountains and the Nuria escarpment. Stretching east and south of the confluence of the Caroní and Orinoco Rivers, these mountains form the northeastern edge of the land system surrounding the ancient Guianan Shield (Paynter 1982) and comprise a unique type of submontane wet forest that rarely exceeds 400 m in elevation (Huber and Alarcon 1988).

In the past 30 yr, a large portion of the Sierra Imataca suffered extensive deforestation (Huber and Alarcon 1988). In the last decade renewed emphasis on ranching, logging and mining activities resulted in a profusion of access roads opening the forests. Mining concessions cover >7% of the Imataca Reserve lands (Luy 1992). New technology, economic markets, and politics promote further degradation of large expanses of these forested lands. Thus, what remains of the original vegetation continues to shrink. Current logging practices and the conversion of forest to pasture and agricultural settlement contribute significantly to the sweeping fragmentation of the landscape.

Recently, the Venezuelan Audubon Society described the Venezuelan System of Forest Reserves, compiling and reviewing more than 400 references on forest management and related investigations (Luy 1992). Facing massive habitat change, current knowledge of the potentially threatened ecosystems is largely inadequate for conservation management (Bisbal 1988, Luy 1992). For example, Venezuelan scientists and resource managers do not have access to basic vertebrate listings for Imataca.

The Rio Grande area of the Imataca Forest Reserve has an international reputation as one of Venezuela's best locations for viewing birds; its rich avifauna exceeds 400 species (S.L. Hilty pers. comm.). This knowledge however, is held almost exclusively by birdwatchers visiting from foreign countries.

A decade ago little was known about the biology of raptors in Latin America, except for broad regional accounts which are now dated (Ellis and Smith 1986). Recently the Peregrine Fund's Maya Project, located in Central America's Peten Region, greatly increased our knowledge of habitat requirements for many Neotropical species of raptor (Burnham et al. 1989, 1990, Whitacre et al. 1991).

Raptor research has received little attention in Venezuela, yet it is a country where at least 66 species of Falconiformes have been recorded (Altman and Swift 1986). Since the late 1970s, E. Alvarez (EA) has compiled records of the birds of prey and other vertebrate fauna of the lowlands of the Guayana Region (Alvarez et al. 1986).

In 1987 the logging companies built a bridge over the Rio Grande (hereafter referred to as the RGB) and began building roads and removing logs from the forest to the east. That year 2500 gold prospectors invaded the reserve (J.R. Blanca, Internal Report, CODEFORSA 1988).

Intending to study the harpy eagle in eastern Venezuela, EA began wildlife inventories in that portion of the Imataca Range in 1988 (Kinzey et al. 1988). In 1989 he combined field observations and information collected mainly by foreign birdwatchers to produce a Preliminary Status Report on Venezuelan Raptors (EA unpubl. report 1989).

STUDY AREA AND METHODS

In the dry season of 1993 (22 February–1 March) we conducted a preliminary survey of the raptor community east of El Palmar, Bolívar. We worked near and inside to two adjacent logging concessions located in the Rio Grande area of the Imataca Reserve and managed by the companies SOMAGUA and CODEFORSA. Alvarez has listed >250 species of birds for these FMUs (EA unpubl. reports 1988–93).

We sampled a road network of 25 km of paved state road (serving agricultural and cattle ranching settlements); 16 km of unpaved state rural road (serving settled agricultural land cleared near the boundary of the reserve); and 35 km of newer and largely unimproved pathways (serving the logging concessions). We usually drove over a transect in one direction in the early morning (starting by 0800 h) and again in the afternoon as we returned to our base camp near the RGB shortly before sunset.

As we drove on the state roads and logging paths, we kept constant watch in all directions for birds and other wildlife. We made additional observations while walking along abandoned logging trails and foot trails. At least two experienced observers worked together at all times during the survey of dirt roads, riding on the back of an open-bed pickup. Because our primary intent was to find out which species of raptors were present, we drove at moderate speed (20–30 km h^{-1}) according to road conditions and stopped as necessary to identify birds. No standard speed was established for the study.

Since no reliable maps were available for these roads, we determined the location of the transects with a hand-held Global Positioning System (GPS) receiver (Magellan GPS-5000). To describe our sampling routes, we assigned a three-letter code to conspicuous features and sites along the transects.

The odometers of our vehicles served to estimate the length of each leg and the total distance travelled observing raptors. We considered each time that we drove a transect as one sample, so that a round trip was counted as two samples amounting to twice the route length.

Species Accounts

Ornithological investigations of the Imataca Region in the past were few and patchy in coverage (Paynter 1982). Because of the exploratory nature of the survey, we felt that it would be most useful to compile a checklist of the diurnal raptors known for the Rio Grande area. We annotated our observations during this survey and the records of EA (EA unpubl. notes and reports 1988–93). We also included unique species sightings credited to other observers whom we considered reliable.

RESULTS

King vulture. This species was observed almost every day in the Imataca area, but was less common than the greater yellow-headed vulture. During the survey, single adults and juveniles and small flocks were often seen soaring high above forested and open landscapes. Birds were also observed on the ground, eating carrion (snakes and mammals). A group of up to five birds was observed twice feeding at the carcass of a tapir left by poachers in the brushy edge of the main logging road east of RGB.

Black vulture. This is a species adapted to towns, cities and cattle ranches throughout Venezuela and elsewhere, concentrating in large numbers around slaughterhouses and garbage dumps. Black vultures were rare in the forest interior but were seen along the paved road where the forest was converted to pasture near the town of El Palmar. On 17 June 1992 EA observed two birds just west of RGB in a clearing along the road.

Turkey vulture. A vulture of varied habitat, this species was rarely seen over the forest landscapes of the Imataca, but was common over adjacent savannas and lands cleared for pasture.

Lesser yellow-headed vulture. These vultures were uncommon in the Imataca forest, but were sometimes seen soaring at low elevation or near the canopy top along logging roads and above clearings.

Greater yellow-headed vulture. Very similar to the lesser yellow-headed vulture in habits, this species was seen commonly inside and near the forest of the Sierra Imataca.

White-tailed kite. These kites were absent from the forest interior of Sierra Imataca, but they were common in the larger cleared areas along the paved roads and near towns.

Pearl kite. Never observed in forest interior, pearl kites were frequently seen perched in trees in pastures and on power lines along the paved roads near El Palmar. One sighting was made during the survey. This was 6 km northwest of RGB on Guanamito Road among forest remnants and pasture land.

Swallow-tailed kite. This species was common in the rain forest interior and in more open areas, pasture lands and settled areas. Swallow-tailed kites were most often seen soaring just above the canopy in groups of 3–30, foraging on the wing.

Gray-headed kite. This species is widespread in Venezuela, in forests, savannas and forest edges. It is occasionally seen in the forest interior in the Sierra Imataca. An adult was seen twice during the survey, 6 km east of RGB (24–25 February 1993). Records of EA indicate that a juvenile was seen at the same site (21 June 1992) and adults were observed vocalizing in display flight (May 1992) north of the SOMAGUA camp.

Hook-billed kite. This is a rare and very local bird in the Sierra Imataca. We saw one bird along the logging road east of RGB during the survey; it was near a water impoundment. One previous record exists for this species (EA unpubl. data).

Double-toothed kite. Although small and perhaps easily overlooked, this species is widespread in the Sierra Imataca. It is seen in partially cleared areas at the edge of logging roads on perches with an open view of canopy height or lower. One sighting occurred during the survey, 6 km northeast of RGB. EA had encountered this raptor in the past. He saw individual birds of this species 2 and 4 km north of RGB on 25 June 1992. On 26 February 1992 he sighted two birds when he was 21 km east of RGB and another one was observed 13 km from the bridge.

Plumbeous kite. This is the most common kite in the forest interior, and one of the more abundant raptors in all the environs of the area. Adult and immature individuals perch on open branches of tall trees adjacent to logging roads and clearings. We observed these birds soaring in large groups over the rain forest; > 50 were seen soaring north of RGB on 25 February 1993.

Snail kite. A specialized raptor of lagoons and impoundments, this species is very abundant to the north along the delta of the Orinoco River. Probably an occasional vagrant into the area, the snail kite is included in the master list made by birdwatchers for the Rio Grande area.

Slender-billed kite. A rarely observed bird, this kite occurs in the forest interior. We have one record of an adult, seen in May 1988 at an impoundment made by a logging road 18 km east of the RGB (EA unpubl. data).

Bicolored hawk. This species is secretive and difficult to see. We have one record. One was seen in 1989 at the RGB (EA unpubl. data) in the human-altered riverine forest.

Tiny hawk. This is an easily missed and secretive small sparrowhawk. It uses forest edge, second growth areas and clearings. We saw it during the survey at the Guanamito Road junction (6 km west of the RGB), perched in a small tree at the edge of the paved road in the remnants of an abandoned orchard.

White-tailed hawk. We never observed this hawk in the Imataca rain forest, but frequently saw it in the larger deforested areas now used for pasture east of El Palmar.

Zone-tailed hawk. This widespread species soars at low altitude, appearing to imitate the turkey vulture. It was frequently seen over lands converted to ranching as well as in the forest interior.

Broad-winged hawk. This is a migratory raptor, and it is rare in this portion of the country. It is listed by birdwatchers for the Imataca area, and one bird was seen by M.R. Cuesta near the RGB in 1989.

Roadside hawk. Said to be absent from dense forest (Brown and Amadon 1968), this species is one of the most common and most visible species in disturbed areas, i.e. in partially deforested areas and along logging roads, in the rain forest of the Sierra Imataca.

Short-tailed hawk. These hawks are occasionally seen over the Imataca forest, but are more common in savannas to the west. During the survey, we saw two light-morph birds 30 km east of the RGB. Records of EA indicate a bird stooping (25 May 1992) 17 km east of the RGB, and another bird was seen 20 km east of the RGB (27 June 1992).

Gray hawk. This is a widespread raptor, but not as common as the roadside hawk in the Sierra Imataca, which uses similar habitat. We had a survey sighting of one bird 10 km east of the RGB. EA saw a pair 11 km east of the RGB and a single bird at 18 km east of the RGB on 27 June 1992. Birds were also seen by EA at 18 and 28 km east of the RGB 10 July 1992.

White hawk. Occasionally seen in the Imataca forest, one was sighted during the survey 5 km north of the RGB by the tree nursery (El Baquiro). EA saw a bird near the SOMAGUA camp on 10 July 1992.

Black-face hawk. A rarely seen but conspicuous hawk, the only records are from other observers in past years. S.L. Hilty saw an adult perched mid-level in forest about 3 km west of the RGB on 6 February 1989. In 1990 M.L. Goodwin also saw this species near the bridge.

Black-collared hawk. A common hawk in marshes and lagoons of open areas and very abundant in the Orinoco River delta, the species was seen repeatedly near El Palmar (5–10 km to the west of our survey area; S.L. Hilty 1991–93).

Savanna hawk. This species is common on pasture land and savannas, sometimes near the forest edge. From the paved road, we saw an individual 14 km west of the RGB.

Great black hawk. This is a large conspicuous hawk usually seen near lagoons and creeks. Sightings during the survey included two birds 11 km east of the RGB on 26 February (one bird was seen at same site by EA on 20 June 1992) and one bird 15 km east of the RGB. Another was seen on a foot path north of SOMAGUA camp.

Crested eagle. A large but very rare and secretive raptor, the light morph is easily confused with a fledgling harpy eagle. The dark morph is similar to the adult harpy eagle. Crested eagles are differentiated by their slender and longer legs, long black line behind the eye and single crest. An adult with normal plumage was seen while walking along the road north of the river crossing, just after finishing a survey from the vehicle. EA observed a dark morph of this species in October 1992. This bird was perched close to the ground in the forest interior south of the main logging road.

Harpy eagle. This very large, rare and secretive predator primarily takes arboreal mammals in the Rio Grande forests. Apparently less likely to soar than the crested eagle, this species is hardly ever seen except at active nests. Sightings have been made by F. Espinoza in 1988 (near the main logging road) and by EA in 1989 (just north of the SOMAGUA camp). Eight nests are currently being monitored in the Sierra Imataca and seven nesting events have been documented between 1989–92 in the general area of this survey (EA unpubl. data).

Black and white hawk-eagle. In October 1992 EA observed a bird soaring over a large clearing on the main logging road. Its propensity to soar and the paucity of records suggests that this species is probably very uncommon in Sierra Imataca.

Ornate hawk-eagle. This is a slender bird of prey, not common but conspicuous when soaring. We had one sighting during the survey north of the RGB. EA saw the species in 1989 in a selectively logged forest, and in 1992 soaring over planted strips.

Black hawk-eagle. This species is similar to the ornate hawk-eagle in size, and is apparently more common in the area. While the observer was driving, one bird was seen in small trees adjacent to the main logging road (21 km east of RGB), and another was observed roosting in the snags of a roadside wetland (21 km east of the RGB). EA observed birds in February 1992 at 19 km east of the RGB and also on a trail north of the SOMAGUA Camp. He saw another in July 1992 on the road north of the RGB.

Crane hawk. A conspicuous hawk, the species is often seen soaring and calling just above the tree canopy. A pair was seen twice during the survey near the RGB 23 and 28 February 1993. S.L. Hilty saw a bird flying about 2 km east of the RGB in January 1991; EA saw one 7 km east of RGB on 20 June 1992.

Laughing falcon. Laughing falcons hunt by waiting almost motionless at a perch until prey, especially snakes, is detected. This species uses open areas, clearings and roadsides, and was seen during the survey 30 km east and 2 km west of the RGB. Between 1989 and 1992, EA saw these falcons quite frequently in the area sampled.

Collared forest falcon. This large forest falcon is rarely seen outside the forest. In December 1992 EA saw one at the clearing near the RGB. This bird was chasing chickens on the ground <100 m from the CODEFORSA camp buildings. The bird was later shot by the loggers, as were other poultry hunting raptors.

Black caracara. The black caracara is a social raptor, and not as common in the Rio Grande area as the red-throated caracara. Groups were observed 30 km east of the RGB on 26–27 February (survey) in selectively logged forest. One adult was seen in open pasture by the paved road, 10 km west of the RGB on 5 March 1992. EA observed this bird perching on the tin roof of a small stable with cattle.

Red-throated caracara. This is a gregarious raptor often seen in large groups and detected by its loud calling. We saw the species almost daily in the forest interior and along roads during the survey.

Yellow-headed caracara. This open field raptor is common in the human-dominated landscape. We had one sighting, which was 10 km west of the RGB in pasture land.

Crested caracara. A bird of open savannas and pastures, the species was commonly seen along the paved road. It was never observed in the forest interior.

Bat falcon. This falcon was a fairly common resident of the forest, nesting in cavities of tall dead snags in forest inundated with water because of road impoundments in 1989, 1991 and 1992 (EA unpubl. data). Bat falcons were detected at least five times during the survey along the main logging road.

Aplomado falcon. This bird occurs in open landscapes, sometimes close to the forest edge. One was seen near El Palmar in 1992 (EA unpubl. data).

American kestrel. A raptor found in open landscapes, this species usually perched on power lines and fence posts along roads. It was common near the Rio Grande area. One bird was seen near El Palmar by EA in 1992.

DISCUSSION

In nine days we drove 310 km looking for raptors during daylight hours (0530–1830 H); >200 km of this driving occurred on dirt roads. Half of our sample area was an open landscape dominated by agriculture and ranching, and only 25 km of this was paved. This area was created by deforestation in the last 30 yr. We covered a small portion (16 km of unpaved road) of the frontier of agricultural lands adjacent to the Forest Reserve. Most of the survey period was spent using the logging roads, penetrating 5 km north and 30 km east into the reserve from the RGB. This land is strictly protected from agricultural settlement, but poaching is common.

Our survey confirms a remarkable diversity in the raptor community of a forest landscape subjected to agricultural expansion and the ongoing disturbance of large-scale logging. According to R. Bierregaard (pers. comm.) about 90 specics of Falconiformes (including the New World vultures) breed in Central and South America. Comparing our results with the raptor communities of the four most thoroughly known forests in Tropical America (Manaus, Brazil; Manu Park, Peru; Barro Colorado Island, Panama; and La Selva, Costa Rica), our poorly studied area of Venezuela ranks on top, with at least 42 species of raptor (Karr *et al.* 1990).

By driving over this relatively small area during a short period of time we were able to detect more than 70% of the species of raptors known for the country, including some which are generally considered rare or are rarely found anywhere in their continental range (Ellis *et al.* 1990).

We agree with Vannini (1989) that more research is needed on habitat use by Neotropical raptors before making generalizations about habitat preferences. We saw raptors that are said to be forest interior species and others that are considered to be open-country species. These birds were roosting, foraging and nesting in man-made openings inside the forest that were less than 3 yr old.

In contrast to our success in surveying from the roads, a previous survey by walking 72 km into the interior of the unlogged forest yielded surprisingly few sightings of raptors. EA (unpubl. data) walked a 3 km long trail under continuous canopy 12 times, with only one raptor sighting (a rare crested eagle). Round trips of 1 day were made in May, June and October, and during eight consecutive days in November of 1992. One additional survey was made in March 1993. This forest tract proved to be otherwise rich in avifauna including birds following army ant (*Eciton* spp.) swarms. On at least two occasions EA heard vocalizations along that trail which were probably made by raptors, but the birds could not be located visually. We suggest that, to sample raptors in this forest, recording and playback, call identification, and observation above the canopy are needed as complementary techniques (Whitacre *et al.* 1991).

Current forestry practices in the FMUs of the Sierra Imataca result in drastic changes in forest structure and composition which signify a wide array of potential interactions with the landscape and the biota occupying it (Franklin and Forman 1987, Luy 1992). The diverse community of raptors presents an outstanding opportunity for field research related to forest management. For example, the role of roads as potential environmental barriers is a research priority of the National Research Council (1990). If conservation is to succeed, future research must be complemented with appropriate regulations and their enforcement, conservation education, sound habitat management, and adequate benefits to the local human population (Ellis and Smith 1986).

In North America, birdwatchers often cooperate with researchers to study birds. Hundreds of volunteers are tracking the annual winter populations of feeder birds in the United States and Canada (*WildBird Magazine*, January 1994:20–21). In a similar way, visitors could provide a valuable contribution to a monitoring program for raptors in the Sierra Imataca. We hope that our preliminary survey will encourage cooperation between ecotourists, scientists, conservationists and managers to continue gathering and disseminating useful information about these little-known habitats and their avifauna.

ACKNOWLEDGMENTS

We are grateful to SOMAGUA and CODEFORSA for their long-term cooperation and logistic support during this survey. We wish to thank John Goodwin, who collaborated as an observer, and Pablo Cardenali and Rafael Alvarez for their participation and motivation to learn about raptors. We recognize Mary Lou Goodwin, Steven L. Hilty and Douglas Mason for contributing observations and comments, and Eduardo Iñigo for reviewing the manuscript.

REFERENCES

ALTMAN, A. AND B. SWIFT. 1986.*Checklist of the Birds of South America*. A. Altman, Simon's Rock of Bard College, Great Barrington, Massachusetts, USA.
ALVAREZ, E., L. BALBAS, I. MASSA AND J.E. PACHECO. 1986. Aspectos ecológicos del Embalse Guri. *Interciencia* 11: 325–333.
BISBAL, E., F.J. 1988. Impacto humano sobre los habitat de Venezuela. *Interciencia* 13: 226–232.
BROWN, L. AND D. AMADON. 1968. *Eagles, hawks and falcons of the world*. 1989 edition, Wellfleet Press, Secaucus, New Jersey, USA.
BURNHAM, W.A., J.P. JENNY AND C.W. TURLEY (eds.) 1989. *Maya Project: Use of raptors as environmental indices for design and management of protected areas and for building local capacity for conservation in Latin America*. Progress Report II, 1989. The Peregrine Fund, Inc., Boise, Idaho USA.
—, D.F. WHITACRE AND J.P. JENNY (eds.). 1990. *Maya Project: Use of raptors as environmental indices for design and management of protected areas and for building local capacity for conservation in Latin America*. Progress Report III, 1990. The Peregrine Fund, Inc., Boise, Idaho USA.
ELLIS, D.H. AND D.G. SMITH. 1986. An overview of raptor conservation in Latin America. *Birds of Prey Bull*. 3: 21–25.
—, R.L. GLINSKI AND D.G. SMITH. 1990. Raptor road surveys in South America. *J. Raptor Res*. 24: 98–106.
FRANKLIN, J.F. AND R.T.T. FORMAN. 1987. Creating landscape patterns by forest cutting: ecological consequences and principles. *Landscape Ecology* 1: 5–18.
GORZULA, S. AND G. MEDINA-CUERVO. 1986. La fauna silvestre de la cuenca del río Caroní y el impacto del hombre: Evaluación y perspectivas. *Interciencia* 11: 317–324.
HUBER, O. AND C. ALARCON. 1988. Mapa de Vegetación de Venezuela. Ministerio del Ambiente y de los Recursos Naturales Renovables. DGIIA, División de Vegetación, Caracas, Venezuela.
KARR, J.R., S. ROBINSON, J.G. BLAKE AND R.O. BIERREGAARD JR. 1990. Birds of four neotropical forests. Pp. 237–269 in A.H. Gentry, ed. *Four Neotropical Forests*. Yale University Press, New Haven, USA.

KINZEY, W.G., M.A. NORCONK AND E. ALVAREZ-CORDERO. 1988. Primate survey of eastern Bolívar, Venezuela. *Primate Conservation* 9: 66–70.

LUY, G.A. 1992. *La investigación en Reservas Forestales y Lotes Boscosos de Venezuela.* Sociedad Conservacionista Audubon de Venezuela. Caracas, Venezuela.

NATIONAL RESEARCH COUNCIL. 1990. *Forestry research: A mandate for change.* Committee on Forestry Research, Commission on Life Sciences, Board on Biology and Board on Agriculture, NRC (U.S.). National Academy Press, Washington D.C., USA.

PAYNTER, R.A. 1982. *Ornithological Gazetteer of Venezuela.* Harvard University, Cambridge, Massachusetts, USA.

VANNINI, J.P. 1989. Neotropical raptors and deforestation: Notes on diurnal raptors at Finca El Faro. Quetzaltenango, Guatemala, *J. Raptor Res.* **23**: 27–38.

WHITACRE, D.F., W.A. BURNHAM AND J.P. JENNY (Eds.). 1991. *Maya Project: Use of raptors and other fauna as environmental indicators for design and management of protected areas and for building local capacity for conservation in Latin America.* Progress Report IV, 1991. The Peregrine Fund, Inc., Boise, Idaho, USA.

26

Value of Nest Site Protection in Ameliorating the Effects of Forestry Operations on Wedge-tailed Eagles in Tasmania

Nick J. Mooney and Robert J. Taylor

Abstract – The Tasmanian wedge-tailed eagle is an endemic subspecies which is considered to be endangered. Threatening processes include disturbance and loss of nests during logging operations. Nests are often found at a stage during logging where it is too late to provide ideal protection for a site. In situations where protection measures can be applied, this is achieved by retention of the nest tree and a surrounding buffer and by distracting disturbance during breeding. The reuse of protected and unprotected nests and their territories was compared with that of a control group where no disturbance had occurred. Disturbance of a nest site led to a dramatic drop in the use of that site and a lowered reproductive rate for a pair within their territory. Implementation of protection measures for nest sites markedly improved the breeding success of the pair subject to the disturbance, either by their remaining at the nest site and successfully raising young or by their being more successful at fledging young elsewhere within their territory. Conservation of wedge-tailed eagles could be improved by ensuring nests are discovered before logging commences. Reuse by eagles of nests where disturbance has occurred may lead to the development of a population more tolerant to such disturbance.

Key words: nesting; conservation; wedge-tailed eagle; forestry; breeding; Tasmania.

Disturbance of nesting birds can lead to a variety of effects, ranging from changes in behaviour to complete desertion of a site (Newton 1979). Many factors may potentially lead to desertion and their effects can be cumulative. Exposure to several disrupting factors, which individually may not be critical, can often cause desertion (Newton 1979). For example, forestry operations can involve catastrophic disturbance and habitat modification but may also potentially be followed by changes in prey, predators, competitors and/or pathogens, exposure to pesticides and herbicides and a permanent increased level of disturbance associated with an increased ease of access by humans. A common result of desertions is for a new nest site to be chosen for subsequent nesting (Newton 1979). However, these alternative nests may be less suitable, leading to an increased chance of failure in the future. A nest failure, including that due to disturbance, can in turn increase the chances of future desertion (Grier 1975).

Some raptors, such as the peregrine falcon, osprey and barn owl, are extremely adaptable nesters and readily use artificial sites in human-altered environments. However, such species are exceptional and the vast majority of taxa do not show such flexibility. The degree of adaptability demonstrated is likely to be a complex function of individual experience, genetic potential, the characteristics of the environmental change and the time and opportunity available for adaptation.

The wedge-tailed eagle in Tasmania is an endemic subspecies classified as endangered on the basis of a small total population (around 100 successfully breeding pairs), continued persecution and decreased productivity owing mainly to disturbance from forestry operations (Mooney and Holdsworth 1991, Garnett 1992). The Tasmanian wedge-tailed eagle nests only in trees in forests composed predominantly of old growth native species (usually *Eucalyptus* spp.; Mooney and Holdsworth 1991). Pairs always choose nest sites in trees that are among the tallest available (Mooney and Holdsworth 1991).

Like most *Aquila* the wedge-tailed eagle is a shy nester. Forestry operations have in the past often led to the destruction of nest trees or the desertion of a nest site owing to disturbance around the nest from logging and/or road construction. In order to reduce these adverse effects, measures for the protection of wedge-tailed eagle nests in Tasmania were formulated (Mooney and Holdsworth 1991) and have largely been accepted by the Forestry Tasmania and private logging companies. These measures are administered through a Forest Practice Code (Anonymous 1993) which requires that adequate protection be given to threatened species. Key aspects of these measures are the immediate reporting of nests when found, restriction of heavy disturbance to >500 m from active nests and a reserved area of habitat of ≥ 10 ha around all nests with habitat concentrated uphill on slopes. Such measures are similar to those applied to nests of some other disturbance-sensitive species (Cline 1990). Results of such nest protection measures are not well known, although there is some optimism (Nelson 1982).

This paper reports on the adaptability of Tasmanian wedge-tailed eagles to nesting in areas subject to forest operations, and investigates the value of prescriptions relating to nest protection in promoting reuse of disturbed areas.

METHODS

The effects of disturbance on wedge-tailed eagles was assessed by comparing breeding success for pairs in relatively undisturbed territories with that of pairs in areas subject to disturbance, either with or without nest protection measures applied. Very few protection efforts involved identical procedures since nests were invariably found at different stages of breeding, at different distances from logging and had suffered disturbance over differing periods. Nevertheless, nests or territories were regarded as protected if at least 8 ha of forest were left associated with the nest and heavy disturbance had not occurred within 500 m of the nest for more than two consecutive days during the breeding season. Eleven relatively undisturbed territories monitored for seven consecutive breeding seasons were used as a control group.

The breeding status of nests was assessed by discreet observation from the ground or from a fixed-wing aircraft (Mooney 1988). Nests found during the breeding season were checked at the first opportunity to determine their breeding status with at least one subsequent visit to confirm that status. A nest was considered to be occupied if fresh faeces or prey remains were under or on it, of if eagles were seen perched in the nest tree. A territory was considered occupied if eagles were seen in the general vicinity of the nest. Nests were considered to be active, i.e. being utilised for breeding, if they were at least adorned with fresh leaves and territories were considered active if they contained at least one such nest.

Pairs were considered to have been successful if they produced at least one fledgling. Very little natural mortality of older nestlings occurs (NM unpubl. data) and so the presence of a large, well-feathered nestling was considered to be a success. If a nest was deserted, i.e. the nest was left resulting in a failed breeding attempt, and not used again, efforts were made to locate other nests in that territory. The breeding season (broadly defined as the period from the beginning of nest building/repair to voluntary fledging of the young) covers about five months, generally commencing in August (Mooney and Holdsworth 1991).

Productivity of a standard sample of 16 territories monitored throughout the study showed great consistency between years (Mooney and Holdsworth 1991). Hence no distinctions between particular years were made in the analysis. At undisturbed nests the first breeding attempt subsequent to the first breeding record was used to compare the effect of nest success on subsequent breeding. For disturbed nests, the first breeding attempt after the breeding season when disturbance occurred was used. Only those nests that were active in the season prior to or at the time of protection measures being applied were used to examine the value of nest protection.

The significance of differences in the frequencies of occurrence were tested using Chi-square tests and differences in the numbers of young produced in different circumstances were examined using t-tests.

RESULTS

Disturbance at a nest site led to a lower number of young being produced at that nest site over the following five years (Table 1). However, if the nest site was protected, the reduction in young produced was significantly less. Nest sites that were disturbed were used significantly less in the year after the disturbance, with the effect being more pronounced for the unprotected nests. Even after five years

Table 1. Comparison of breeding of wedge-tailed eagles in Tasmania at nest sites and in the territories that contain them after the first breeding season that they were found. Comparisons are between undisturbed sites and disturbed sites with and without protection measures. Data recorded over five years were used.

	Disturbed		Undisturbed	Undisturbed versus disturbed/protected	Undisturbed versus disturbed/unprotected
	Protected	Unprotected			
First subsequent use of same nest					
The next season	8	1	8	$\chi^{2a} = 7.996$	$\chi^{2a} = 3.886$
After 2–5 yr	9	4	2	$P = 0.018$	$P = 0.1432$
No use after 5 yr	5	10	1		
Number of nests monitored	22	15	11		
Mean number of fledglings/nest over 5 yr	2.0	0.4	3.5	<0.01[b]	<0.001[b]
First subsequent use of same territory					
The next season	15	9	8	$\chi^{2a} = 3.094$	$\chi^{2a} = 0.084$
After 2–5 yr	5	5	3	$P = 0.21$	$P = 0.77$
No use after 5 yr	0	2	0		
Number of territories monitored	20	15	11		
Mean number of fledglings/nest over 5 yr	2.9	2.2	3.5	N.S.[b]	<0.05[b]

[a] chi-square test
[b] t-test

a high proportion of nest sites that had not been protected were still unused (Table 1). Once deserted during breeding, disturbed nest sites were likely to remain unused for a longer period than for those deserted in undisturbed areas. Thus, for disturbed nest sites, of the 18 nests deserted during breeding, eight were subsequently used within seven years, missing an average of 3.8 seasons (range 1–6). For the five undisturbed nest sites the number of seasons missed was 1.8 (range 1–3).

When the effects of disturbance are viewed in terms of the production of young by breeding pairs, i.e. at the territory-wide scale, not just at the original nest site, the impacts are reduced. In this scenario, only pairs originally from unprotected nest sites suffered a significant drop on the number of young produced (Table 1). Number of young produced per successful nesting was very similar for all three groups: 1.06 (undisturbed), 1.08 (disturbed and protected) and 1.04 (disturbed and unprotected). The reduced breeding rate for territories with unprotected nest sites is probably related to the increased rate of change to these sites. Thus, territories where changes of nest had occurred relatively often (at least once in five years) produced fewer young over a five year period than did those territories where no changes occurred (2.2 compared with 3.1; $t = 2.1$, $df = 25$, $P < 0.05$).

Change of use of nest site was highly influenced by the outcome of breeding at that nest. For the 11 undisturbed pairs monitored, there were an average of 1.3 changes of nest site per year. All but one of these changes followed breeding failure at the nest. For all nests monitored, there was a significantly higher chance ($P < 0.05$) of a nest being reused if breeding was successful (23 of 36) than if breeding had failed (7 of 20). There was no significant difference between disturbed and undisturbed sites in the rate of reuse of nests for those where breeding was successful (10 of 19 and 13 of 17, respectively). A higher rate of change of nest site after failure of breeding did appear to occur at disturbed sites (11 of 15) compared with undisturbed sites (2 of 5). This difference however, was not significant and probably was due to the small sample size. Failure of a breeding attempt was also more likely to lead to a failure by that pair in the next breeding season, regardless of where they nested in a territory, i.e. whether they stayed at a nest or moved to a new site; 28 successes of 33 attempts followed a successful breeding whereas 9 successes of 20 attempts followed a breeding failure; $P < 0.01$. Failure at a disturbed nest site appeared more likely to lead to subsequent failure of a pair whenever they nested in a territory than did failure for undisturbed pairs (10 of 15 compared with one of four, respectively). This difference was not significant either, and was probably due to the small sample size.

DISCUSSION

Disturbance of a nest site led to a dramatic drop in the use of that nest site and lowered reproductive rate for a pair within their territory. Implementation of protection measures for nest sites greatly improved the breeding success of the pair subject to the disturbance, either by their remaining at the nest site and successfully raising young or by their being more successful at fledging young elsewhere within their territory. The adage "success breeds success" applies well to wedge-tailed eagles and a strategy of keeping a pair at a successful nest site appears to be the best course of action for conservation purposes and for industry.

Most territories have multiple nests and the eagles were able to move to other nest sites within their territory and compensate, to some extent, for the disturbance. However, a reliance on the use of alternative nests is not a particularly viable long-term strategy. If eagles are not encouraged to stay where they are, then the process of nest discoveries in the midst of logging with consequent disruption to the birds and to the timber industry, will continue. Eagles may move from disturbance only to be disturbed again, and sometimes this is repeated. Unless nests are protected as they are found, loss of nests could well continue until many territories have no natural nest sites. Wedge-tailed eagles nest only in old growth trees. Since the length of the cutting cycle will be shorter than the time required for trees of old growth form to develop, a serious shortage of nest sites could develop without conservation of nest sites. It could also well be the case that not all nest sites are equal and eagles may have to use less productive nest sites as the extent of disturbance increases. To date, about 40% of Tasmania's wedge-tailed eagle territories have been exposed to serious disturbance (Mooney and Holdsworth 1991).

The time interval to reuse of conserved nests after desertion (about four years) may be a function of several factors. It may take years for birds to habituate to drastic disturbance at a favourite nest, or other nests may be less successful, promoting moves between nests until the original nest is reused. After about four years there may be a turnover of the original pair, and then newcomers, not having experienced the disturbance, may occupy the original nest.

Implementation of protection measures, even in "rescue" situations where the nest is deserted, may have positive consequences in the longer term. Arguably, these conserved sites are used by birds that are naturally more tolerant of, and have at least partly adapted to, considerable human disturbance. They may also produce offspring that are similarly tolerant and are therefore of special value in populating an environment which will be increasingly disturbed. Given that forestry operations are likely to continue, there may be some long-term, habituative advantage to the eagles in continuing, sub-critical exposure to humans (Parsons and Burger 1982). At least some wedge-tailed eagles obviously have the ability and/or adaptability to nest in sites subject to some disturbance, so there is long-term potential to change or expand their nesting habits in Tasmania.

Although all successful nestings were similarly productive, even conserved nests had lower productivity for years after disturbance. Obviously, protection efforts could be improved. One way to ensure this is for nests to be located before heavy disturbance occurs. Even the best "rescues" are not as good as planned protection. Ideally, nests should be found well ahead (in time and space) of logging and not disturbed during breeding. Disruption of logging, with its consequent bad will toward the birds, such as occurred with attempts at protection of the spotted owl (Gup 1990), would also decrease. This in turn promotes cooperation, encourages the rapid reporting of nests and decreases vandalism such as shooting or malicious felling of nest trees. Measures are presently being implemented to try to ensure that pre-logging surveys of areas likely to contain suitable nesting sites are routinely undertaken. Continuing surveys of breeding success may allow the identification of especially productive territories or pairs and enable their special protection.

Despite the sometimes small samples in this study, there was an obvious beneficial effect shown for efforts to protect nests of the Tasmanian wedge-tailed eagle. The use of buffer zones around nest sites has also been recommended for the protection of other raptors (Olendorff *et al.* 1980). However, the use of buffers may not necessarily be sufficient for species that need old growth forest for foraging. Fragmentation of forest may well improve food availability for the wedge-tailed eagle, which is a generalist hunter (Marchant and Higgins 1993), as distinct from species that are forest specialists. For the latter, protection on a nest-by-nest basis may be ineffectual considering the large areas of forest that may be needed for foraging (Thiollay 1989). However, given time, sufficient food and a more sensitive attitude by people, it may be possible that some apparently forest-dependent raptors can survive in what has been presumed to be unsuitable habitat (Kenward and Widén 1989). This potential is vital for the future survival of many raptors and must be explored.

ACKNOWLEDGMENTS

We would like to thank the many people that reported nests and the results of breeding attempts. Principal among them were the staff of the Forestry Tasmania, notably Bob Hamilton, Sue Jennings, Barry Crawford, Peter Duckworth, Paul Smith, Brett Warren, Mick Miller, Ross Lucas, Steve Davis, Mike Boyden, Peter Watson, Barry Hunt, David Tucker, Alan Futcher and Mike Smith and of private forestry companies, especially Richard Hart, Geoff Wilkinson, Tony Price, Daryl Clark, Peter Naughton, Mike Mangum, Mike Warner and Barry Burns. Various private landowners were very generous in their help and we particularly thank Henry Foster, Tom Dunbabin and Rex Kemp. Our colleague, Mark Holdsworth, and many volunteers provided various assistance over many years. Among these, members of the Australian Raptor Association were outstanding, with special thanks to Chris Spencer, Peter Tonelley, Norma Buttle, Tom Ralph, Simon Plowright and Jason Wiersma. Penny Olsen, David Baker-Gabb and a third anonymous reviewer gave useful comments on the manuscript.

REFERENCES

ANONYMOUS. 1993. Forest Practices Code. Forestry Tasmania, 199 Macquarie St., Hobart 7001, Australia.

CLINE, K. 1990. Bald eagles in the Chesapeake: a management guide for landowners. Instit. Wildlife Res., Natl. Wildl. Fed., Washington, D.C. USA.

GARNETT, S. 1992. The action plan for Australian Birds. Aust. Nat. Parks Wildl. Serv., Canberra, Australia.

GRIER, J.W. 1975. Patterns of bald eagle productivity in northwest Ontario, 1966–72. Pp. 103–108 in J.R. Murphy, C.M. White and B.E. Harrell, eds. *Population status of raptors*. Raptor Res. Rep. 3, Fort Collins, Colorado USA.

GUP, T. 1990. Owl vs man. *Time* magazine. June.

KENWARD, R. AND P. WIDÉN. 1989. Do goshawks *Accipiter gentilis* need forests? Some conservation lessons from radio tracking. Pp. 561–567 in B.-U. Meyburg and R.D. Chancellor, eds. *Raptors in the modern world*. Lentz Druck, Berlin, Germany.

MARCHANT, S. AND P. HIGGINS. 1993. *Handbook of Australian, New Zealand and Antarctic birds*. Vol. 2. Oxford Univ. Press, Melbourne, Australia.

MOONEY, N.J. 1988. Efficiency of fixed-wing aircraft for surveying eagle nests. *Aust. Raptor Assoc. Newsl.* **9**: 28–30.

— AND M. HOLDSWORTH. 1991. Effects of disturbance on nesting wedge-tailed eagles *Aquila audax fleayi* in Tasmania. *Tasforests* **3**: 15–29.

NELSON, M.W. 1982. Human impacts on golden eagles: A positive outlook for the 1980s and 1990s. *J. Raptor Res.* **16**: 97–106.

NEWTON, I. 1979. *Population ecology of raptors*. Buteo Books, Vermillion, South Dakota USA.

OLENDORFF, R.R., R.S. MOTRONI AND M.W. CALL. 1980. Raptor management: the state of the art in 1980. U.S. Dept. Int. Tech Note No. 345, Bureau of Land Manage., Denver, Colorado USA.

PARSONS, K.C. AND J. BURGER, 1982. Human disturbance and nestling behaviour in black-crowned night herons. *Condor* **84**: 184–187.

THIOLLAY, J.M. 1989. Area requirements for the protection of rainforest raptors and game birds in French Guiana. *Conserv. Biol.* **3**: 128–137.

Raptors in Industrial Landscapes

27

Use of Reservoirs and other Artificial Impoundments by Bald Eagles in South Carolina

A. Lawrence Bryan, Jr., Thomas M. Murphy,
Keith L. Bildstein, I. Lehr Brisbin, Jr.
and John J. Mayer

> Abstract – Active bald eagle nest territories in South Carolina increased from 12 in 1977 to 84 in 1993. Nest territories associated with reservoirs increased from one in 1982 to 29 in 1993. This was a significantly faster rate of increase than was the rate for territories not associated with reservoirs. Reservoir territories also produced significantly more fledglings per nest than a sample of non-reservoir territories in the ACE basin. Eagle sightings on a newly constructed reservoir (L-Lake) increased steadily throughout the study, while sightings on the 33-yr old Par Pond reservoir were minimal until a partial drawdown of that site created more favorable foraging conditions. Bald eagles appear to be able to rapidly find and use both new reservoirs or newly conducive conditions at older reservoirs. Eagle use of these reservoirs did not appear to be linked to densities of waterfowl and marsh birds (as potential prey) or other fish-eating birds (as indicators of abundant fish).
>
> Key words: bald eagle; breeding territory; foraging habitat; impoundments; reservoirs.

Historically, nesting bald eagles were common throughout the coastal plain of the southeastern United States (US), as well as along major river drainages and the few large lakes found in the region (US Fish and Wildlife Service (USFWS) 1984, Wood et al. 1990). Southeastern eagle populations were greatly reduced in the 1970s by the same factors (shooting, habitat alteration, and most recently organochlorine pesticides) that depleted other populations of this species (USFWS 1984). The increased protection later afforded to eagles and their habitats has resulted in a recovery of bald eagle populations, and several southeastern states have reached or exceeded recovery goals as laid out in the Bald Eagle Recovery Plan (USFWS 1984, Wood et al. 1990).

At the same time that population expansions were taking place, large amounts of new aquatic habitat were being created in the southeastern US in the form of man-made impoundments of many sizes, and many of these are now receiving significant use by both breeding and non-breeding eagles. Three major reservoirs (comprising lakes Marion, Moultrie and Wateree) in South Carolina's coastal plain were completed in the 1940s and were colonized by eagles as the reservoirs matured and waterbird use intensified. Some use of these reservoirs by nesting eagles was documented, but this was all but eliminated as eagle populations declined in the 1970s. Since that period, as the South Carolina bald eagle population has recovered, eagle use of reservoirs has also expanded.

South Carolina supports a bald eagle population which has been growing steadily since the late 1970s and contains the third largest population in the southeastern states (USFWS 1984, Wood et al. 1990). Population growth has occurred in the absence of a reintroduction program in the state, although such programs exist in adjacent North Carolina, Tennessee and Georgia.

Bald eagles have been observed at reservoirs on the US Department of Energy's Savannah River Site, in southwestern South Carolina, since the late 1950s (Norris 1963, Mayer et al. 1985). An eagle nest was discovered adjacent to Par Pond in 1986 (Mayer et al. 1988) and a second nest was discovered near L-Lake in 1990 (Wike et al. 1993).

Using South Carolina as an example, our goal is to describe the use of reservoirs and other freshwater, man-made impoundments >40 ha in size by breeding bald eagles. We will also compare the reproductive successes of nests associated with reservoirs to those of nests from the Ashepoo-Cumbahee-Edisto River (ACE) Basin in South Carolina, where eagles feed in coastal and riverine habitats. In order to more thoroughly describe the actual patterns of impoundment use by eagles in this region, the reservoir system on the Savannah River Site was selected for detailed surveys designed to examine the spatial and temporal patterns of eagle use of such aquatic systems.

METHODS

Study Area

South Carolina has >1600 man-made freshwater impoundments >4.05 ha in size, covering >210 000 ha (SCWRC 1991). Nineteen impoundments are >405 ha. The function of most of these impoundments is recreational, but 15 of the 19 larger impoundments are used primarily to produce electric power.

South Carolina also has >200 000 ha of coastal marshes, including approximately 28 500 ha of shallow, man-made coastal impoundments (Tiner 1977). Most of these coastal impoundments were built in the eighteenth and nineteenth centuries for the purpose of growing rice (Rogers 1970). The majority are currently managed to attract waterfowl for hunting (Tiner 1977).

The ACE Basin occupies approximately 142 000 ha of largely undeveloped land and water areas, including coastal impoundments, within the boundaries of Beaufort, Charleston, and Colleton counties of South Carolina (NOAA

1991). The basin is bounded by the Atlantic Ocean (southeast), the North Edisto River (northeast), the Coosaw River (southwest), and extends inland to include most of the Ashepoo and Combahee River drainages.

The 77 701 ha Savannah River Site (SRS), which is located along the north shore of the Savannah River in southwestern South Carolina, has been used for the production of plutonium and tritium for nuclear weapons since its closure to the public in 1952 (Fig. 1). L-Lake (405 ha), Par Pond (1100 ha), and Pond B (87 ha) are three man-made impoundments (Fig. 2) on the SRS that were constructed to serve as cooling reservoirs for thermal effluent from nuclear reactors. Par Pond was formed in 1960 and maintained a constant water level until mid-July of 1991, when the discovery of structural anomalies in its dam required the lowering of its water level by 6 m, reducing its volume and surface area by 65% and 50%, respectively. Pond B was formed in 1961 and L-Lake was formed in late 1985, and both have maintained constant water levels since the time of their construction.

Productivity

Active breeding territories of South Carolina eagles were classified as either reservoir (man-made) or non-reservoir (riverine-coastal systems), depending on the primary feeding areas used by the breeding birds. Eagles using shallow coastal impoundments (rice fields) were classified as non-reservoir. Occupation of nest sites and reproductive success of eagles were monitored by annual aerial surveys and ground observations. Numbers of active breeding territories and numbers of young fledged were recorded and related to primary feeding areas from 1977 through 1993. Rates of population increase (numbers of breeding territories) from 1982 to 1993 for reservoir and non-reservoir (statewide) eagle populations were compared by the application of a homogeneity-of-slopes model (Proc GLM, SAS 1988) to log-transformed data. The reproductive success (number of young fledged per breeding territory) of eagle populations associated with reservoirs from 1982–93 was compared with the reproductive success of ACE basin nests during the same period.

Use of SRS Impoundments

Reservoir surveys were conducted on the SRS from the fall of 1987 through the summer of 1993. They were continued on a seasonal basis four times per year (except for the winter of 1989) and from the fall of 1989 through the spring of 1991. All birds, including eagles, were counted from a small boat cruising the reservoir shoreline on three consecutive days (one day at each of the three reservoirs) at two-week intervals, three times each season (for details, see Bildstein et al. 1994). All birds sighted on or flying over the water, as well as those flying over land ≤20 m from the shoreline, were counted from a stationary boat at approximately 50-m intervals. Numbers of eagles, other members of the

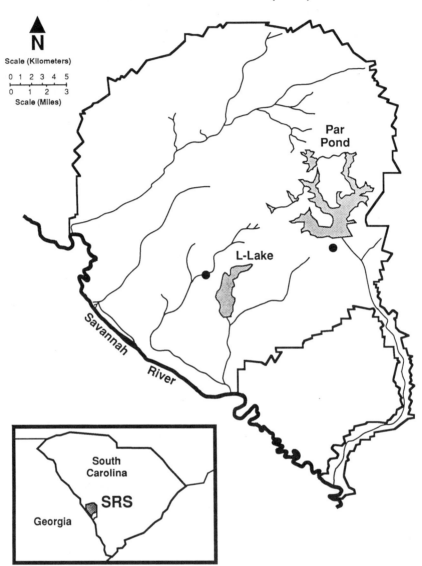

Figure 1. US Department of Energy's Savannah River Site, showing locations of the L-Lake and Par Pond reservoir systems. Solid circles indicate locations of bald eagle nests.

open-water fish-eating guild, marsh birds, primarily American coots and common moorhens, and waterfowl were censused to determine whether eagle densities were related to densities of other fish-eating birds or potential prey species (marsh birds and waterfowl; see Appendix 1 for avian species observed during censuses). Densities were quantified as birds per km of shoreline for each

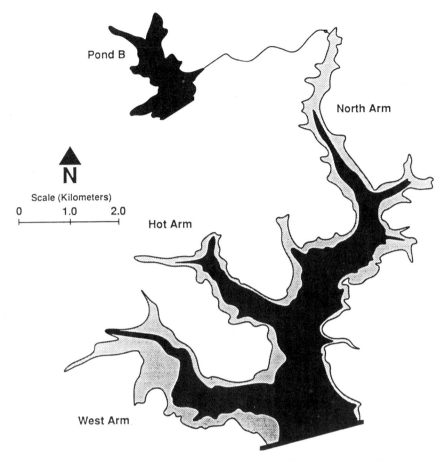

Figure 2. The Par Pond reservoir on the US Department of Energy's Savannah River Site. Blackened areas represent open water, and stippled areas indicate mud flats exposed following partial drawdown of the reservoir in 1991. Boat surveys routes were confined to the reservoir's Hot Arm.

reservoir: L-Lake 20.2 km; Par Pond 7.9 km; and Pond B 9 km. Only the 7.9 km of shoreline comprising Par Pond's Hot Arm was surveyed.

RESULTS

Numbers of Nests

The number of known active bald eagle nest territories in South Carolina grew rapidly from 1977 ($N = 12$) to 1993 ($N = 86$). This included exponential increases in both reservoir and non-reservoir territories (Fig. 3). Comparative

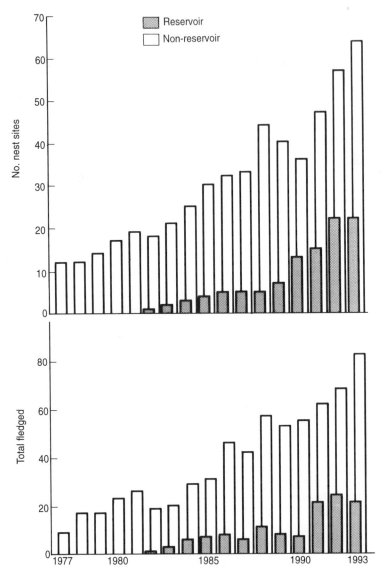

Figure 3. Number of active bald eagle nesting territories and fledglings produced on reservoir and non-reservoir sites in South Carolina: 1977–93.

analysis of the increase in numbers (*log-transformed*) of both types of territories indicated that the population of eagles associated with reservoirs increased at a faster rate (slope = 0.11) than the population not associated with reservoirs (slope = 0.04; Student's $t = -8.0$, Prob $>/t/ = 0.0001$). The first South Carolina eagle nest associated with a reservoir since the 1970s was reported in 1982, when a nest was found on the Lake Marion reservoir. By 1993, 26% of the state's total of 88 nests was associated with reservoirs.

Table 1. Comparison of the number of fledglings produced by South Carolina bald eagle nests associated with man-made reservoirs and non-reservoir habitats between 1982 and 1993.

Number of fledglings	Eagle territory type	
	Reservoir N (%)	Non-reservoir N (%)
0	29 (28.4)	65 (29.5)
1	22 (21.6)	87 (39.6)
2	50 (49.0)	65 (29.5)
3	1 (1.0)	3 (1.4)
Total	102	220

Reproductive Success

The average annual productivity of statewide eagle territories ranged from 0.69 fledglings per territory in 1977 to 1.39 in 1986. Total production of fledglings statewide increased to a high of 103 in 1993.

Approximately 72% of both reservoir and ACE basin territories fledged young from 1982 through 1993. Eagle territories associated with reservoirs ($N = 102$) produced an average of 1.23 ± 0.88 (SD) fledglings per nest during this period, while ACE Basin territories ($N = 220$) produced 1.03 ± 0.80 (SD) fledglings. Reservoir nests produced significantly more fledglings per nest than ACE Basin nests ($\chi^2 = 14.07$, df = 3, $P = 0.003$), with a higher percentage of nests fledging a second nestling (Table 1).

Avian Use of SRS Impoundments

Avian use of the SRS reservoirs was dominated by migratory waterfowl, primarily ring-necked ducks and lesser scaup and marsh birds (primarily American coots) during winter and spring seasons (Fig. 4). Fish-eating birds, primarily double-crested cormorants, were abundant in spring, and long-legged waders were most common in summer and fall (Fig. 4).

Bald eagles were observed on L-Lake throughout the study and their use of that area appeared to increase during the study period (Fig. 5). Eagles were not observed on Par Pond during the surveys until the winter of 1992 (Fig. 6). Observations of eagles on Pond B, the smallest of the three reservoirs on the SRS, were infrequent. Numbers of bald eagles observed during surveys of the L-Lake and Par Pond reservoirs did not appear to be directly associated with seasonal variations in the densities of any of the other guilds observed (Figs. 5 and 6). However, the densities of both eagles and other fish-eating birds increased on Par Pond after the drawdown of the reservoir during summer of 1991.

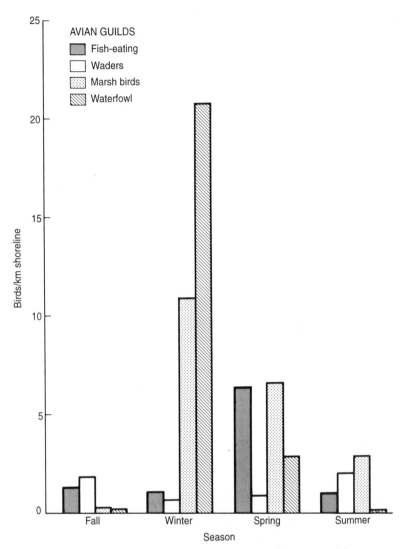

Figure 4. Seasonal use of the L-Lake Reservoir by avian guilds.

DISCUSSION

As of 1993, the South Carolina population of breeding bald eagles had surpassed its recovery goal of 40 occupied breeding territories (USFWS 1984) by 120%. Many breeding areas have been and still are associated with reservoirs. The percentage of eagle breeding territories associated with reservoirs has increased steadily into the 1990s and linear models (this study) suggest that the rate of increase in territories is greater in reservoir habitats than in non-reservoir

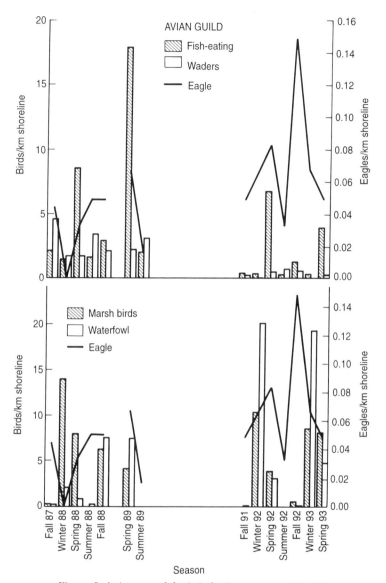

Figure 5. Avian use of the L-Lake Reservoir: 1987–93.

habitats. Continued study of these populations and their habitats is needed to accurately estimate population growth and possible carrying capacities.

The dispersal of eagles inland from historic (coastal) areas could be a response to the availability of food resources at these reservoir sites. However, the continued growth of ACE Basin and other non-reservoir eagle populations during the last 17 yr suggests that eagles moving to reservoirs were not compelled to do so due to a lack of available habitat in this region. Dead and injured fish associated with hydro-electric dams have been documented as

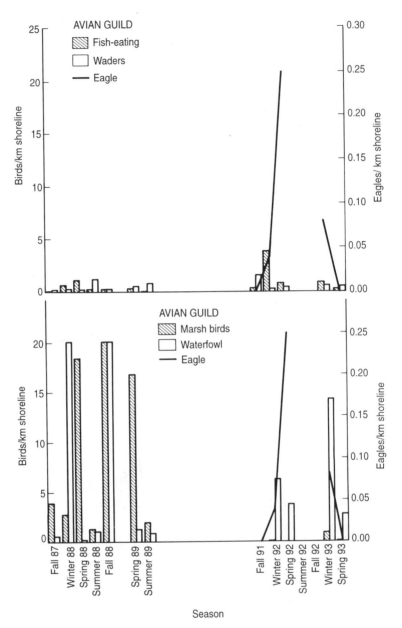

Figure 6. Avian use of the Par Pond Reservoir: 1987–93.

attracting foraging bald eagles (Stalmaster 1987). Fish may be more abundant in recently constructed reservoirs because of the "trophic upsurge" associated with internal and external nutrient loading (see Kimmel and Groeger 1986) and more available because of reduced littoral vegetation at these sites (Bildstein et al. 1994). Reservoir drawdowns, such as the Par Bond reservoir in 1991, can make prey more available by concentrating fish in a reduced area, as well as by reducing protective cover for fish if water levels drop below the level of littoral vegetation. These factors may have been responsible for the increase in numbers of eagles and other piscivorous birds at Par Pond in the winter of 1992 (Fig. 6).

Recent eagle colonization and increased numbers of other species in the fish-eating guild at the recently constructed SRS reservoirs suggest that these sites are capable of providing sufficient food resources. Bald eagles appear to be able rapidly to find and use both new reservoirs (L-Lake) or newly conducive conditions at older reservoirs (Par Pond drawdown). The timing of the breeding chronology of southeastern bald eagles is such that they are incubating and hatching eggs when peak numbers of overwintering waterfowl and marsh birds (potential prey) are present on southeastern reservoirs. The dispersal of eagles inland (see Mayer *et al.* 1988) from their coastal natal areas may have resulted from the availability of food at reservoirs, combined, perhaps more importantly, with the lack of competition from other eagles for resources and space.

Breeding success of eagle pairs associated with reservoirs was slightly higher than that of ACE basin eagles, although other non-reservoir breeding territories in South Carolina were as successful as reservoir territories (T.M. Murphy unpubl. data). Regardless, reservoir eagles are producing at least as well as eagles using more natural habitats and do not appear to be paying a cost in reproductive output as a result of occupying these more recent man-made habitats.

An early concern for bald eagle recovery was that populations in the southeast were largely disjunct, with isolated remnant populations in Florida, South Carolina, and the Chesapeake Bay region (USFWS 1984). If the eagles' association with reservoirs continues, the fact that impoundments and reservoirs are fairly widespread geographically in the state will reduce the likelihood of a single catastrophic event destroying either the state's eagle population or eagle production, i.e. nestlings, for a year. A prime example of the potential for such an event is the impact of Hurricane Hugo in 1989, which resulted in the loss of nest trees from 25 of South Carolina's 54 breeding areas (Murphy 1991).

Potential costs to breeding eagles using impoundments include higher rates of disturbance, notably from development of reservoir shorelines and recreational activities (Buehler *et al.* 1991, Smith 1988). Habitat alteration, including disturbance at nest sites, has been suggested as the single most important factor inhibiting eagle recovery in the southeastern region (USFWS 1984). Recreational boating was reported by Wood *et al.* (1990) to reduce the numbers of eagles utilizing particular areas of reservoirs in Florida and North Carolina, though whether this affected foraging behavior or productivity was unknown.

Contaminants often found in reservoirs, such as mercury, may also be of concern in certain situations. In general, mercury levels are higher in fish from younger oligotrophic reservoirs and lower in fish from older eutrophic reservoirs (Eisler 1987). Eagle prey species, i.e. American coot and largemouth bass (see Mayer *et al.* 1988) collected from reservoirs on the SRS for example, were found to contain levels of mercury >0.05 ppm (Clay *et al.* 1979, Pinder and Giesy 1981). This level has been associated with adverse effects in sensitive avian species (Eisler 1987). While productivity of SRS eagle nests has been reasonably high, no data are available on the mercury levels in these birds. Furthermore, the survivorship of the fledglings produced at this site is unknown. Grier (1980) maintained that survival rates may be a more important limiting factor to bald eagles than productivity.

The recovery of eagles and their association with reservoirs in South Carolina may be a model for how other states with low initial numbers of breeding eagles might eventually effect recovery, i.e. partial build-up in historic areas, followed by expansion into newly available habitats: reservoirs. States that hack eagles on reservoirs (Odum 1980, Wood *et al.* 1990) may experience a fast rate of recovery as this species rapidly adapts to new foraging opportunities in reservoir habitats.

ACKNOWLEDGMENTS

R. A. Kennamer and T. V. Youngblood assisted with data analyses and other aspects of manuscript preparation. J. W. Coker assisted with data collection concerning eagle nest sites and their productivity. Nest site monitoring was partially funded by a cooperative grant (Sect. 6) between the US Fish and Wildlife Service and the South Carolina Wildlife and Marine Resources Department. D. E. Gawlick, C. Golden, C. Viverette and D. P. Ferral conducted most of the avian surveys of SRS reservoirs. This is Hawk Mountain Sanctuary Contribution Number 13. J. Frazier, J. Gessaman, G. Hunt and D. Varland improved earlier drafts of this manuscript. This research was supported by the United States Department of Energy, Savannah River Operations contract DE-AC0976SROO-819 with the University of Georgia's Institute of Ecology, Savannah River Ecology Laboratory.

REFERENCES

BILDSTEIN, K.L., D.E. GAWLICK, D.P. FERRAL, I.L. BRISBIN, JR. AND G.R. WEIN. 1994. Wading bird use of established and newly created reactor cooling reservoirs at the Savannah River Site, near Aiken, South Carolina, USA. *Hydrobiologia* 279/280: 71–82.

BUEHLER, D.A., T.J. MERSMANN, J.D. FRASER AND J.K. SEEGAR. 1991. Effects of human activity on bald eagle distribution on the northern Chesapeake Bay. *J. Wildl. Manage.* 55: 282–290.

CLAY, D.L., I.L. BRISBIN, JR., P.B. BUSH AND E.E. PROVOST. 1979. Patterns of mercury contamination in a wintering waterfowl community. *Proc. Ann. Conf. S.E. Assoc. Fish and Wildl. Agencies* 32: 309–317.

EISLER, R. 1987. *Mercury hazards to fish, wildlife, and invertebrates: a synoptic review.* US Fish Wildl. Serv. Biol. Rep. 85(1.10).

GRIER, J.W. 1980. Modeling approaches to bald eagle population dynamics. *Wildl. Soc. Bull.* 8: 316–322.

KIMMEL, B.L. AND A.W. GROEGER. 1986. Limnological and ecological changes associated with reservoir aging. Pp. 103–109 in G.E. Hall and M.J. Van Den Avyle, eds, *Reservoir fisheries management: strategies for the 80's.* American Fisheries Society, Bethesda, Maryland.

MAYER, J.J., R.T. HOPPE AND R.A. KENNAMER. 1985. Bald and golden eagles on the Savannah River Plant, South Carolina. *Oriole* 50: 53–57.

—, R.A. KENNAMER AND F.A. BROOKS. 1988. First nesting record for the bald eagle on the Savannah River Plant. *Chat* 52: 29–32.

MURPHY, T.M. 1991. *The effects of Hurricane Hugo on nesting bald eagles in South Carolina.* South Carolina Wildl. and Marine Resources Rep. Work Order 89-1.

NATIONAL OCEANIC AND ATMOSPHERIC ADMINISTRATION. 1991. Ashepoo-Combahee-Edisto (ACE) Basin National Estuarine Research Reserve in South Carolina: final environmental impact statement and draft management plan. NOAA Office of Ocean and Coastal Res. Manage. Washington, D.C.

NORRIS, R.A. 1963. Birds of the AEC Savannah River Plant area. *Contrib. Charleston Mus. Bull.* 14: 1–78.

ODUM, R.R. 1980. Current status and reintroduction of the bald eagle in Georgia. *Oriole* 45, 1–14.

PINDER, J.E. AND J.W. GIESY. 1981. Frequency distributions of the concentrations of essential and nonessential elements in largemouth bass (*Micropterus salmoides*). *Ecology* 62: 456–468.

ROGERS, G.C. 1970. *The history of Georgetown county, South Carolina.* Univ. South Carolina Press. Columbia, South Carolina.

SMITH, T. 1988. The effect of human activities on the distribution and abundance of the Jordan Lake–Falls Lake bald eagles. M.Sc. thesis, Virginia Polytechnic Inst. and State Univ., Blacksburg, Virginia.

SOUTH CAROLINA WATER RESOURCES COMMISSION. 1991. *Inventory of lakes in South Carolina: ten acres or more in surface area.* South Carolina Water Res. Comm. Rep. No. 171.

SAS INSTITUTE, INC. 1988. SAS/STAT User's Guide, 6.03 Edition. SAS Institute, Cary, North Carolina.

STALMASTER, M.L. 1987. *The Bald Eagle.* Universe Books, New York, New York.

TINER, R.W. 1977. *An inventory of South Carolina's coastal marshes.* South Carolina Wildl. Marine Res. Dep. Tech. Rep. No. 23.

US FISH AND WILDLIFE SERVICE. 1984. *Southeastern states bald eagle recovery plan.* US Fish and Wildl. Serv., Southeast Region. Atlanta, Georgia.

WIKE, L.D., R.W. SHIPLEY, J.A. BOWERS, A.L. BRYAN, C.L. CUMMINS, B.R. DEL CARMEN, G.P. FRIDAY, J.E. IRWIN, J.J. MAYER, E.A. NELSON, M.H. PALLER, V.A. ROGERS, W.L. SPRECHT AND E.W. WILDE. 1993. *SRS ecology: environmental information document.* Rep. WSRC-TR-93-496. Westinghouse Savannah River Co., Savannah River Site, Aiken, South Carolina.

WOOD, P.B., D.A. BUEHLER AND M.A. BYRD. 1990. Bald Eagle. Pp. 13–21 in B.A. Giron Pendleton, ed. *Proc. southeast raptor management symposium and workshop.* Natl. Wildl. Fed., Washington, D.C.

APPENDIX 1

Avian guild components on Savannah River Site Reservoirs.

Guild

Open-water fish-eating:
Common loon
Horned grebe
Double-crested cormorant
Anhinga
Osprey
Bald eagle
Bonaparte's gull
Ring-billed gull
Herring gull
Caspian tern
Forster's tern
Least tern
Belted kingfisher

Long-legged Wading Birds:
Great blue heron
Great egret
Snowy egret
Wood stork
Little blue heron
Tricolored heron
Green-backed heron
Black-crowned night heron
American bittern
Least bittern

Guild

Marsh Birds:
Pied-billed grebe
American coot
Common moorhen
Purple gallinule
Lesser yellowlegs
Killdeer
Spotted sandpiper

Waterfowl:
Mallard
American black duck
Blue-winged teal
Ruddy duck
Bufflehead
Wood duck
Ring-necked duck
Lesser scaup
Northern pintail
Gadwall
American widgeon
Red-breasted merganser
Hooded merganser

28

Attraction of Bald Eagles to Habitats just below Dams in Piedmont North and South Carolina

Richard D. Brown

Abstract – Midwinter surveys of bald eagles were conducted annually from 1983 to 1987 and 1989 to 1993. Flights were made along the Yadkin, Pee Dee, and Catawba Rivers in Piedmont North and South Carolina in fixed-wing planes or helicopters. Of the 117 eagles seen, more were found just below dams (103; 88%) than in other parts of the reservoirs and rivers (14; 12%). Additional year-round observations support these findings. Dams serve as good "eagle feeders" so long as other habitat features are present. While fish may go through turbines, increased water turbulence and the current below dams may attract feeding fish, hence the eagles. Perch sites below dams and other features of the habitat may also be important attraction factors. Of 18 dams in the study area, only six (33%) had consistent eagle use. Studies are needed to determine why some dams are more attractive to eagles than others. Habitat management should increase the numbers of bald eagles at dams and the use of tailwater habitats.

Key words: bald eagle; dams; midwinter; survey; Carolinas.

Bald eagles are known for their attraction to bodies of water (Snow 1973). Often man-made reservoirs are used (Busch 1981, Murphy et al. 1984, McClelland 1992). Eagles frequently are seen below dams in winter (Imler and Kalmbach 1955, Spencer 1976, Steenhof 1978) and have been reported from Arizona (Brown et al. 1989), California (Hunt et al. 1992), Connecticut (Russock 1979), Illinois (Musselman 1949), Iowa (Hodges 1959), Maine (Cammack unpubl. rep.), North Dakota (Splendoria 1973), South Dakota (Grewe 1966), Washington (Wood 1979), and Wisconsin (Ross and Follen 1988).

During the 1983 Christmas Bird Count, six bald eagles were observed immediately below Narrows Dam on the Falls Reservoir of the Yadkin River (Greene 1983). This is 10 km northeast of Albemarle, Stanly County, in Piedmont North Carolina (Fig. 1).

Because of the numbers seen and the fact that bald eagles had never been reported from this area, a more thorough census was established for the region. The purpose of this paper is to report on 10 year of censuses and the finding that bald eagles are attracted to habitats just below dams in Piedmont North and South Carolina. I compared numbers of eagles seen near dams with those seen in other parts of the reservoirs and rivers.

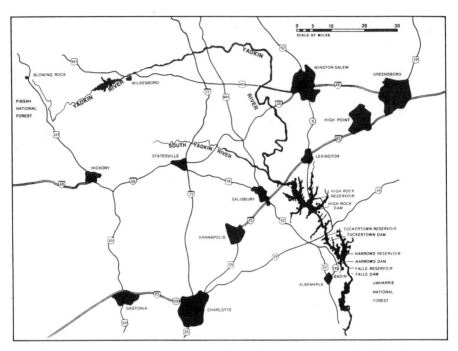

Figure 1. Major reservoirs and dams used by bald eagles along the Yadkin River and upper Pee Dee River in Piedmont North Carolina.

STUDY AREA

During 1983 through 1987 and 1989 through 1993, volunteers from the Carolina Raptor Center, Inc., and I conducted aerial counts of bald eagles in connection with the National Midwinter Bald Eagle Survey. This survey was started by the National Wildlife Federation in 1979 for the lower 48 conterminous states.

My study area included two drainage systems: the Catawba River west of Charlotte, and the Yadkin/Pee Dee Rivers to the east. The Catawba River originates in the North Carolina mountains, flows east for about 70 km, and then heads generally south on the west side of Charlotte. The Catawba River was surveyed from Lake James near Marian, McDowell County, North Carolina, to I-20 at Camden, Kershaw County, South Carolina. The approximately 280 km of river surveyed included 11 reservoirs and hydroelectric dams. After 5 yr of survey with no eagles seen along the Catawba River north of Charlotte, flights were restricted to the portion of the Catawba River south of Charlotte.

The Yadkin River originates in the mountains north of the Catawba River (Fig. 1). It flows easterly to Winston-Salem, Forsyth County, North Carolina, then south-southeasterly to just east of Albemarle. Here the Yadkin is joined by the Uwharrie River and together they form the Pee Dee River. The survey included the Yadkin River from W. Kerr Scott Reservoir west of North Wilkesboro, Wilkes County, North Carolina, to the Pee Dee River at US 1, Cheraw, Chesterfield County, South Carolina. This approximately 300 km stretch included seven reservoirs and hydroelectric dams. In 1987 the upper portion of the Yadkin River was also eliminated from the survey because no eagles had been seen there in 5 yr.

METHODS

Surveys were made in a fixed-wing Helio Courier, fixed-wing Cessna, or by helicopter between 6–16 January annually. Flights occurred when weather conditions permitted, and covered all hours of the day. Most flights began shortly after sunrise, depending on fog conditions at the reservoirs, and lasted until early afternoon. None of the waters were iced over during the surveys.

Rivers and shorelines of reservoirs were surveyed as completely as possible. Flights occurred so observers could cover both sides of the rivers. Four or more observers, including the pilot, participated in each flight, except in 1993 when two observers participated. Larger reservoirs were flown up and/or down one side rather than in the middle to maximize coverage. Time and financial restraints usually prevented covering all shorelines of the larger bodies of water. It is possible that incomplete coverage of the larger reservoirs resulted in some eagles being missed.

All locations of eagle sightings were recorded on data sheets. Once eagles were seen, they were usually circled to confirm the sighting and to look for markers. Eagles seen were aged as adults or immatures. Adult eagles were easily seen at a distance because of their white heads. Because of this, I believe that few, if any, were missed that were associated with the larger bodies of water. Because of their cryptic coloration, immature eagles were hard to see even when flights were close to them. I am confident that eagles seen in one location were not counted in another because the aircraft flew considerably faster than the eagles.

RESULTS

During the 10-yr survey, a total of 117 bald eagles was seen. Of these, 67 (57%) were adults and 50 (43%) were immatures. The data show that the majority of eagles (103; 88%) were found along tailrace and riverine sections less than 750 m (and usually much closer) below dams. Fourteen (12%) were found much farther than 750 m from the dams on reservoirs and rivers. The data show a highly significant attraction of bald eagles to habitats created along the tailwater (section below the dam that is affected by reservoir releases). There were more eagles below dams than expected by chance alone ($P < 0.0001$; $\chi^2 = 67.701$, df $= 1$). Non-winter censusing and year-round observations from biologists and the general public who use these rivers and reservoirs confirm that eagles are found mostly below the dams even in summer.

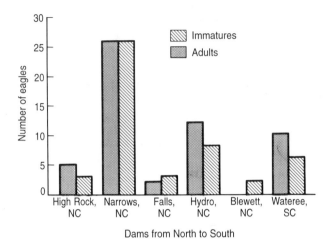

Figure 2. Comparison of total number of immature and adult bald eagles seen at dams along the Yadkin, upper Pee Dee, and Catawba Rivers in Piedmont North and South Carolina during 1983 to 1987 and 1989 to 1993 January surveys.

Table 1. Age class of bald eagles seen per year at dams along the Catawba and Yadkin/Pee Dee Rivers in Piedmont North and South Carolina.

Year	Adults	Immatures	Totals
1983	0	6	6
1984	5	2	7
1985	9	7	16
1986	7	7	14
1987	6	6	12
1988	—	—	—
1989	7	3	10
1990	7	6	13
1991	4	6	10
1992	3	1	4
1993	7	4	11
Totals	55	48	103

Only six (33.3%) of the 18 dam tailrace areas had eagles (Fig. 2). Wateree Lake dam in South Carolina is the only dam along the Catawba River that had eagles during yearly surveys. The other five dams are along the Yadkin/Pee Dee River (Fig. 2). Table 1 shows the age class of eagles seen per year at the dams.

Table 2. Age class of bald eagles seen per year at locations other than at dams along the Catawba and Yadkin/Pee Dee Rivers in Piedmont North and South Carolina.

Year	Adults	Immatures	Totals
1983	2	0	2
1984	0	0	0
1985	0	0	0
1986	3	1	4
1987	3	0	3
1988	–	–	–
1989	0	0	0
1990	1	0	1
1991	1	0	1
1992	1	1	2
1993	1	0	1
Totals	12	2	14

The numbers of adults ($N = 55$) and immatures ($N = 48$) were similar. Table 2 shows an age class breakdown of eagles seen in areas other than near dams. The number of adults and immatures was 12 and two, respectively.

DISCUSSION

The dams along the Yadkin/Pee Dee River drainage, from High Rock Reservoir to Blewett Falls Lake, have been the most consistently used locations by bald eagles in North Carolina since surveys began in 1983 (T. Henson, pers. comm.). Eagles have been seen in varying numbers for at least 15 yr and in recent years eagles have been reported year-round. At present, no nests have been found, but are highly suspected. Gerrard et al. (1992) showed with radiotracking studies that two pairs of eagles had a home range of 4 km^2 and 7 km^2, respectively. The presence of adult eagles year-round below dams might indicate nesting in suitable habitat or at least intent to nest in the future. Russock (1979) mentioned the wintering of eagles below the Shepaug Hydroelectric Dam in Connecticut and the possible existence of an active eagle's nest in the area. Year-round observations in the study area seem to indicate that eagle numbers diminish during the spring and summer, and that immigration occurs in the fall. More studies are needed to determine the amount of immigration, emigration and permanent residency.

South Carolina has a well-established coastal breeding population of eagles (T. Murphy, pers. comm.) as does Virginia within the Chesapeake Bay. North Carolina, sandwiched between these two states, has only recently reported nine pairs of nesting eagles (T. Henson, pers. comm.). From 1984, when the first nest in recent times was found, to 1992, five of the seven coastal pairs have produced 37 eaglets. Two nests are from Chatham and Wake Counties near Jordan Lake and Raleigh in the Piedmont. Jordan Lake was filled in 1982 and from that time eagles started using the lake. The first nest was reported in 1989, but no young have been produced. Jordan Lake has very little shoreline development and the dam is not hydroelectric.

Traditionally, greater numbers of eagles have been seen along the coast of North Carolina, and now they are being found inland. The data in this study show that more eagles are attracted to habitats immediately below dams in January than to other aquatic habitats.

Eagles below dams were frequently seen feeding. Tailrace sections below dams may be providing foraging opportunities (Russock 1979, Steenhof et al. 1980, Brown et al. 1989), attracting eagles like bird feeders do for other birds. Hunt et al. (1992) showed that numbers of bald eagles wintering along a northwestern US river were correlated with numbers of salmon. As the fish passed through the turbines, they became stunned, killed, or cut up depending on the type of turbines and the type and size of fish (S. Johnson pers. comm.). Russock (1979) and Steenhof et al. (1980) also reported fish passing through generators and turbines. Certainly any dead or injured fish would be easy prey to eagles and would comprise an attractive foraging location.

Discharge of water from reservoirs during hydroelectric generation can result in the transport of reservoir fish, plankton and aquatic insects (Hudson and Cowell 1966, Benson and Cowell 1968, Walburg 1971, Matter et al. 1983). Discharges can also scour tailwater substrates (Armitage 1977, Brooker and Hemsworth 1978, Matter et al. 1983), thus increasing food availability to fish (Barwick and Oliver 1982, Barwick and Hudson 1985). Observations of fishermen show that they usually do not fish when the hydroelectric facilities are not generating power. Barwick and Hudson (1985) showed that fish at Hartwell Reservoir on the Savannah River between Georgia and South Carolina fed primarily on aquatic insects, crayfish, and terrestrial organisms originating from the tailwater. Apparently more aquatic food attracts more fish, and more fish may attract the eagles.

The Narrows Dam contained the largest number of eagles (Fig. 2). It is located along the Yadkin River that is largely remote and bordered by the Uwharrie National Forest on the east, and to the west by property owned by Alcoa and the Morrow Mountain State Park. All six areas where eagles were found below dams contained mature trees, good perch sites, and protection from wind (Steenhof et al. 1980). The surrounding land and shoreline is generally remote and undeveloped. Other dams lacking bald eagles tended to be void of perch trees and/or were considerably developed along the tailwater. Differences in the head and length of the dam, and the amount of water released in the tailwater may account for some dams being more attractive than others. However, good habitat seems to be most important.

As a result of the surveys, Harris (1994) studied the ecology of bald eagles on the Yadkin and Pee Dee Rivers and gave habitat management recommendations. However, additional studies are needed to determine the year-round availability of food at hydroelectric dams along the Yadkin/Pee Dee Rivers and to examine why some dams are better "eagle feeders" than others. Various characteristics of the tailwater habitats, once identified, could be managed at other dams to attract and support eagles.

ACKNOWLEDGMENTS

I am grateful to George Gowen for financing the surveys, WSOC-TV Charlotte for use of "Chopper 9", and the many survey participants, especially M. Engelmann, D. S. Griffin, S. Harris, M. McGrady, B. O'Leary, D. Roberts and S. White. Many thanks to pilots B. Griffin, S. Holmberg, J. Callihan, R. Gilliland, M. Thorpe and J. Franklin for safe flights. I thank J. W. Grier, P. Nye and J. J. Negro for helpful manuscript reviews.

REFERENCES

ARMITAGE, P.D. 1977. Invertebrate drift in the regulated River Tees, and an unregulated tributary Maize Beck, below Cow Green Dam. *Freshwater Biol.* 7: 167–183.
BARWICK, D.H. AND J.L. OLIVER. 1982. Fish distribution and abundance below a southeastern hydropower dam. *Proc. Ann. Conf. Southeast Assoc. Fish and Wildl. Agencies* 36: 135–145.
— AND P.L. HUDSON. 1985. Food and feeding of fish in Hartwell Reservoir tailwater, Georgia–South Carolina. *Proc. Ann. Conf. Southeast Assoc. Fish and Wildl. Agencies* 39: 185–193.
BENSON, N.G. AND B.C. COWELL. 1968. The environment and plankton density in Missouri River reservoirs. Pp. 358–373 in *Reservoir fishery resources symposium*. Reservoir Comm., Southern Div. Am. Fish. Soc., Bethesda, MD.
BROOKER, M.P. AND R.J. HEMSWORTH. 1978. The effect of the release of an artificial discharge of water on invertebrate drift in the R. Wye, Wales. *Hydrobiologia* 59: 155–163.
BROWN, B.T., R. MESTA, L.E. STEVENS AND J. WEISHEIT. 1989. Changes in winter distribution of bald eagles along the Colorado River in Grand Canyon, Arizona. *J. Raptor Res.* 23: 110–113.
BUSCH, D.E. 1981. Wintering bald eagles at southwest Nebraska reservoirs. *Neb. Bird Rev.* 49: 34–35.
GERRARD, J.M., A.R. HARMATA AND P.N. GERRARD. 1992. Home range and activity of a pair of bald eagles breeding in Northern Saskatchewan. *J. Raptor Res.* 26: 229–234.
GREENE, J.W. 1983. Six for the new year. *Chat* 47: 63–65.
GREWE, A.A., JR. 1966. Some aspects in the natural history of the bald eagle, (*Haliaeetus leucocephalus*), in Minnesota and South Dakota. Ph.D. thesis, Univ. of South Dakota, Vermillion.
HARRIS, S.D. 1994. Ecology and habitat management of bald eagles on the Yadkin and Pee Dee Rivers of North Carolina. MS thesis, Univ. of N.C. at Charlotte.
HODGES, J. 1959. The bald eagle in the Upper Mississippi Valley. *Iowa Bird Life* 29: 86–91.

HUDSON, P.L. AND B.C. COWELL. 1966. Distribution and abundance of phytoplankton and rotifers in a main stem Missouri River reservoir. *Proc. S. D. Acad. Sci.* **45**: 84–106.

HUNT, W.G., B.S. JOHNSON AND R.E. JACKMAN. 1992. Carrying capacity for bald eagles wintering along a northwestern river. *J. Raptor Res.* **26**: 49–60.

IMLER, R.H. AND E.R. KALMBACH. 1955. The bald eagle and its economic status. US Dept. of Interior, Fish Wildl. Serv. Circular 30.

MATTER, W.J., P.L. HUDSON AND G.E. SAUL. 1983. Invertebrate drift and particulate organic material transport in the Savannah River below Lake Hartwell during peak power generation cycle. Pp. 357–370 in T.D. Fontaine III and S.M. Bartell, eds. *Dynamic of lotic ecosystems.* Ann. Arbor. Sci., Ann Arbor, Mich.

McCLELLAND, P.T. 1992. Ecology of bald eagles at Hungry Horse Reservoir, Montana. MS thesis, Univ. of Montana.

MURPHY, T.M., F.M. BAGLEY, W. DUBUC, D. MAGER, S.A. NESBITT, W.B. ROBERTSON, JR., AND B. SANDERS. 1984. Southeastern states bald eagle recovery plan. US Fish Wildl. Serv. Atlanta, Ga.

MUSSELMAN, T.E. 1949. Concentrations of bald eagles on the Mississippi River at Hamilton, Illinois. *Auk* **66**: 83.

ROSS, D.A. AND D.G. FOLLEN, SR. 1988. Bald eagles wintering at the Pentenwell Dam, Wisconsin. *Passenger Pigeon* **50**: 99–106.

RUSSOCK, H.I. 1979. Observations on the behavior of wintering bald eagles. *Raptor Res.* **13**: 112–115.

SNOW, C. 1973. *Habitat management series for endangered species.* Rep. 5: Southern bald eagle and northern bald eagle. US Dept. Interior, Bur. Land Manage.

SPENCER, D.A. 1976. Wintering of the migrant bald eagle in the power 48 states. Nat. Agric. Chem. Assn., Washington, D.C.

SPLENDORIA, F.A. 1973. Observations on wintering bald eagles at Garrison Dam, North Dakota. *Prairie Nat.* **5**: 6.

STEENHOF, K. 1978. *Management of wintering bald eagles.* US Dept. Interior, Fish Wildl. Serv. Publ. FWS/OBS-78/79, Washington, D.C.

—, S.S. BERLINGER AND L.H. FREDRICKSON. 1980. Habitat use by wintering bald eagles in South Dakota. *J. Wildl. Manage.* **44**: 798–805.

WALBURG, C.H. 1971. Loss of young fish in reservoir discharge and year-class survival, Lewis and Clark Lake, Missouri River. Pp. 441–448 in G.E. Hall, ed. *Reservoir fisheries and limnology.* Am. Fish. Soc. Spec. Pub. 8.

WOOD, B. 1979. *Winter ecology of bald eagles at Grand Coulee Dam, Washington.* Appl. Res. Stations, Habitat Manage. Div., Washington Dept Game, Olympia, Washington.

29

Reclaimed Surface Mines: An Important Nesting Habitat for Northern Harriers in Pennsylvania

Ronald W. Rohrbaugh, Jr. and Richard H. Yahner

Abstract – Northern harriers traditionally nest in agricultural habitats, emergent wetland areas, abandoned fields, and natural grasslands throughout their geographical range. These four habitats, however, are declining in area and quality throughout North America. In Pennsylvania, preliminary research has shown that harriers are nesting in reclaimed grassland surface mines. Grassland surface mines are a relatively new habitat type in the state, and little is known about their use by harriers. Based on Pennsylvania Breeding Bird Atlas data, we compared the number of probable and confirmed breeding attempts by harriers among the six physiographic sections of the Appalachian Plateau Province, each containing varying numbers of surface mines. Seventy-eight breeding attempts by harriers were documented in the six sections of the Appalachian Plateau Province. A significantly greater number ($P < 0.005$) of breeding attempts than expected occurred in the Pittsburgh Plateau Section of the Appalachian Plateau Province compared with the remaining physiographic sections. In addition, the Pittsburgh Plateau Section contained a significantly greater number ($P < 0.001$) of surface mines than expected compared with the remainder of the Appalachian Plateau Province. We have obtained evidence that northern harriers are associated more frequently than expected with open grassland habitat created by reclamation of surface mines during the breeding season. If managed properly, reclaimed grassland surface mines will provide important nesting habitat for northern harriers in Pennsylvania and help supplement a net loss of suitable harrier habitat throughout North America.

Key words: northern harrier; nesting; reclaimed grassland; surface mines; Pennsylvania.

Northern harriers breed in open habitats from northern Alaska east to Newfoundland, south to southeastern Virginia, and west to northern Baja, California (American Ornithologists' Union 1983). The number of breeding harriers declined annually by 1.1% from 1966 to 1987 in the US (Robbins et al. 1986). The primary reason for the widespread decline in breeding harriers is habitat loss, particularly loss of nonforested habitat in the northeast US (Gill 1985, Serrentino and England 1989, Serrentino 1992). Harriers are a "Species of

Special Concern" in Pennsylvania and listed as "at risk" by the Pennsylvania Game Commission (Gill 1985, Pennsylvania Game Commission 1991).

Harriers traditionally nest in agricultural habitats, emergent wetland areas, abandoned fields, and natural grasslands throughout their geographical range (Hamerstrom and Kopeny 1981, Gill 1985, Hamerstrom 1986, Serrentino 1987, Christiansen and Reinert 1990). These habitats are declining in area and quality throughout North America. The value of agricultural habitats, particularly in the northeastern US, to breeding harriers has been reduced as a result of earlier and more frequent hay cropping and conversion of hay crops to row crops (Toland 1985, Hamerstrom 1986, Andrle and Carroll 1988, Herkert 1994, Warner 1994). Wetland habitats are being eliminated or degraded by commercial and residential development and intensive agricultural practices (Tiner 1984). Another type of wetland, Atlantic coast saltmarshes, are suitable nesting habitat for harriers but are being altered by ditching to control mosquitos (Clarke *et al.* 1984, Serrentino and England 1989). Abandoned fields are being converted to residential and commercial development or to secondary forests (Brooks 1989) and natural grasslands are yielding to development and agriculture (Herkert 1994, Warner 1994).

Little quantitative information is available on the historic or current breeding distribution and abundance of harriers in Pennsylvania. From 1983 to 1989 the Pennsylvania Breeding Bird Atlas (PBBA) (Brauning 1992) reported possible, probable, and confirmed breeding attempts by harriers in 334 (7%) breeding bird atlas blocks (24.3 km^2) throughout the state (Fig. 1). Data obtained by the PBBA and preliminary results generated from a long-term research project (Yahner and Rohrbaugh 1993, 1994) confirm that harriers are nesting in reclaimed surface mines in northcentral and northwestern Pennsylvania.

Current reclamation procedures for surface mines in Pennsylvania have created nonforested habitats dominated by herbaceous plants interspersed with some woody vegetation (Fig. 2) (see Rafaill and Vogel 1978 and Samuel *et al.* 1978 for information regarding surface-mine reclamation). Prominent herbaceous plants include fescue (*Festuca* spp.), orchard grass (*Dactylis* spp.), timothy (*Phleum* spp.), birdsfoot trefoil (*Lotus* spp.), clover (*Trifolium* spp., *Melilotus* spp.), and goldenrod (*Solidago* spp.). Deciduous and coniferous trees, such as black locust and Austrian pine, are planted in small plantations (<5 ha). In addition, these mined lands are associated with numerous human-created wetlands (cattail [*Typha* spp.] marshes) that are designed to leach metals, e.g. iron, from water and soil. The average size of surface-mined areas in north-

---

Figure 1. The six sections of the Appalachian Plateau Province and the number of Breeding Bird Atlas blocks (Brauning 1992) reporting breeding attempts by northern harriers, 1983–89 (Modified from Brauning 1992)

1 = Pocono Plateau Section
2 = Glaciated Low Plateau Section
3 = Allegheny High Plateau Section
4 = Allegheny Mountain Section
5 = Pittsburgh Plateau Section
6 = Glaciated Section

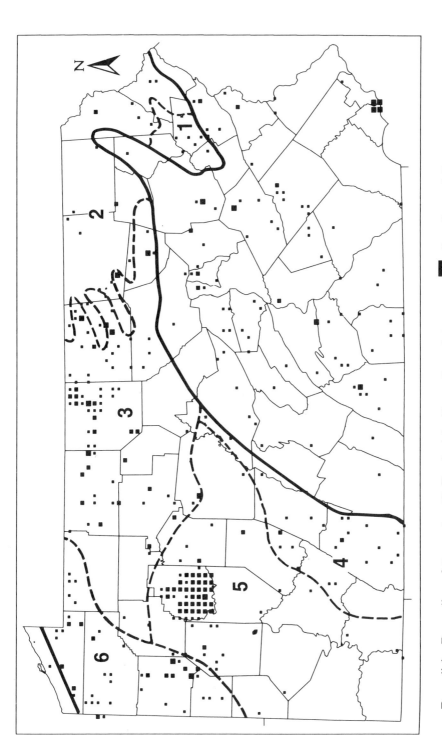

Figure 2. Surface-mine reclamation procedures create open grasslands dominated by herbaceous plant species interspersed with some woody vegetation that occurs principally in small (≤5 ha) plantations.

central and northwestern Pennsylvania is 157 ha ($N = 17$ mine sites, $SE = \pm 32$) and there are over 300 surface mines in these regions (Energy Information Administration 1989).

These open grassland surface mines are a relatively new habitat type in Pennsylvania, and little is known about their influence on raptor populations. Grassland surface-mine habitats are particularly abundant in the Pittsburgh Plateau Section of Pennsylvania, which historically had nesting harriers (Todd 1940). Our objective was to compare the number of probable and confirmed breeding attempts by northern harriers among the six physiographic sections of the Appalachian Plateau Province each containing varying numbers of surface mines, and to correlate the number of breeding attempts in each physiographic section to the number of surface mines in each section.

METHODS

We determined the number of probable and confirmed breeding attempts by harriers and the number of surface mines in each of the six physiographic sections of the Appalachian Plateau Province, using PBBA data (Brauning 1992) and data from the Energy Information Administration, respectively. The PBBA reports possible, probable, and confirmed breeding attempts; however, we excluded possible breeding attempts from our analyses to reduce the risk of including questionable observations. A breeding attempt was defined as probable if characterized by behavioral evidence, e.g. courtship displays or territorial behavior, and confirmed if physical evidence, e.g. occupied nest or fledged young, was obtained. We restricted our analyses to the Appalachian Plateau Province because we had few data on harriers and surface mines outside of this region. We assessed the amount of regional variation in the number of probable and confirmed breeding attempts by harriers and in the relative number of surface mines among the six physiographic sections of the Appalachian Plateau Province using a G-test for goodness-of-fit (Sokal and Rohlf 1981). Expected values for the goodness-of-fit tests were equal to the total number of breeding attempts or surface mines in all physiographic sections of the Appalachian Plateau Province multiplied by the percent land area of a given physiographic section. We related the number of breeding attempts by harriers in each physiographic section to the number of surface mines in each section using simple linear regression.

RESULTS

Probable and confirmed breeding attempts by northern harriers were recorded in 98 (2%) of PBBA blocks from 1983 to 1989 (Brauning 1992). Of the 98 blocks reporting breeding attempts, 78 (80%) were located in the Appalachian Plateau Province. The Pocono Plateau, Glaciated Low Plateau, Allegheny High Plateau, Allegheny Mountain, Pittsburgh Plateau, and Glaciated Sections each contained 1 (1%), 2 (3%), 22 (28%), 3 (4%), 38 (49%) and 12 (15%) breeding bird atlas blocks with probable and confirmed breeding attempts, respectively.

A significantly greater number ($G = 10.56$, df $= 1$, $P < 0.005$) of breeding bird atlas blocks contained probable and confirmed breeding attempts than expected in the Pittsburgh Plateau Section of the Appalachian Plateau Province compared with the remaining physiographic sections. We noted 38 and 23 observed and expected breeding attempts in the Pittsburgh Plateau Section and 40 and 55 observed and expected attempts in the remaining sections of the Appalachian Plateau Province, respectively. In addition, the Pittsburgh Plateau Section contained a significantly greater number ($G = 234.64$, df $= 1$, $P < 0.001$) of surface mines than expected compared with the remainder of the Appalachian Plateau Province. We noted 207 and 83 observed and expected surface mines in the Pittsburgh Plateau Section and 70 and 194 observed and expected mines in the remaining sections of the Appalachian Plateau Province, respectively. The Pittsburgh Plateau Section represented approximately 30% of the land area of the Appalachian Plateau Province and contained 49% of the harrier breeding attempts and 75% of the Province's 277 surface-mines. The number of breeding attempts in each physiographic section was correlated positively ($r = 0.81$, df = 5, $P = 0.05$) with the number of surface mines in each section.

Within the Pittsburgh Plateau Section, 44 (64%) of the 69 breeding attempts were recorded in Clarion County. Of the 13 counties contained at least partially within the Pittsburgh Plateau Section, Clarion County contains the second largest area (km^2) of surface mined land (Pennsylvania Coal Association 1992).

CONCLUSIONS

We have obtained evidence that during the breeding season northern harriers are associated more frequently than expected with open grassland habitat created by reclamation of surface mines in Pennsylvania. Nest and roost sites of harriers have been observed in wetland areas contained within these grassland surface mines (Yahner and Rohrbaugh 1993, 1994). However, the extent to which harriers use reclaimed surface mines for breeding is unknown. Tentative results from an ongoing research project suggest that harriers are approximately five times more abundant in grassland surface mines than agricultural habitats from April through July in Pennsylvania (Yahner and Rohrbaugh 1993, 1994). Unlike agricultural habitats, reclaimed surface mines are relatively undisturbed, which may benefit nesting harriers. Unfortunately, no quantitative data are available on the historic density or relative abundance of harriers breeding in the state prior to mine reclamation.

We are currently studying the use of reclaimed grassland surface mines by harriers to determine long-term trends and ecological relationships of harrier populations occupying surface mines in Pennsylvania. There are approximately 800 000 ha of surfaced-mined land throughout the state. Results of our studies will be necessary to develop management guidelines for harriers using surface-mine habitat. The development of management guidelines is timely because harrier populations and traditional habitats are declining. Furthermore, ecological succession is changing the plant species composition and structure of reclaimed surface mines, which may affect use of these habitats by harriers. If managed properly, reclaimed grassland surface mines will provide important nesting habitat for northern harriers in Pennsylvania and help supplement a net loss of suitable harrier habitat throughout North America.

ACKNOWLEDGMENTS

We thank the Pennsylvania Game Commission and the Max McGraw Wildlife Foundation for providing funding for this study. We also appreciate cooperation from the Pennsylvania Breeding Bird Atlas Project for sharing breeding data on northern harriers in the state. In addition, we thank D. Brauning, R. Simmons, and B. MacWhirter for reviewing the manuscript.

REFERENCES

AMERICAN ORNITHOLOGISTS' UNION. 1983. *Checklist of North American birds* 6th ed. Allen Press, Lawrence Kansas.

ANDRLE, R.F. AND J.R. CARROLL, eds. 1988. *The atlas of breeding birds in New York State.* Cornell Univ. Press, Ithaca, New York.

BRAUNING, D.W., ed. 1992. *Atlas of breeding birds in Pennsylvania.* Univ. of Pittsburgh Press, Pittsburgh, Pennsylvania.

BROOKS, R.T. 1989. Status and trends of raptor habitat in the Northeast. Pp. 123–132 in B.G. Pendleton, ed. *Proc. Northeast Raptor Management Symposium and Workshop.* Nat. Wildl. Fed., Washington, D.C.

CHRISTIANSEN, D.A., JR. AND S.E. REINERT. 1990. Habitat use of the northern harrier in a coastal Massachusetts shrubland with notes on population trends in southeastern New England. *J. Raptor Res.* **24**: 84–90.

CLARKE, J.A., B.A. HARRINGTON, T. HRUBY AND W.E. WASSERMAN. 1984. The effect of ditching for mosquito control on saltmarsh use by birds in Rowley, Massachusetts. *J. Field Ornithol.* **55**: 160–180.

ENERGY INFORMATIOJN ADMINISTRATION. 1989. *Coal production 1989.* DOE/EIA-0118(89). Distribution Category UC-98. US Dept. Energy, Washington, D.C.

GILL, F.B., ed. 1985. Birds. Pp. 259–351 in H.G. Genoways and F.J. Brenner, eds. *Species of special concern in Pennsylvania.* Carnegie Museum of Natural History. Special Publication, no. 11. Pittsburgh, Pennsylvania.

HAMERSTROM, F. 1986. *Harrier, hawk of the marshes, the hawk that is ruled by a mouse.* Smithsonian Inst. Press, Washington, D.C.

— AND M. KOPENY. 1981. Harrier nest-site vegetation. *J. Raptor Res.* **15**: 86–88.

HERKERT, J.R. 1994. The effects of habitat fragmentation on midwestern grassland bird communities. *Ecol. Appl.* **4**: 461–471.

PENNSYLVANIA COAL ASSOCIATION. 1992. *Pennsylvania coal data 1992.* Pennsylvania Coal Association, Harrisburg, Pennsylvania.

PENNSYLVANIA GAME COMMISSION. 1991. *Pennsylvania's wild birds and mammals: special concern species, extirpated species, established exotics.* Pennsylvania Game Commission, Harrisburg, Pennsylvania.

RAFAILL, B.L. AND W.G. VOGEL. 1978. A guide for vegetating surfaced-mined lands for wildlife in eastern Kentucky and West Virginia. US Fish Wildl. Serv, FWS/OBS-78/84.

ROBBINS, C.S., D. BYSTRAK AND P.H. GEISSLER. 1986. *The breeding bird survey: the first 15 years, 1965–1979.* US Fish Wildl. Serv. Res. Publ, no. 157. Washington, D.C.

SAMUEL, D.E., J.R. STAUFFER, C.H. HOCUTT AND W.T. MASON. 1978. *Surface mining and fish/wildlife needs in the eastern United States.* US Fish Wildl. Serv, FWS/OBS-78/81.

SERRENTINO, P. 1987. The breeding ecology and behavior of northern harriers in Coos County, New Hampshire. M.S. thesis, Univ. Rhode Island.

— 1992. Northern harrier, *Circus cyaneus*. Pp. 89–117 in K.J. Schneider and D.M. Pence, eds. *Migratory nongame birds of management concern in the Northeast.* US Dept. Interior, US Fish and Wildl. Serv., Newton Corner, Massachusetts.

— AND M. ENGLAND. 1989. Raptor status reports: northern harrier. Pp. 37–46 in B.G. Pendleton, ed. *Proc. Northeast Raptor Management Symp. and Workshop.* Nat. Wildl. Fed., Washington, D.C.

SOKAL, R.R. AND F.J. ROHLF. 1981. *Biometry.* W.H. Freeman and Co., San Francisco, California.

TINER, R.W., JR. 1984. *Wetlands of the United States: current status and recent trends.* US Dep. Interior, US Fish Wildl. Serv., Nat. Wetlands Inventory, Washington, D.C.

TODD, W.E.C. 1940. *Birds of western Pennsylvania*. Carnegie Mus., Univ. Pittsburgh Press, Pittsburgh, Pennsylvania.

TOLAND, B. 1985. Nest site selection, productivity, and food habits of northern harriers in southwest Missouri. *J. Nat. Areas* 5: 22–27.

WARNER, R.E. 1994. Agricultural land use and grassland habitat in Illinois: Future shock for midwestern birds? *Conserv. Biol.* 8: 147–156.

YAHNER, R.H. AND R.W. ROHRBAUGH, JR. 1993. Long-term status and management of northern harriers, short-eared owls, and associated wildlife species in Pennsylvania, Annual Report, Pennsylvania Game Commission.

— AND — 1994. Long-term status and management of northern harriers, short-eared owls, and associated wildlife species in Pennsylvania, Annual Report, Pennsylvania Game Commission.

30
Raptors Associated with Airports and Aircraft

S. M. Satheesan

Abstract – An ecological study of bird hazards at 30 Indian aerodromes was conducted between 1980 and 1993. Analyses from 552 aircraft strike remains identified through microscopic and macroscopic methods between 1966 and July 1993 revealed that 55.4% of the incidents involved 18 species of raptors. Vultures (*Gyps* and *Neophron* spp.) and kites (*Milvus* spp.), the commonest of the raptors on the Indian subcontinent were involved in 48.4% of them. The economic loss due to their collision with aircraft ranged from over a dozen air crashes to severe damage to engines and other airplane parts. The superabundant food in urban areas available at garbage and carcass dumps as well as sanitary landfills has attracted multitudes of raptors. High numbers of scavenging raptors in cities have led to increased collisions with aircraft. Aerodromes provide open, vast and relatively tranquil areas with natural plant and animal food and thus favourable habitat for feeding, resting, roosting and nesting. The attraction of raptors to aerodromes proves catastrophic to raptors and to humans.

Key words: raptors; human-altered environment; airports; aerodrome; aircraft strikes; India.

The rapid growth of the human population and the drive to increase the quality of life is progressively depleting natural resources. Over-exploitation affects raptorial birds and their natural habitats. Hence, raptors adapt to seemingly inhospitable and hostile habitats. An airport is a human-altered environment that several species of raptors utilize, despite the high risk of encounters with aircraft.

The heavy financial loss incurred by Indian aviation due to bird hits to aircraft was the motivation for this study. The Bombay Natural History Society (BNHS) has been identifying species from bird–aircraft strike remnants since 1966. Field research at 30 Indian airports started in 1980 with the basic objective being to save aircraft and birds by developing conservation-oriented methods (Grubh *et al*. 1988, 1989).

STUDY AREAS

The association of raptors with airports and aircraft was studied in 30 civil and military aerodromes in India between 1980 and 1993, namely: Jammu,

Srinagar, Delhi, Hindan, Agra, Gorakhpur and Patna in the north; Ambala, Sirsa, Jodhpur, Jaislamer and Utterlai in the northwest; Calcutta, Kalaikunda, Tezpur and Chabua in the northeast; Naliya, Jamnagar, Bhuj, Bombay, Juhu, Pune and Dabolim (Goa) in the west; Gwalior, Nagpur, Dundigal and Hyderabad in central India; and Bangalore, Madras and Trivandrum in south India.

METHODS

Several methods were employed to study the intensity and extensiveness of raptor associations with airports and aircraft.

(1) Field work consisted of: (a) tower-top or hill-top observations from lofty positions including the control tower of the aerodrome; (b) infield observations while walking the entire area of the aerodrome; (c) buffer zone observations within a 25 km radius of the aerodrome midpoint to find out the attractions for birds inside and adjacent to the aerodrome; and (d) collection of specimens of all fauna and flora of the aerodrome area, including carcasses of birds struck by aircraft.

(2) Laboratory work comprised: (a) the preservation and identification of all biological specimens collected from the aerodrome, including aircraft-struck specimens; and (b) analyses of the stomach contents of all animals to study food habits. Bird specimens received from aerodromes all over India were identified at the BNHS using macroscopic and microscopic methods.

(3) To identify the problem birds, problem aerodromes, and to evolve practical measures to minimize bird–aircraft collisions, bird strike data sent to the BNHS from various Indian aerodromes were analysed.

RESULTS

Association of Raptors with Aircraft

Analysis of bird–aircraft strike data gathered from 1966 to 1993 revealed that 77 species of birds and three species of bats were involved in 612 collisions with aircraft in India. Eighteen species of raptors were involved in 54.3% of the collisions in 56 Indian aerodromes (55.4% of 552 incidents analysed up to July 1993). Vultures (*Gyps* and *Neophron* spp.) and kites (*Milvus* spp.) were responsible for 46.9% of bird hits in India (48.4% out of 552 incidents up to July 1993). Of these incidents, 7.4% were caused by 13 species of birds of prey other than vultures and kites (Table 1).

Table 1. Raptors sighted during surveys and hit by aircraft at airports in India between 1966 and 1993.

Raptor species	Airports where sighted ($N = 30$)	Airports recording strikes ($N = 56$)	Percent of incidents[a] (body weight g)
Black-shouldered kite	18	7	1.1 (270)
Honey buzzard	12	0	0
Pariah kite	21	29	25.5 (680)
Black-eared kite	1	1	0.3 (900–1000)
Brahminy kite	8	6	2.6 (600)
Shikra	2	0	0
Sparrowhawk	3	1	0.16 (200)
Long-legged buzzard	4	0	0
White-eyed buzzard	9	0	0
Tawny eagle	7	0	0
Steppe eagle	1	0	0
White-bellied sea-eagle	2	0	0
Pallas's fish-eagle	1	0	0
Red-headed vulture	2	0	0
Long-billed vulture	19	2	0.3 (5000)
Indian white-backed vulture	27	32	17.32 (4500)
Egyptian vulture	18	2	0.33 (2000)
Hen harrier	4	0	0
Pale harrier	7	1	0.16 (300)
Montagu's harrier	6	2	0.33 (250)
Pied harrier	5	0	0
Marsh harrier	15	1	0.16 (400)
Short-toed eagle	5	1	0.16 (1500–2000)
Shaheen falcon	1	0	0
Laggar falcon	1	0	0
Red-headed merlin	1	2	0.33 (225)
Common kestrel	17	3	0.49 (140)
Hobby	1	0	0
Barn owl	5	3	0.49 (300)
Eagle owl	9	3	0.49 (1100)
Spotted owlet	19	1	0.16 (120)
Short-eared owl	0	1	0.16 (366)
Brown hawk owl	1	0	0
Unidentified raptors			0.16%
Unidentified harriers			0.33%
Unidentified *Gyps* vultures			3.30%

[a] Number of raptor hits (332) in total number of bird hits (612).

Raptor strikes. Most of the incidents were caused by pariah kites (25.5%) and vultures (21.1%; Grubh et al. 1988, 1989, Satheesan 1991, 1992; Satheesan and Grubh 1992; Table 1). Among other raptors, the more predominant species were brahminy kite, black-shouldered kite, kestrel, harriers (three species) and owls (four species). They were involved in only 1.3% of all incidents.

Month and time of raptor strikes. The majority of raptor strikes occurred during their breeding season (September–March) or when the juveniles fledged (January–April; Fig. 1). Juvenile birds (unfamiliar with the airport environment) and nesting adults (hunting for young) seemed especially susceptible to aircraft strikes (Satheesan 1991, 1992).

The temporal distribution of raptor strikes showed that 84% of the incidents ($N = 275$) occurred between 0700 h and 1500 h (Fig. 2). This is related to the availability of thermals needed for soaring and foraging, as well as resting and sunning behaviour of raptors in the aerodrome area. There was a positive correlation ($\dot{r} = 0.614$) among kite activity, kite aircraft hits and aircraft movements at Bangalore aerodrome in south India (Satheesan 1991, 1992).

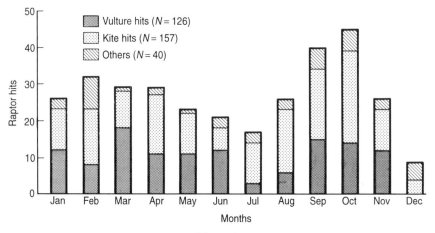

Figure 1.

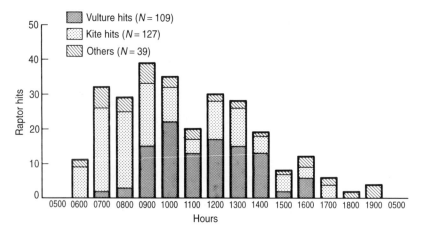

Figure 2.

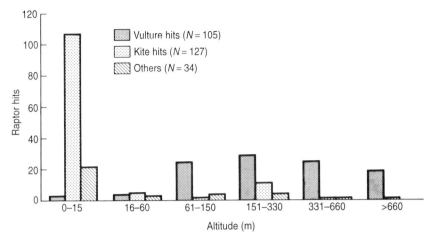

Figure 3.

Altitudinal distribution of raptor strikes. About 46% of raptor strikes ($N = 226$) occurred >60 m above the ground (Fig. 3). About 47% of incidents involved soaring raptors ($N = 247$) such as vultures, pariah kites and brahminy kites. About 50% of the incidents involved hovering species ($N = 10$) such as kestrels and black-shouldered kites and occurred at altitudes >60 m and outside the airport area. About 54% of all raptor strikes as well as the majority of hits by pariah kites (88%) and brahminy kites (81%) occurred at altitudes ≤60 m and on or very close to the runway. These birds were likely either resting, feeding or flying low within the airport area (Satheesan 1991).

Locality of raptor strikes. About 90% of kite hits ($N = 133$), 46% of vulture hits ($N = 107$) and 61% of strikes due to other raptors ($N = 42$) occurred in urban areas. About 94% of vulture hits were outside the airports, whereas 82% of kite hits as well as 69% of other raptor hits were inside the airport area. Only bird hits at altitudes of aircraft flight ≤60 m during climb or approach were classified as occurring inside an airport. Military aircraft suffered more due to vulture hits during low-level exercises outside the aerodrome whereas the civil aircraft suffered more due to kite hits mainly inside the aerodrome because immediately after take-off, civil aircraft climb well above the bird layer (Table 2). In general, the occurrence of about 51% of raptor hits ($N = 275$) inside aerodrome areas and 70% of raptor hits ($N = 324$) in cities and towns indicates that urban habitats and airports attract raptors.

Geography of raptor strikes. Vulture hits were more common in northwest, northeast, west, central and southcentral India, while kite hits occurred mainly in south India and in larger cities. Collisions due to other raptors occurred more or less uniformly on the subcontinent.

Table 2. Raptor hits inside and outside Indian airports at various flight phases of military and civilian aircraft between 1966 and 1993 (N = 275).

Raptor hits			Flight phases				
			Inside airport			Outside airport	
		Taxi	Take-off run and landing roll	Initial climb and final approach	Climb and approach	Descent and level	Total hits by aircraft
Vulture hits to aircraft	Military	1	1	4	34	53	93
	Civil	0	1	0	12	1	14
Kite hits to aircraft	Military	0	24	0	9	9	42
	Civil	1	80	4	5	1	91
Other raptor hits to aircraft	Military	0	9	3	5	5	22
	Civil	0	10	2	1	0	13
Total raptor hits		2	125	13	66	69	275[a]

[a] 140 of these were inside airports and 135 were outside.

Economic loss as a result of raptor–aircraft strikes. The financial loss caused by raptor hits, i.e. millions of rupees was due to: (a) effects on flights, such as aborted takeoffs (26 cases), precautionary landings (72 cases), engine shutdowns (18 cases), delays of departure for engine checks or repair, and loss of aircraft from crashes or crash-landings (12 due to vulture hits and one due to a kite hit); (b) damage to aircraft engines and other parts due to collisions with vultures (55 incidents), kites (6 incidents) and other raptors (12); and (c) damage to property on the ground, such as destruction of standing crops, vehicles or buildings (three cases). Raptor hits also caused loss of life and injury to pilots, crew and passengers in three cases. Vultures weighing over 4500 g caused most of the economic losses to aviation in India.

DISCUSSION

Association of Raptors with Airports

The field surveys carried out by the BNHS revealed that 33 species of raptors used airport areas for one or more of their basic behaviours such as feeding, roosting, nesting or resting. The attractions for raptors to and near aerodromes appear to be: food, resting places, roosting sites, nesting sites and water-logged areas.

Food. Observation of birds of prey and analyses of stomach contents of raptors revealed that they feed on various species of insects, worms, snails, amphibians, reptiles and rodents. Raptors are especially attracted to grass during thinning operations or mowing to feed on insects that were being flushed (Grubh et al. 1989, Satheesan and Rao 1990, Satheesan et al. 1990, Satheesan 1991, 1992). Raptors like pariah kites make regular searches over and around runways during morning hours for carcasses of bats, small birds and other animals killed by aircraft or vehicles. Garbage such as kitchen refuse and food waste from restaurants, canteens, other eating houses and aircraft catering services deposited in open garbage bins, pits or huts inside aerodrome areas is readily accessible to and fed on by scavenging raptors, e.g. pariah kites. Runways and taxi tracks are literally used as a dining table by pariah kites, harriers, kestrels and black-shouldered kites to feed on live prey (grasshoppers or crabs) or offal (obtained from a nearby garbage dump; Satheesan and Rao 1990, Satheesan 1991, 1992).

Resting places. Aerodromes provide safe communal resting places for raptors such as pariah kites. An aerodrome is used by kites as a communal resting place if food is available regularly within the aerodrome area or if a regular feeding site is present within 100–200 m. This is the case at Trivandrum, Bombay, Bangalore and Hyderabad airports where regular garbage dumps exist very close to the airports (Satheesan 1989, 1991, 1992). Utility structures like power pylons, floodlight towers, other electrical installations, radars, control towers, fire towers, hangars and blast pens in airport areas are used by raptors as resting areas and sites from which to search for prey. In desert environments where tall trees are not available, utility structures are always used by raptors for locating and hunting prey.

Roosting sites. Harriers roost in the shoulder grass of runways (Satheesan and Rao 1990, Satheesan et al. 1990), whereas owls roost in aircraft hangars and in blast pens. Roost trees are available for black-shouldered kites and king vultures.

Nesting sites. Nesting sites are available for raptors in vegetation and on utility structures inside aerodromes. King (red-headed) vultures and pariah kites nest in trees (Satheesan 1991, 1992) and owls nest in aircraft hangars.

Water-logged areas. Pools, puddles, ponds and lakes attract raptors like kites for bathing during the heat of summer (Satheesan 1991, 1992).

In general the openness, vastness, tranquillity and the safety afforded by the protected nature of the airports coupled with the availability of food, resting, roosting and breeding facilities available there attract raptors to this new habitat. Compared with other open areas such as sport stadiums, playgrounds, race courses or highways, the main attraction for raptors in an airport area is the safety afforded by areas with less disturbance. Hence, several species of raptors apparently identify the airport environment as a favourable habitat. The adaptation of some raptors to this human-altered environment has even changed their food habits in such a way that they prefer to scavenge rather than to hunt live prey. Once a reliable food source is available, they roost and nest in unnatural and even dangerous sites, e.g. a vulture nest and a kite nest were located close to active runways at Agra and Bombay airports, respectively.

Mitigation and Management Techniques

In order to save raptors and their natural environments one should work to:

(1) identify and protect the remaining natural habitats from use for agriculture, industry, mining or construction;
(2) provide artificial perching, roosting and nesting sites (including nest boxes) in natural or natural-looking sites away from aerodromes and utility structures;
(3) establish sites to feed endangered raptors, especially vultures, but located away from aerodromes and utility structures;
(4) alter flight schedules to avoid areas where birds fly in concentrated flocks and peak hours of bird activity; and
(5) deny food, resting, roosting and nesting sites to raptors and other birds in the vicinities of aerodromes. The last can be accomplished by making the aerodrome area unattractive in the followings ways: (a) level airport grounds and institute drainage systems to avoid waterlogging; (b) cover open drains with wire mesh or perforated concrete slabs; (c) integrate into the vegetation a species of short grass, that does not produce seeds attractive to birds and other animals. This would reduce the vegetation cover and hence its carrying capacity for insects, birds and other animals preyed upon by raptors; (d) mow grass regularly with lightweight grass cutting machines during the night so as not to attract birds with insects flushed out of vegetation; and (e) adopt proper animal garbage disposal by using covered garbage bins, pits and huts, and dry rendering plants.

Animal garbage should also be denied to raptors by: (a) replacing primitive slaughtering places with modern abattoirs; (a) bird-proofing meat, fish and poultry markets, bonemills, piggeries, carcass processing plants; (c) covering dumps and sanitary landfills; (d) introducing legislation to ban indiscriminate dumping of animal carcasses and setting up carcass processing centres in urban areas to convert meat into chicken feed and by-products such as fertilizer, animal fat and glue for use in the textile industry; and (e) establishing dry rendering plants to process waste food from restaurants, large eating houses, residential buildings and kitchens.

CONCLUSION

The basic philosophy for protecting raptors, aircraft, and humans is to manage raptor habitats so that natural habitats are more attractive to them than human-altered environments.

ACKNOWLEDGMENTS

The data presented in this paper were collected as a part of bird strike remnant identification carried out since 1966, field projects undertaken since 1980 as well as the Bird Hazard Research Cell started in 1987. These BNHS research projects were funded by the Operational Problems Panel of Aeronautics, R & D Board of Defence Ministry, Government of India. The presentation of this paper was made possible by the travel assistance from the Director General of Civil Aviation and Air India. I am grateful to the Director of BNHS, Dr. J. S. Samant, the late Dr. Salim Ali, Dr. Robert B. Grubh and all the scientists who worked with BNHS and contributed towards the bird strike prevention research.

REFERENCES

GRUBH, R.B., S.M. SATHEESAN AND G. NARAYAN. 1988. Ecological study of bird hazards at Indian aerodromes with special reference to Bombay, Delhi, Agra and Ambala. Pp. 32–38 in *Proc. of First Natl. Seminar on Bird Hit Prevention*. Inst. Aviation Manage. New Delhi, India.

—, —, —, P. RAO, H. DATYE, R.B. SINGH, C. KARMORKAR, M. RAJASEKARAN, M. AYYADURAI AND S. MURALIDHARAN. 1989. Ecological study of Bird Hazard at Indian Aerodromes: Phase 2. Final Report. Part Two. Vols. 1–17. Bombay Natural History Society, Bombay, India.

SATHEESAN, S.M. 1989. Communal resting behaviour of black (pariah) kites in India. *Gabar* **4**: 16.

— 1991. Ecology and behaviour of the pariah kite *Milvus migrans govinda* (Sykes) as a problem bird at some Indian aerodromes. Ph.D. diss., Univ. of Bombay, Bombay, India.

— 1992. Solutions to the kite hazard at Indian airports. *Internat. Civil Aviation Organiz. J.*, Feb.: 13–15.

— AND P. RAO. 1990. Roosting and feeding of harriers in Secunderabad, Andhra Pradesh. *J. Bombay Natl. Hist. Soc.* **87**: 13.

—, — AND H. DATYE. 1990. Biometrics and food of some harriers. *Pavo. J. Indian Ornithol.* **28**: 75–76.

— AND R.B. GRUBH. 1992. Bird-strike remains identification in India. Pp. 5–41 in *Proc. 21st meeting of bird strike committee Europe (BSCE)*. Jerusalem, Israel, Working Paper 6.

Raptors at Large

31

The Effect of Altered Environments on Vultures

David C. Houston

Abstract – Vultures are specialised birds of prey that obtain their food largely by scavenging from carcasses, rather than killing prey themselves. The aim of this paper is firstly to consider the impact that human activity has had on this group of birds, and the causes of population decline, concentrating on the Old World species. Secondly, I discuss some of the management techniques that can be used to support vulture populations, and consider the success of these methods in sustaining viable vulture populations in industrialised countries. In particular, I emphasise the ways in which the conservation of vultures differs from that of other predatory birds.

Key words: vultures. Old World; impact; human activity; management.

THE STATUS OF VULTURE SPECIES

There are a total of 22 species of vultures in the world, which belong to two quite unrelated groups. The 15 species of Old World vultures are closely related to the eagles and buzzards in the Falconiformes. In the New World however, the seven species of Cathartid vulture are descended from the same line of ancestry as the storks (Sibley and Ahlquist 1990). The superficial appearance of birds from the two groups is extremely similar. They are text-book examples of convergent evolution. The two groups of vultures are so close in their ecology that the conservation problems they face are almost identical. The New World group contains the most endangered of all vulture species, the California condor. This is now one of the rarest birds in the world and is receiving perhaps the most intensive conservation management ever given to any endangered bird species. The conservation problems faced by this species are reviewed by Snyder and Snyder (1989). There are active conservation programmes also for the Andean condor (Wallace and Temple 1987), and in the Old World for the European griffon; (Terrasse 1983), Cape griffon (Mundy *et al.* 1992), cinereous vulture (M. Bijleveld pers. comm.), lappet-faced vulture (Meretsky and Lavee 1989), Egyptian vulture (R. Nadi pers. comm.) and bearded vulture (Frey and Walter 1989). Proportionally, the vultures must be receiving among the most active conservation action of any bird group, with over one-third of species currently the subject of active management and reintroduction programmes.

It might be thought that scavenging birds would be highly vulnerable to human disturbance. These birds evolved to feed on the carcasses of large ungulates, such as deer and antelope, which are among the first species to be lost when natural ecosystems are converted to agricultural land. Most of these changes occurred before historical records. The only documented case (Brooke 1984) recorded the distribution of the Cape vulture in South Africa from sight records made between 1800 and 1969 and contrasted this with records made from 1970 to 1983. This shows a clear but small contraction in range. If we consider the extreme ecological changes that have occurred in the whole of southern Africa since 1800, it is perhaps surprising that this species has adapted so well. Today the Cape griffon feeds largely on domestic stock, as is the case for most vulture populations in developed countries.

It is a feature of most forms of extensive livestock farming that cattle, sheep or goats are kept at densities close to, or exceeding, those that the habitat can sustain. The provision of artificial water supplies and supplementary feeding means that some areas given over to free-ranging livestock today probably support higher biomass figures than would have been found under natural wildlife ecosystems. To a vulture, a dead cow or sheep is just as acceptable as a dead antelope. This move to domestic animals as their primary food source has undoubtedly, for some species, led to increased food supplies and caused some vulture populations to expand. India is perhaps the best example, where those vulture species which have adapted to living near cities, such as the Indian white-backed vulture and Egyptian vulture, are now so abundant within the sub-continent that it is difficult to believe that they have not benefited enormously from human activity and are today far more abundant than would have been the case under natural wildlife systems.

CAUSES OF POPULATION DECLINE

Vultures are unlike almost all other birds of prey in that we can quantify their food supply with considerable accuracy. For most other predatory animals we can obtain estimates for prey density within their foraging range. But this cannot be assumed to represent the food supply because of all the problems of availability and whether prey can be captured. For many birds of prey the favoured hunting areas are not where prey is most abundant, but areas where prey can be caught most efficiently. In the case of vultures this problem does not arise. We need only to estimate carcass numbers. If we know the density of livestock and wild ungulates in an area, and we have estimates of the mortality rates for each species, we can estimate the number of animals which will die each year. In most forms of extensive livestock farming any animals that die are left for natural scavengers to deal with – either they are not found by the owner or it is impractical to bury or dispose of the carcass. Three studies in man-dominated environments have estimated the food supply available for griffon vulture populations in this way, and all have concluded that the potential food available is likely to be sufficient to supply the food requirements for the breeding vulture population, e.g. Richardson 1980, Robertson 1983, Komen 1986. However, some other studies have shown that changes in livestock farming methods, and reduction in stock densities, may have resulted in insufficient food being available for vulture populations. This has been suggested for cinereous vultures in the Evros region in Greece (Grant 1985) and for lappet-faced and other vultures in Israel following the reduced number of Bedouin goat herds, largely because of the political problems of this region (Meretsky and Lavee 1989). It is important to note that even where we estimate that the food supply may be abundant, the seasonal variation must be considered. Ungulate mortality is highly seasonal, and so a calculated annual mean food surplus for the vulture population may be meaningless if it consists of a short period of super-abundance followed by a long period of food shortage.

If we look at studies of habitat utilisation, we again find that some vulture species appear to benefit from human activity, while others have obviously suffered. A recent study carried out by Xirouchakis (1993) in the Evros region of Greece showed griffon, cinereous and Egyptian vultures there spent more time searching over disturbed habitats than over undisturbed natural woodland, presumably because more food was available in these areas, and that the Egyptian vulture had different habitat preferences from the other two species, favouring arable land.

We can therefore conclude that human activity need not cause a decline in the potential food supply for vultures, and in many situations may lead to a substantial increase. Why then should some species such as the California condor have shown such dramatic contraction in their range within historical times?

Good information on the causes of these population changes is needed. This sounds a simple question, but it is far from easy to answer. The California condor has received intensive research to determine the cause of population decline, and we still know comparatively little about the factors responsible. For other species we are lacking almost any reliable data.

Vultures clearly evolved in conditions in which they had very low annual mortality and also a very low rate of reproduction. Most species lay only a single egg, and larger species may not start breeding until they are six to eight years old. The large vulture species have among the slowest reproductive rates of any bird in the world. It follows that if mortality factors start to increase and the birds are unable to adjust their reproductive rate to compensate, the population will inevitably start to slide towards extinction. Two recent studies have been made to model the dynamics of griffon vulture populations, and such models clearly show that small increases in mortality rates alone can cause populations to become extinct (F. Sarrasin and S. Piper pers. comm.).

When we consider the mortality factors influencing vultures, it is difficult to assess their relative importance. Few birds are found dead. Those which are located are likely to form a highly biased sample, e.g. a bird found drowned in a farmer's cattle trough is far more likely to be reported than a bird dying from starvation.

Irresponsible shooting is undoubtedly a problem for most vulture populations, as for many other raptors. Some species get blamed (usually unjustly) for killing farm livestock, but vultures are usually not shot because they are perceived as pests. It is their large size, slow flight, and tendency to fly close to hillsides that makes them particularly vulnerable. The fact that they form such an easy target, together with their impressive size, seems to present a challenge to the uncivilized hunter.

Pesticides and the accumulation of other persistent toxic chemicals, which had such serious impact on other raptor populations, were probably a factor in the decline of griffon vultures in a number of countries, such as Israel, in the past (Leshem 1985). But today, in countries where there is still widespread use of these persistent chemicals, there is no evidence that this is an important factor. Mundy et al. (1982) studied levels of DDT in the eggs of Cape griffons and white-backed vultures in southern Africa, and showed comparatively low levels and only slight eggshell thinning. By comparison, pesticide levels in America led to a 20–32% decrease in California condor eggshell thickness (Kiff et al. 1979) and an 11–18% decrease in that of turkey vulture eggs (Wilbur 1978), and may have contributed to the decline of the condor.

Poisoning is a problem with many birds of prey, but in the case of vultures it is among the most important causes of population decline. Vultures are peculiarly vulnerable because, unlike most other raptors, they feed on large mammal carcasses and these are easily poisoned. In addition, vultures often feed in large social groups, birds from a wide area being drawn towards a single food source. A single poisoned carcass will not only kill local birds, but may eliminate a substantial proportion of the total population in a region.

There are two forms of poisoning. The first is the deliberate use of poison by farmers to control predatory mammals. Earlier this century it was common practice in all mountain areas of Europe to use strychnine to control wolves and foxes, which led to heavy mortality of vultures (Hiraldo et al. 1979). In much of Africa it is still routine to kill hyaenas, jackals, lions, wild dogs and other potential predators on stock in this way. A survey suggested that 10 of the 21 farmers interviewed in Transvaal in South Africa still use strychnine in predator control. It seems remarkable that such lethal poisons are still routinely used in an uncontrolled manner even in advanced agricultural communities. Carcasses so poisoned are, of course, highly dangerous to vulture populations. On one occasion in Cape Province in South Africa a single poisoning led to 42 birds being killed, this being one-tenth of the total population of the species in this province of South Africa. Another incident concerning a dead cow in Botswana, laced with strychnine, led to the death of about 100 vultures (Mundy et al. 1992). The number of such events which went, and which continue to go, unrecorded must be substantial. Strychnine poisoning was thought to have been the cause for decline in vulture numbers in Kenya (Jackson 1938), and is one of the causes of mortality identified today for the California condor (Snyder and Snyder 1989).

The second form of poisoning is deemed unintentional. The most serious example is lead poisoning. This can arise when hunters wound an animal, which later dies with lead shot or bullets in the tissues. If such a carcass is found by a vulture, the lead may be ingested and can rapidly lead to death. Lead is highly toxic when ingested into the alimentary tract. It is not known how significant this cause of mortality is for Old World vultures, because birds killed in this way are unlikely to be discovered or diagnosed. However, the research on California condors has shown this to be an important cause of mortality in that species (Snyder and Snyder 1989), and it is likely this is true for all vulture species living in areas where firearms are widely used. Attempts are underway to introduce steel shot or other less toxic metals into the ammunition used within the condor range.

Another factor to which vultures are peculiarly susceptible is electrocution on electricity pylons. They are more vulnerable than other raptors because of a combination of their liking for tall perch sites for roosting, and a massive wingspan which may short-circuit the current. Such incidents usually also lead to a breakdown in the electricity supply. This has the fortunate consequence that electricity companies have a vested interest in trying to minimize this problem. In Israel during a 23-month period in the early 1980s one short stretch of power line with three pylons was responsible for killing 43 griffon vultures; this represented a quarter of the total number of birds in the northern Israel population (Leshem 1985). Almost all of the birds killed were first year or immature birds, and this seems a feature of this cause of mortality that it is mostly inexperienced birds that suffer. They probably have poor flying skills and are more likely to lose their balance when trying to perch on electric pylons. In parts of southern Europe, America and South Africa power lines have claimed the lives of large numbers of birds. Ledger and Annegarn (1981) and Mundy *et al.* (1992) report on a highly successful programme, in collaboration with an electricity company, to modify pylon design in South Africa to overcome this problem. Mortality can also be caused by birds flying into tower-support wires or power lines. A single television transmission tower in South Africa, because of its supporting guy ropes, was responsible for killing 55 Cape vultures in this way (Benson and Dobbs 1984).

In some areas, drowning in vertically sided cattle water troughs is also a major cause of mortality. Vultures probably do not drown when they come to drink. They have such long necks that they can usually reach the water when perched on the side of a water trough. They drown when they try to bathe. Vultures are fastidious about their plumage, and Old World vultures regularly bathe after feeding, often by almost totally submerging themselves in deep pools. But in cattle troughs with vertical sides they are unable to escape after the plumage has become waterlogged. Vultures cannot take off vertically from water, and, although they can swim effectively across wide bodies of water, a confined cattle trough will trap them until they drown. A recent survey of causes of mortality in Cape vultures in southern Africa by Piper (1993) showed this to be the second most important cause of death, following electrocution. The solution is a simple one: plastic crates that are widely used in bakeries for carrying bread can be wired together upside-down to form a raft, which covers part of the water surface. Cattle can easily drink by pushing the plastic crate down with their noses to get at the water, but the raft provides sufficient buoyancy to allow vultures to escape after bathing (Piper 1993).

Surveys of causes of mortality, from recovered dead birds, have been carried out for several vulture species (Cape griffon, Piper 1993; California condor, Snyder and Snyder 1989; lappet-faced vulture in Israel, Meretsky and Lavee 1989), and they all identify poisoning, electrocution, drowning and shooting as the main identified factors. All of these are factors directly caused by human activity, and vulture populations will only recover when we can minimize these risks to a level that the reproductive output of the population can sustain.

CONSERVATION MANAGEMENT

A number of Old World vulture species are now receiving active conservation management to reduce the impact of these factors. Fortunately, these birds respond readily to a number of simple techniques. One of the first to be established was the use of "vulture restaurants" or feeding stations, in which carcasses can be provided regularly for the birds. In this respect vultures are again distinct from most other predatory birds. We can supplement their food supply easily, without needing to modify their normal feeding behaviour substantially. The advantages of feeding stations are numerous. First, they provide additional food, and a reliable source that birds can depend upon. This is important because of the considerable seasonal variation in the mortality of ungulates, with the consequence that the food supply for vultures also varies greatly in abundance. Periods of plenty when vultures rarely need to visit a feeding station could be followed by periods of extreme food shortage when the feeding station may be vitally important in maintaining the birds in the area. Coupled to this variation in food availability, there are also great variations in the food requirements of birds. When breeding, the daily food requirements of griffon vultures are up to four times greater than outside the breeding season (Houston 1976). The important feature of these sites is that the birds soon learn that they can rely on food being available there even if they may not always choose to take advantage of it, and it is known that there is seasonal variation in the use of feeding areas (Brown 1990). But from some restaurant sites birds can obtain a substantial proportion of their total food requirements (Brown and Jones 1989). The secondary advantages of restaurants are also important. The detailed radio-telemetry studies on condors in the Americas have shown that the foraging behaviour of vultures is influenced by the distribution of food sources. This is hardly surprising, but it does mean that if a reliable source of food is provided at one fixed site, birds will be restricted in their range of movements compared with individuals that are having to forage for widely dispersed natural food resources. By being encouraged to remain in protected sites, birds are less likely to wander into areas where strychnine baits are in use, or carcasses containing lead shot can be found, or where direct persecution occurs and birds may be shot or trapped. Vulture feeding stations started in South Africa in 1966 and in the Pyrenees in 1969 (Terrasse 1985), and a good account of their success is given in Mundy *et al.* (1992). They are now used to sustain vulture populations in Europe, Israel, southern Africa and California. Although there have been no detailed studies on the impact of these sites on vulture populations, it is likely that they have contributed substantially to vulture conservation in these areas. The European griffon populations in parts of the Pyrenees have been increasing by 8% per year, probably owing to a combination of reduced persecution and the widespread use of vulture restaurants (Donazar 1987).

A second management technique that has proved remarkably successful is the use of captive-bred birds to supplement or reintroduce natural populations. The California condor programme is the best-known example, where the whole survival of the species is now dependent on captive breeding. But prior to this work starting there were very successful reintroductions in Europe of griffon vultures into the Cévennes region of France (Terrasse 1983), and bearded vultures into the Alps (Frey and Walter 1989). The success of these projects rests on a number of features that are common to all vultures species. First, it appears that most species settle into captivity well, and given sufficient attention will breed readily. This is probably a consequence of the numerous adaptations they have developed to reduce energy demand. The need to reduce energy costs permeates every aspect of the activities of vultures. One of the results is that they show a relatively inactive behaviour pattern. Provided they are fed regularly, vultures will spend most of the day resting. Vultures show none of the nervous hyperactive behaviour patterns of some other raptor species in captivity that make them so difficult to establish and breed in aviaries. In captivity, by using double-clutching and artificial incubation, vultures can be made to produce far more young than would be achievable from the same birds in the wild. The release of young in this way can either reintroduce birds to an area from which they have become extinct, or help to support a natural population by offsetting the raised mortality rates and buying time while methods are investigated to remove the dangers in the environment.

When released, the birds do need to learn flying skills, which take a considerable time to develop in vultures. But they do not need to show complicated prey capture behaviour like most other predatory birds, and so a different form of "hacking back" is needed to establish vultures (Wallace and Temple 1983, 1985). Provided that carcasses can be made available to support the released birds, Sarrazin *et al.* (1994) have shown that the survival rate of introduced birds can be far higher than would be expected in a natural population. This is assisted by the fact that most vultures help each other in food searching, and so if birds are released as a group and at the same site each year they can learn together as they explore their new environment, and young birds can gain experience from older individuals. This reduces the probability of weak or immature birds dying before they have gained sufficient survival skills. Most other raptors are solitary, and at release sites need to be cared for individually. This is demanding on facilities, staff time and logistic support. Vulture releases are comparatively easy and best carried out with the birds in a group, and efforts should be made to retain a gregarious foraging behaviour. The large size of vultures also means that they have almost no natural predators, which is an obvious advantage when comparatively inexperienced captive-bred birds are released into the wild.

Finally, and importantly, vultures are spectacular birds, highly visible and ideal species around which to base public education programmes to highlight conservation problems. Everyone has heard of vultures, and their rather macabre popular reputation helps to attract attention. The experience in South Africa and Europe has been that the publicity given to vulture conservation projects has acted to raise the general public attitude to much wider wildlife conservation and environmental issues, and so has made a major contribution to the preservation of wildlife systems even in heavily industrialised countries.

ACKNOWLEDGMENTS

I am very grateful to Peter Mundy and Ed Henckel for their improvements to the manuscript.

REFERENCES

BENSON, P.C. AND J.C. DOBBS. 1984. Causes of Cape Vulture mortality at the Kransberg colony. Pp. 87–93 in J.M. Mendelsohn and C.W. Sapsford, eds. *Proc. 2nd symp. Afr. Predatory Birds*. Natal Bird Club, Durban, South Africa.

BROOKE, R.K. 1984. *South African Red Data Book: Birds*. South African National Scientific Programmes Report No 97. CSIR: Pretoria, South Africa.

BROWN, C.J. 1990. An evaluation of supplementary feeding for Bearded Vultures and other avian scavengers in the Natal Drakensberg. *Lammergeyer* **41**: 30–36.

— AND S.J.A. JONES. 1989. A supplementary feeding scheme in the conservation of the Cape Vulture at the Waterberg, South West Africa/Namibia. *Madoqua* **16**: 111–129.

DONAZAR, J.A. 1987. Apparent increase in a Griffon Vulture *Gyps fulvus* population in Spain. *J. Raptor Res.* **21**: 75–77.

FREY, H. AND W. WALTER. 1989. The reintroduction of the Bearded Vulture to the Alps. Pp. 341–344 in B.-U. Meyburgh and R.D. Chancellor, eds. *Raptors in the modern world*. World Working Group on Birds of Prey, Berlin & London, UK.

GRANT, M.C. 1985. *Black vultures in Evros: feeding and food supply*. Report to Edinburgh University, Scotland, UK.

HIRALDO, F., M. DELIBES AND J. CALDERON. 1979. *El Quebrantahuesos* Gypaetus barbatus *Monograph 22*. Agriculture Ministry & National Institute for Conservation, Madrid, Spain.

HOUSTON, D.C. 1976. Breeding of the White-backed and Rüppell's Griffon Vultures, *Gyps africanus* and *Gyps rueppellii*. *Ibis* **118**: 14–40.

JACKSON, F.J. 1938. *The birds of Kenya Colony and the Uganda Protectorate*. Gurney & Jackson, London, UK.

KIFF, L.F., D.B. PEAKALL AND S.R. WILBUR. 1979. Recent changes in California Condor eggshells. *Condor* **81**: 166–172.

KOMEN, J. 1986. Energy requirements and food resource of the Cape Vulture in the

Magaliesberg, Transvaal. M.Sc. thesis, Univ. of the Witswatersrand, Witswatersrand, South Africa.

LEDGER, J.A. AND H.J. ANNEGARN. 1981. Electrocution hazards to the Cape Vulture *Gyps coprotheres* in South Africa. *Biol. Conserv.* **20**: 15–24.

LESHEM, Y. 1985. Vultures under high tension. *Israel: land and nature* **10**: 149–153.

MERETSKY, V.Y. AND D. LAVEE, 1989. *Conservation and management of the Negev Lappet-faced Vulture in Israel.* Report to Israel Nature Reserves Authority, Tel Aviv, Israel.

MUNDY, P.J., K.I. GRANT, J. TANNOCK AND C.L. WESSELS. 1982. Pesticide residues and eggshell thickness of Griffon Vulture eggs in southern Africa *J. Wildl. Manage.* **46**: 769–773.

—, D. BUCHART, J. LEDGER AND S. PIPER. 1992. *The Vultures of Africa.* Acorn Books and Russel Friedman Books, Randburg, South Africa.

PIPER, S.E. 1993. Mathematical Demography of the Cape Vulture. Ph.D. thesis, Univ. of Cape Town, Cape Town, South Africa.

RICHARDSON, P.R.K. 1980. The natural removal of ungulate carcasses and the adaptive features of the scavengers involved. M.Sc. thesis, Univ. of Pretoria, Pretoria, South Africa.

ROBERTSON, A.S. 1983. The feeding ecology and breeding biology of a Cape Vulture colony in the south-western Cape Province. M.Sc. thesis, Univ of the Witwatersrand, Witwatersrand, South Africa.

SARRAZIN, F., C. BAGNOLINI, J. L. PINNA, E. DANCHIN AND J. CLOBERT. 1994. High survival estimates of Griffon Vultures *Gyps fulvus fulvus* in a reintroduced population. *Auk* **111**: 853–862.

SIBLEY C.G. AND J.E. AHLQUIST. 1990. *Phylogeny and classification of birds.* Yale Univ. Press, New Haven and London, UK.

SNYDER, N.F.R. AND H.A. SNYDER. 1989. Biology and conservation of the California Condor. *Current Ornithology* **6**: 175–267.

TERRASSE, M. 1983. The status of vultures in France. Pp. 81–85 in S.A. Wilbur and J.A. Jackson, eds. *Vulture Biology and Management.* Univ. of California Press, Berkeley, California.

— 1985. The effects of artificial feeding on Griffon, Bearded and Egyptian Vultures in the Pyrenees. Pp. 429–430 in I. Newton and R.D. Chancellor, eds. *Conservation Studies on Raptors.* International Council for Bird Preservation, Cambridge, UK.

WALLACE, M.P. AND S.A. TEMPLE. 1983. An evaluation of techniques for releasing hand-reared vultures to the wild. Pp. 400–426 in S.A. Wilbur and J.A. Jackson, eds. *Vulture Biology and Management.* Univ. of California Press, Berkeley, California.

— AND — 1985. A comparison between raptor and vulture hacking techniques. In D.K. Garcelon and G.W. Roemer, eds. *Proceedings of the International Symposium on Raptor Reintroduction.* Institute for Wildlife Studies, Arcata, California.

— AND — 1987. Releasing captive-reared Andean Condors in the wild. *J. Wildl. Manage.* **51**; 541–550.

WILBUR, S.R. 1978. Turkey Vulture eggshell thinning in California, Florida and Texas. *Wilson Bull.* **90**: 642–643.

XIROUCHAKIS, S. 1993. Habitat Utilisation by predatory birds at Evros, Greece. M.Sc. thesis, Univ. of Reading, Reading, UK.

32

The Impact of Man on Raptors in Zimbabwe

Ron R. Hartley, Kit Hustler and Peter J. Mundy

Abstract – Habitat destruction has had a negative effect on 31 of 73 species of raptors in Zimbabwe, but land clearance, dam construction and artificial nest sites have assisted in 18 other species. DDT spraying ceased in 1990, but indirect poisoning of vultures has continued. The human population growth rate has been >3.0% per year, and 60% of communal lands (42% of Zimbabwe) are considered to be overpopulated. While most ranch lands in commercial farming areas (31%) provided suitable habitats, intensive cropping areas have been sprayed with agrochemicals, and cultivation has been increasing. Protected areas (15%) have provided secure habitats for most species. Ayres' and martial eagles, Cape griffon, and peregrine and taita falcons need careful monitoring.

Key words: exotic trees; habitat; organochlorines; poisoning; utilization; conservation; Zimbabwe.

References to the impact of man on raptors in Zimbabwe have focused mainly on five topics:

(1) pesticides (Whitwell et al. 1974, Tannock et al. 1983; Thomson 1984a, Douthwaite 1992, Hartley and Douthwaite in press);
(2) poisoning (Mundy et al. 1992);
(3) incidental reports on the use of artificial nest sites (Harwin 1972, Brooke 1974, de la Harpe 1990);
(4) collisions (Mundy 1988, 1990, 1993a,b); and
(5) electrocutions (Hartley 1991a, Mundy 1993a).

This study reviews the impact of man on raptors in Zimbabwe against a background of >3.0% population growth per annum (Central Statistical Office 1984), changing land-use patterns (Zinyama and Whitlow 1986, Sill 1992) and a national wildlife policy based on utilization (Parks and Wild Life Act 1975; Department of National Parks and Wild Life Management 1991). It includes information on the use of exotic trees and man-made structures for nesting, the increase and loss of habitat, pesticide and other chemical poisoning, direct persecution, drowning, electrocution, collisions with a variety of man-made objects, and human utilization.

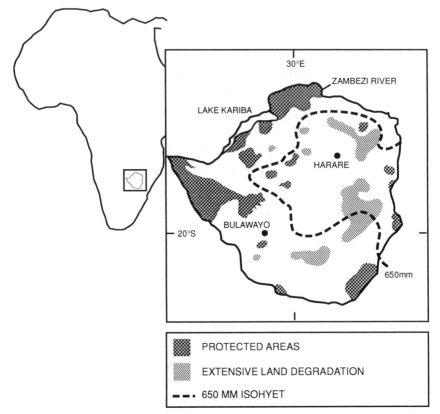

Figure 1. Location of Zimbabwe, showing protected areas and extensive land degradation (after Whitlow 1988).

METHODS

Information was obtained from published literature and unpublished records of the Zimbabwe Falconers' Club (Hartley 1993a), Natural History Museum of Zimbabwe (Cotterill and Hustler 1993), and Ornithological Unit of the Department of National Parks and Wild Life Management. Breeding records are from Zimbabwe Falconers' Club and Ornithological Association of Zimbabwe nest record cards (each card = one breeding attempt), the latter maintained at the Natural History Museum of Zimbabwe. Nomenclature mostly follows Maclean (1993) and all scientific names of birds are given in the Appendix of this book.

STUDY AREA

Zimbabwe covers an area of 390 460 km^2 (Zinyama and Whitlow 1986) within the Afro-tropical region (Fig. 1). The physical background has been described by Irwin (1981). Mean national annual rainfall is about 647 mm with a range of 335–1193 mm, varying from 157–2640 mm in different parts of the country in different years (Department of Meteorological Services *in lit.*). Although the principal vegetation type is *Brachystegia* (Miombo) woodland, 24 sub-types have been described (Wild 1965 in Child and Heath 1992).

The quality and abundance of natural habitat is related to three land-use categories. General Land (32%) is privately owned and has a density of 7.6 persons/km^2, with ranch lands still intact and relatively undamaged (Whitlow 1988). Some areas have had densities of breeding raptors that have compared favourably with protected areas (National Parks Estate = 12.7%; Forestry Areas = 2.3%). Habitat quality has declined rapidly in Communal Lands (42%) which average 25.5 persons/km^2 (Zinyama and Whitlow 1986) and 60% of the areas are either overpopulated or grossly overpopulated (Whitlow 1988).

There are 61 species of diurnal and 12 species of nocturnal raptors (Irwin 1981; Table 1) in a country where rural areas contain 60% of the population (mainly peasant farmers occupying 42% of the country; Sill 1992). The granite dwala and kopje terrain of the Matopos Hills supports the greatest diversity and abundance of raptors recorded in Africa (Gargett 1990). This habitat type extends over 20% of the country, but it has suffered the greatest human population pressure (Whitlow 1988). A similar diversity and abundance of raptors has been recorded in the Batoka Gorges (Hartley 1993b), in protected areas of the Zambezi Valley (RRH unpubl. data), on Debshan Ranch (P. J. Mundy and J. Dale unpubl. data) and for eagles and vultures in Hwange National Park (Howells and Hustler 1984).

RESULTS AND DISCUSSION

Use of Exotic Trees

Nesting in eucalyptus and pine trees has been recorded for 19 and four species, respectively. Grown mainly on General Land in wetter areas (>650 mm annual rainfall; Fig. 1) as stands of 1–100 ha in size, eucalyptus trees have been adopted most extensively by accipiters (Table 2) and cuckoo hawks. While gabars and little banded goshawks, Ovambos and little sparrowhawks have been found nesting close together in stands of smaller eucalyptus, the larger black sparrowhawk has usually replaced these species in the stands of larger trees (RRH unpubl. data). Red-breasted sparrowhawks nest only in exotic pines in the Eastern Highlands (also see Steyn 1982), where they are sometimes replaced by the African goshawk (RRH unpubl. data) which is much larger. As in South Africa (Tarboton and Allan 1984), it is likely that all of these species have extended their range as a result of exotic tree plantations.

The bat hawk, Wahlberg's, long-crested, martial, and crowned eagles and the lanner falcon also nest in eucalyptus, while pines have been used by black-breasted snake eagle. Pearl-spotted owls have used nest holes in jacaranda and

Table 1. A summary of human impacts on raptors in Zimbabwe. b = breeding; m = migrant; i = intra-African; r = resident; p = palearctic; v = vagrant; T = tree; C = cliff; G = ground; F = falconry; EM = ethno-medicine; × = recorded; ? = likely, but data lacking.

Species	Status	Natural nest site	Positive		Negative						
			Nest on exotic trees	Nest on man-made structures	Increased habitat	Habitat destruction	Pesticide poisoning	Direct persecution	Drowning	Electrocution	Collisions	Utilization
Secretary bird	b,r	T			×			×		×	×	EM
Hooded vulture[a]	b,r	T				×						EM
Cape griffon[b]	r	C					×				×	EM
White-backed vulture[a]	b,r	T			×		×					EM
Lappet-faced vulture[a]	b,r	T			×	×	×					EM
White-headed vulture[a]	b,r	T				×	×					
Yellow-billed kite	b,i,m	T			×					×	×	
Black kite	p,m				×						×	
Black-shouldered kite	b,r	T	×		×						×	
Cuckoo hawk	b,r	T	×									
Bat hawk	b,r	T				×					×	
Black eagle[a]	b,r	C,T				×		×		×	×	
Tawny eagle	b,r	T				×	×	×		×	×	
Steppe eagle	p,m					×				×	×	
Lesser spotted eagle	p,m					×				×	×	
Wahlberg's eagle	b,i,m	T	×			×		×		×	×	
African hawk eagle[a]	b,r	T	×			×		×		×	×	F
Ayres' hawk eagle[a]	b,r	T				×		×		×	×	F
Long-crested eagle	b,r	T	×		×			×		×		
Martial eagle[b]	b,r	T	×			×		×	×	×		
Crowned eagle[a]	b,r	T	×			×		×		×		F
Brown snake eagle	b,r	T	×		×	×						
Black-breasted snake eagle	b,r	T				×				×	×	
Southern banded snake eagle	r	T				?						
Bateleur	b,r	T				×						EM?
African fish eagle	b,r	T	×		×		×	×		×	×	
Common buzzard	p,m				×		×					
Augur buzzard	b,r	C,T									×	
Lizard buzzard	b,r	T	×							×	×	
Red-breasted sparrowhawk	b,r	T	×		?							F

Species	Status	Habitat							
Ovambo sparrowhawk	b,r	T	×						F
Little sparrowhawk	b,r	T	×						F
Black sparrowhawk[a]	b,r	T	×					×	F
Little banded goshawk	b,r	T	×			×		×	F
African goshawk	b,r	T	×			×		×	F
Gabar goshawk	b,r	T				×		×	F
Dark chanting goshawk	b,r			×		×		×	
European marsh harrier	p,m								
African marsh harrier	b,r	G							
Montagu's harrier	p,m								
Pallid harrier	p,m								
Gymnogene	b,r	T	×		×				
Osprey	p,m				×				
Siberian peregrine falcon[a]	p,m				?	×	×	×	F
African peregrine falcon[a]	b,r	C	×	×	?	×	×	×	F
Lanner falcon[a]	b,r	C,T	×	×	?	×	×	×	F
Hobby falcon	p,m				?		×		
African hobby falcon	b,r	C			?		×		
Taita falcon[a]	b,r	C					×		
Red-necked falcon	b,r	T			×		×		
Eastern red-footed kestrel	p,m								
Rock kestrel	b,r	C		×					
Greater kestrel	b,r	T	×	×					
Lesser kestrel	p,m								
Dickinson's kestrel	b,r	T	×	×	×		×		F
Barn owl	b,r	C,T	×	×	×		×		
Grass owl	b,r	G			×				
Wood owl	b,r	T		×					
Marsh owl	b,r	G		×			×		
Scops owl	b,r	T					×		
White-faced owl	b,r	T			×		×		
Pearl-spotted owl	b,r	T	×				×		
Barred owl	b,r	T					×		
Cape eagle owl	b,r	C/G					×	×	
Spotted eagle owl	b,r	G	×	×				×	
Giant eagle owl	b,r	T			×		×	×	
Pel's fishing owl	b,r	T					×		

No impacts are recorded for: Egyptian vulture[b] (v); Honey buzzard (p,m); Booted eagle (i/p,m); Western banded snake eagle (b,r); Palmnut vulture[b] (v); Pale chanting goshawk (b,r); Sooty falcon (v); Western red-footed kestrel (p,m).

Order of species and nomenclature almost entirely follow Maclean (1993).

[a] Restricted (Department of Natural Parks and Wild Life Management 1991).

[b] Specially protected (Department of National Parks and Wildlife Management 1991).

Table 2. Raptors that have made the greatest use of exotic trees for nesting.

Species	N	% Exotic Eucalyptus	Pine	Jacaranda	% Indigenous
Red-breasted sparrowhawk	22		100		
Black sparrowhawk	288	60.4			39.6
Cuckoo hawk	17	52.9			47.0
Ovambo sparrowhawk	119	42.8	0.8		56.3
Little sparrowhawk	62	22.6	1.6	11.3	64.5
Greater kestrel	48	29.2			70.8
African goshawk	100	8.0	13.0		79.0
Little banded goshawk	139	20.9			79.1
African fish eagle	117	12.8			87.2
Gabar goshawk	152	7.9			92.1
African hawk eagle	328	6.4			93.6
Lizard buzzard	113	6.2			93.8
Gymnogene	99	6.1			93.9
Black-shouldered kite	107	3.7			96.3

syringa trees. Flocks of eastern red-footed kestrels and individual Ayres' eagles (Lendrum 1982) have used eucalyptus as roosts.

Use of Man-made Structures

Of 134 separate African peregrine falcon breeding sites, four have been in quarries, one on a dam wall (also recorded by Aspinwall 1975) and another on a high-rise building (de la Harpe 1990). A pair also occupied a cooling tower at a thermal power station, but breeding was not confirmed. Of 226 separate lanner falcon breeding sites, 20 have been in quarries (one of which was subjected to regular blasting; see White et al. 1988 and Holthuijzen et al. 1990), one on a dam wall, two on buildings (Dalling 1975) and one on a power pylon. It is likely that more use has been made of power pylons than suggested by these data, as there have been few investigations in Zimbabwe. Lanner falcons use power pylons often in South Africa (Tarboton and Allan 1984). A pair of rock kestrels bred regularly on a large building at a mine. Two pairs of greater kestrels have used old pied crow nests on power pylons, while another pair bred on a tower in Harare (Borrett 1970). A pair of Dickinson's kestrels used holes within the boxed steel members of a bridge (Brooke 1974). A gymnogene bred on a low tower at an abandoned mine. Barn owls have commonly used buildings, mine shafts, quarries, and mine dumps (Steyn 1982; RRH pers. obs.).

Although there were no systematic investigations, these incidental records indicate that less use has been made of man-made structures by raptors in Zimbabwe than in South Africa (Tarboton and Allan 1984, Allan 1987, Mundy *et al.* 1992), probably because in Zimbabwe there are still adequate natural nesting sites.

Increasing Habitat

Where woodlands are converted to cultivation and grazing (in General Land especially), species like the secretary bird, vultures, black-shouldered, yellow-billed and black kites, black-breasted snake eagle, Montagu's and pallid harriers, lanner falcon and lesser kestrel may have benefited.

Lanner falcons appear to be more common in deforested Communal Lands than in neighbouring protected areas. Twice as many breeding lanners were observed in the Communal Lands of the Matobo Hills than in adjacent Matobo National Park (N. Greaves and RRH unpubl. data). Parts of cleared intensive farming regions appear to have suited the black-shouldered kite, which in the Transvaal, has increased its range (Tarboton and Allan 1984). Increased availability of rodents around cultivated areas may have benefited the barn owl which has probably extended its range since the advent of modern settlement.

Over 8000 dams have been built (Sayce 1990) ranging from Kariba (one of the largest in the world) to small masonry weirs. Over 100 of these are large dams which have provided much additional habitat for fish eagles in particular, but also for osprey, grass and marsh owls, and the harriers. Along the upper and middle Zambezi River, respectively, fish eagles occurred as one pair per 3 km (Bell-Cross and Vernon 1984) and one pair per 2.8 km (Francis *et al.* 1992). Along the whole Lake Kariba, treated as a straight-line, there was one pair per 2 km (R. J. Douthwaite pers. comm.), but the density would be much less than this if one took into account all the shoreline indentations (Francis *et al.* 1992). This equated to one pair per 9 km for Kariba ($N = 158$ pairs; Douthwaite 1992), compared with one pair per 13.5 km for Lakes Darwendale and McIlwaine (P. J. Mundy unpubl. data). Conversely, the proposed Batoka Gorge project will affect the rare taita falcon (Hartley 1993a, b).

The bat hawk (Cooper 1976, Hartley 1988), Ovambo, little and black sparrowhawks, little banded and African goshawks, gymnogene, African peregrine and lanner falcons, rock and greater kestrels, and barn and spotted eagle owls have nested in urban areas. Ayres' hawk eagles have been attracted to towns for food, mainly outside of their breeding season (Lendrum 1982, Hartley 1982a, RRH and P. J. Mundy unpubl. data).

Intensive cereal cultivation on the edge of Harare, and cattle feed-lots on the edges of Bulawayo and Triangle, where prey is abundant, have attracted Ovambo and black sparrowhawks, little banded and Gabar goshawks, tawny and Wahlberg's eagles, African and Ayres' hawk eagles, and peregrine and lanner falcons. Croplands have also attracted black-shouldered kite, barn owl and African marsh harrier.

Telephone and electricity poles (and lines), and road posts are used extensively as hunting perches by black-shouldered kites, long-crested and black-breasted snake eagles, common, augur and lizard buzzards, little banded and dark chanting goshawks, lanner and hobby falcons, eastern red-footed, rock, greater, lesser and Dickinson's kestrels and spotted eagle owls.

Habitat Destruction

Peregrines have vacated sites in heavily degraded Communal Lands (Thomson 1984b, RRH unpubl. data). Gargett (1990) described the reduced density of breeding black eagles in the Communal Lands of the Matobo Hills, compared with the adjacent Matobo National Park. Douthwaite (1992) described smaller inter-nest distances for fish eagles in protected areas compared with Communal Lands along Lake Kariba. Outside of the National Parks Estate, the martial eagle is probably under the greatest threat from habitat loss, as it lives at low densities throughout its range (235–990 km^2/pair, see Howells and Hustler 1984). However, the average territory size for martials in Hwange National Park is thought to be 133 km^2 (Hustler and Howells 1987).

The Communal Lands encompass 68% of the country's rugged granite terrain, most of it under severe human population pressure (Fig. 1; Whitlow 1988). Thus, it is most likely that a considerable area of Matopos-type terrain has failed to support previously high populations of raptors. For example, bateleur eagles have declined in these areas, especially around Harare, where they bred until the late 1970s. Bateleurs seen in these areas now are mainly juvenile individuals with the breeding adults confined mainly to protected areas. The cause of the decline is not clear, but a combination of poisoning and habitat destruction has probably been responsible.

Protected areas and most General Lands still provide suitable habitats. However, current pressures for additional land for resettlement have targeted >50% of General Lands for compulsory acquisition (Sill 1992). So far the resettlement program has resulted in accelerated land degradation (Whitlow 1988). However, in some Communal Lands, the Communal Areas Management Programme for Indigenous Resources (CAMPFIRE) has the potential to arrest this process, because it is based on wildlife utilization and benefit (Murombedzi 1992). So far 26 District Councils have "Appropriate Authority" status which permits them to manage their areas for wildlife utilization. About 19 200 km^2 of habitat (4.9%; Department of National Parks and Wild Life Management 1989) have been secured in this way. More than any other single factor, the success of land redistribution and the associated systems of land husbandry will probably affect raptor populations.

DDT and Other Chlorinated Hydrocarbons

Organochlorines such as DDT have been used extensively in Zimbabwe (Thomson 1984a, Douthwaite 1992) since registration for agriculture use in 1946, extending to tsetse fly (*Glossina* spp.) control in 1967 and mosquito (*Anopheles* spp.) control in 1972. About 40 000 km^2 (10.2%) of woodland have been sprayed with DDT alone for tsetse fly control (Hartley and Douthwaite in press). Levels of chemicals and methods of application have been described by Douthwaite (1992). Banned from domestic use in 1973 and agricultural use in 1985, DDT can now be used only for research purposes connected with tsetse fly and mosquito control. More than 50% of all DDT used has been applied to catchments draining into Lake Kariba (Douthwaite 1992).

Public concern over possible effects on wildlife led to detailed investigations in 1982–83 (Matthiessen 1985a) and in 1987–90 (Hartley 1991b, Douthwaite 1992). As a result, DDT use for tsetse fly control will cease at the end of 1995 (Douthwaite 1992). There has been no ground spraying since 1990. A synthetic pyrethroid, deltamethrin, has been substituted successfully for DDT. The effects on non-target organisms should be less persistent and there is less risk to vertebrates (Lambert et al. 1991). However, the much higher costs for this chemical may result in the resumption of DDT use (Douthwaite 1993).

By the 1970s DDT contamination was widespread in Zimbabwe (Whitwell *et al.* 1974) and levels exceeding 8 ppm (wet weight) were reported in the eggs of nine of 14 species of raptors; fish and bird-eating species were at greatest risk [Thomson 1981; Ratcliffe Index (Ratcliffe 1967) presented in Table 3]. In the Zambezi Valley levels of contamination in fish eagles (Douthwaite 1992), African goshawks (Hartley and Douthwaite in press) and the African peregrine falcon have matched the frequency of spraying (Hartley in press a). Although some residue levels and eggshell thinning are synonymous with threshold levels at which equivalent species declined elsewhere, it appears that only African goshawks and fish eagles in heavily sprayed areas have experienced a significant decline in number of breeding pairs (Hartley and Douthwaite in press) and hatching success (Douthwaite 1992). Levels of contamination have declined at sites where DDT use ceased 5–10 yr ago (Douthwaite 1992). DDT residues dissipate more quickly from the physical environment in tropical than in temperate regions (Matthiessen 1985b). The prognosis is good for raptors in Zimbabwe. In about 10 yr (see Newton and Wyllie 1992) more complete recoveries are expected (Douthwaite 1993).

Although pesticides have been very much in use, very few raptor deaths have been attributed to them, despite their known killing power. For example queletox (active ingredient: fenthion) has been sprayed against the red-billed quelea at its roost sites (Mundy and Jarvis 1989). The quelea is a huge pest of small grain crops. A few giant eagle owls have been killed after feeding on dying queleas (P. J. Mundy unpubl. data). A captive black-shouldered kite nearly died after feeding on sprayed queleas (R. W. Querl pers. comm.). In spite of the potential for killing (see Thomsett 1987, Hunt et al. 1992, Yeld 1993), no diurnal raptors have been reported dying in this way, although it is likely that this has happened.

Dieldrin in a poisoned cape turtle dove killed a pair of captive peregrines (RRH and C. M. Foggin unpubl. data). However, other commonly used chemicals, such as azodrin and anti-coagulants, have not been recorded as causes of raptor mortality. In particular, barn owls have not been reported succumbing to rodents that have been affected by anti-coagulants. However, no

Table 3. Frequency of samples showing total DDT concentrations (ppm wet weight), Ratcliffe Index (Ratcliffe 1967) and percent thinning of eggs in Zimbabwe, 1970–90.

Species	N	<5.0[b]	5.0–9.9	10.0–19.9	>20.0	Mean	Range	N	X	Range	Thinning[c]	Reference[d]
		Total-DDT[a]						Ratcliffe Index				
Secretary bird	1	1				3.61						1
Hooded vulture	1	1				0.6						2
White-backed vulture	4	4				0.03	0.02–0.04	4	3.46	3.43–3.55	−0.6	3
Lappet-faced vulture	2	2				1.1	0.5–1.7					2
Yellow-billed kite	1	1				0.14						1
Black eagle	3	t										2
Crowned eagle	1	t										2
Wahlberg's eagle	5	3,n.d.	1			1.71	n.d–7.64					1,2,4
African hawk eagle	1	1				0.5						2
Black-breasted snake eagle	1	1				0.7						2
African fish eagle	53	19	19	12	3	10.0	2.0–69.42	19	2.43	2.07–2.57	−11.8	4,5
Ovambo sparrowhawk	3	1	2			5.54	4.09–6.44					4
Black sparrowhawk	10	1	4	1	4	17.92	4.50–63.1					2,4
African goshawk	9	3	1		5	23.83	0.2–65.3	9	1.34	1.20–1.46	−8.2	2,4,6
Dark chanting goshawk	2	1,n.d				0.22	n.d–0.44					2
Peregrine falcon	17	12	2	2	1	5.62	0.24–23.94	13	1.68	1.34–1.84	−10.2	1,4,7,8
Lanner falcon	4	1	3			7.76	3.95–9.66					1,4
Spotted eagle owl	2	1,t				0.1	t–0.2					2
Giant eagle owl	1	1				2.1						2
Pel's fishing owl	1				1	24.15						4
Total	122	69	32	15	14			45				

[a] All dry weights converted to wet weights by × 0.2 (after Peakall and Kiff 1979), and clutch mean was used when more than one egg.
[b] t = trace; n.d. = not detected.
[c] Baseline for: White-backed vulture = 3.48 (Mundy et al. 1982); African fish eagle = 2.76 (Douthwaite 1992); African goshawk = 1.46 (Hartley and Douthwaite in press); Peregrine falcon = 1.87 (Peakall and Kiff 1979).
[d] Data sources: 1 = Tannock et al. 1983, 2 = Whitwell et al. 1974, 3 = Mundy et al. 1982, 4 = Thomson 1981, 5 = Douthwaite 1992, 6 = Hartley and Douthwaite (in press), 7 = RRH (unpubl. data), 8 = Peakall and Kiff 1979.

"poison network" has been established to receive and deal with carcasses of poisoned birds.

Poisoning

The deliberate poisoning of raptors is probably slight. Species most at risk are scavengers, particularly vultures. Over the last 10 yr, at least 434 vultures of five species have been poisoned (Mundy et al. 1992). Almost all of these have been accidental, and the practice continues. On one occasion, however, five dead vultures were thought to have been poisoned deliberately by buffalo poachers. On rare occasions tawny, bateleur and fish eagles have been found poisoned alongside vultures. Unfortunately, little is known about the poisons used, because of the inability to test the carcasses. Usually organophosphate cattle dips have been suspected.

Chemicals such as strychnine, telodrex (telodrin) and toxaphene have been used against carnivores, e.g. black-backed jackal and spotted hyena. These species are perceived as "problem animals" for livestock. These chemicals are assumed to cause poisoning of scavenging birds that may have consumed the baits, and also secondary poisoning when the raptors are assumed to eat the dead carnivores. Indeed, the disappearance of the bateleur eagle (adept at finding small baits) from much of Zimbabwe is thought to have been caused by the careless poisoning of jackals in anti-rabies campaigns (W. R. Thomson in Foggin 1980).

Direct Persecution

This has taken the form of shooting, trapping, nest destruction, and illegal egg collecting. Shooting does not appear to have had a significant impact since it is restricted mainly to a small number of the country's 4000 commercial farmers. Targeted species have been martial and African hawk eagles and black sparrowhawks, usually when individual birds (mainly immatures) have resorted to hunting poultry. A few pigeon fanciers in towns have shot Ayres' eagles (Lendrum 1982, Hartley 1982a, RRH and P. J. Mundy unpubl. data), as well as lanner and peregrine falcons. Aviculturalists have sometimes shot African, little banded and Gabar goshawks. Sometimes small raptors have been shot with catapults (Hartley 1982b), which are common in the African population.

More recently, peasant farmers have used *dho-gazzas* to trap mainly lanner falcons in three widely separated areas (RRH unpubl. data). Lanners occur commonly in most Communal Lands, where they eat young chickens especially (Steyn 1982, Thomson 1984b, Zimbabwe Falconers' Club records, RRH pers. obs.).

Nest destruction has reduced the productivity of certain localized raptor populations, including fish eagles (Douthwaite 1992) and black sparrowhawks

(RRH unpubl. data). Stones have been tossed onto nests, some nests have been pulled down, and chicks have been killed and eaten. Most tree-nesting raptor species have been affected, while occasionally cliff-dwelling lanner (RRH unpubl. data) and peregrine falcons (RRH unpubl. data) have suffered when their nests have been accessible. A successful conservation technique has been to reward peasant farmers with beer or cash when they allow the completion of a successful nesting cycle by raptors (Gau 1990, RRH and J. Hough unpubl. data). Sometimes nest trees (eucalyptus and pines especially) have been felled for commercial purposes. Occasionally individuals such as falconers have managed to intervene, delaying the felling until the completion of a nesting cycle. A more formal program of extension work is needed.

Drowning

On one ranch with a breeding pair of martial eagles, two immatures drowned in reservoirs within 14 mo (W. King pers. comm.). A juvenile peregrine was rescued from certain drowning in another reservoir (W. Gau pers. comm.). These reservoirs were steep-walled concrete tanks. Accidental drowning is probably more common than believed. There is a need to promote a simple design modification for reservoirs, which could then be promoted through the national farmer's magazine, local conservation journals and newsletters.

Electrocution

Many raptor species have at some time been electrocuted in Zimbabwe (Table 1). Recent records include black (Mundy 1993a), martial and African fish eagles. Four trained peregrines, a crowned and an African hawk eagle were electrocuted over the past 17 yr (RRH unpubl. data). Systematic work in South Africa has revealed the extent of electrocutions in large raptors such as vultures (Mundy et al. 1992), plus the relevant conservation remedies. This has not been done in Zimbabwe and it should become a priority.

Collisions

Collisions have occurred with vehicles, airplanes, fences, power and telephone lines, a radar screen, buildings, gin traps and minefields. The most frequent casualties have been the ubiquitous spotted eagle owl, killed while feeding on insects on roads at night. *Milvus* kites and white-backed vultures have been struck often when scavenging carrion from roadways (Mundy 1993b, G. Enslin pers. comm.). Air strikes have included yellow-billed kite, spotted eagle owl, and secretary bird (Mundy 1988).

Two taita falcons struck telephone lines (Hartley 1982b, RRH unpubl. data);

four African peregrines (Hartley 1982b, RRH unpubl. data) and a bat hawk (Mundy 1990) struck buildings in Harare alone. Six African peregrines (including three trained hawks) struck fences, all but one (which broke its neck) breaking wings (RRH unpubl. data). A Siberian peregrine was caught in the protective screening of a radar mast (Hartley 1982b), while two black eagles were caught in gin (jaw) traps set for carnivores (Gargett 1990, Hartley 1994). White-backed vultures (Mundy et al. 1992) and an augur buzzard (Hartley 1982b) have been killed in minefields.

Utilization

Ethno-medicine and falconry (Hartley 1993a) are principal utilizations, while the Natural History Museum of Zimbabwe has collected specimens under special circumstances. Zimbabwe has many traditional healers called n'angas with a thriving national association for them (ZINATHA). They use animal and plant products in their cures. Among raptors, vultures are well used as it is believed that their brains are clairvoyant. The brains of a freshly killed vulture are eaten in a "porridge" in order to foretell some aspect of the future, or divine something that is hidden from sight. Africans are also very superstitious about the bateleur eagle and owls, but it is not known if these birds and their parts are used in medicine. One strong belief is that if the nesting tree of a bateleur is climbed the adults will attack with such vigour that the climber will be dislodged and killed. Owls have been cursed and feared because they are believed to be witches' birds, but they are neither persecuted nor utilized as a result. Sadly, the usefulness of owls as rodent killers has not been recognized.

About 30% of the 50 active falconers per annum use peregrines, nearly all of them captive-bred (Hartley 1993a). The remainder use lanner falcons, Ovambo and black sparrowhawks, African hawk eagles, and African and Gabar goshawks which have been wild-taken. There has been no commercialization of raptors in Zimbabwe; trained hawks that are no longer needed for hunting or captive breeding have been released. Surplus captive-bred peregrines have been hacked back to the wild ($N = 39$; Hartley in press b), and exported to start other similar captive breeding programs in Botswana (1 pr) and South Africa (6 pr).

Lanners have been used temporarily because falconers are required first to fly these successfully in order to get a permit to use the peregrine (Hartley 1993a). Nearly all of the lanners have been trapped as immatures or adults and they have been released at the end of the season. Several used for falconry have later been observed at nest sites (Zimbabwe Falconers' Club records).

Egg collections were made legal in 1987 (Mundy et al. 1992), although each needs a permit if the eggshells of specially protected species are possessed. The impact of legal egg collectors ($N = 15$) is not considered significant, and has been far outweighed by the scientific value of these collections. Occasionally Natural History Museum of Zimbabwe personnel have collected specimens, still considered a necessary component in biological research (Cotterill and Hustler 1993). Although illegal egg collecting has occurred, especially in the

Matopos (Gargett 1990), it has not been possible to gauge the extent of this problem.

Conservation Priorities

Of 46 positive impacts (Table 1), exotic trees provided 20 and habitat increase 19. The most significant negative factor was habitat destruction, which has affected 31 species (42.5%). Species scoring five negatives or more were peregrine and lanner falcons, African hawk, Ayres' and tawny eagles. Each scoring four negatives were Wahlberg's, martial, crowned and bateleur eagles, and black sparrowhawks and African goshawks. The relative importance of negative factors on each species has varied, and has been balanced against positive impacts and abundance and distribution of the species.

For example, peregrine falcons (350–400 pairs estimated for Zimbabwe) occur throughout the country and generally in rugged, relatively inaccessible habitats. Some key zones are in protected areas, especially in the Zambezi Valley. Principal threats have been pesticides and habitat destruction, but pesticides seem not to have had a significant effect on productivity (RRH unpubl. data). Increasing deforestation has probably caused a decline in the population (Thomson 1984b, RRH pers. obs.), and a recent wave of uncontrolled settlement in parts of the Zambezi Valley may affect some populations. In contrast, lanner falcons (1400–1500 pairs estimated) can survive in open and even degraded habitats and are probably extending their range at the expense of peregrines. The rare taita falcon (50–70 pairs estimated) can ill afford any negative impacts, especially as this species most likely competes with the peregrine falcon for nest sites and prey (Hartley et al. 1993).

Black (Gargett 1990), tawny, African hawk (Howells and Hustler 1984) and Wahlberg's eagles occur in relatively high densities in protected and some ranching areas. The much scarcer Ayres' eagle (unaccountably rare throughout its African range; Brown 1974) appears to be holding its own in well-wooded areas, especially in zones too rugged for settlement. In towns however, the impact of persecution needs assessment. Despite their demise in highveld areas, bateleur eagles are still encountered commonly in lowveld regions, where there are extensive protected areas (including private game farms). Crowned eagles are found throughout the country, especially in riparian systems where nesting densities can be remarkably high (RRH unpubl. data). The black sparrowhawk and African goshawk (Hartley and Douthwaite in press) are common, and sometimes occur at relatively high densities, even in urban areas.

Most of these have been specially protected (Parks and Wild Life Act 1975), and 16 species will continue to benefit from legal protection in the revised recommendations (Department of National Parks and Wild Life Management 1991) in which eight species will be delisted. Only the martial eagle and the Cape griffon have been placed in the highest category of specially protected. Cape griffons are vulnerable in southern Africa (Brooke 1984), and the population in Zimbabwe is small and localized. Martial eagles are a k-selected species that require extraordinarily large territories and they are particularly sensitive to habitat destruction.

ACKNOWLEDGMENTS

Attendance at the symposium by R. R. Hartley was made possible by The Peregrine Fund Inc., the Endangered Wildlife Trust of southern Africa, the Raptor Research Foundation, the Zimbabwe Falconers' Club Raptor Conservation Fund (administered by the Zimbabwe National Conservation Trust), the Zimbabwe Association of Tour and Safari Operators, and the Biodiversity Foundation for Africa, whose assistance is gratefully acknowledged. Thanks are also due to R. Watson, D. Varland, J. Ledger and A. Sparrow. Falcon College kindly gave R. R. Hartley leave of absence to attend the symposium. T. J. Cade, C. M. White and D. Varland commented on the manuscript.

REFERENCES

ALLAN, C. 1987. Raptors nesting on transmission pylons. *African Wildlife* **42**: 325–326.
ASPINWALL, D.R. 1975. The middle Zambezi and lower Luangwa Valleys. *Honeyguide* **83**: 19–21.
BELL-CROSS, G. AND C.J. VERNON. 1984. Relationship between fish eagle breeding and environmental factors – a general observation. Pp. 95–96 in J.M. Mendelsohn and C.W. Sapsford, eds., *Proc. 2nd Symp. Africa Predatory Birds*. Natal Bird Club, Durban, South Africa.
BORRETT, R. 1970. Birds in the centre of Salisbury city. *Honeyguide* **64**: 11–12.
BROOKE, R.K. 1974. Birds and bridges in Rhodesia. *Honeyguide* **80**: 42–45.
— 1984. *South African red data book—birds*. South African nat. scientific programmes rep. No. 97. Foundat. for Research Devel. Council for Scientific and Industrial Research, Pretoria, South Africa.
BROWN, L. H. 1974. Is poor breeding success a reason for the rarity of Ayres' hawk-eagle? *Ostrich* **45**: 145–146.
CENTRAL STATISTICAL OFFICE. 1984. *1982 population census: a preliminary assessment*. C.S.O., Harare, Zimbabwe.
CHILD, G. AND R.A. HEATH. 1992. Are Zimbabwe's major vegetation types adequately protected? *GeoJournal* **23**: 20–37.
COOPER, P.J. 1976. Bat Hawks in Gwelo. *Honeyguide* **87**: 29–30.
COTTERILL, W. AND K. HUSTLER. 1993. Documenting and conserving biodiversity in Central Africa. *Zimbabwe Wildlife* **70**: 20–22.
DALLING, J. 1975. Lanners in central Salisbury: the first four years. *Honeyguide* **84**: 23–26.
DE LA HARPE, D. 1990. The Harare peregrines – cause for celebrations? *Honeyguide* **36**: 163–164.
DEPARTMENT OF NATIONAL PARKS AND WILD LIFE MANAGEMENT. 1989. *The status of projects involving wildlife in rural development in Zimbabwe*. Dep. Rep., Harare, Zimbabwe.

— 1991. *Protected species of animals and plants in Zimbabwe.* Dep. Rep. Harare, Zimbabwe.
DOUTHWAITE, R.J. 1992. Effects of DDT on the fish eagle *Haliaeetus vocifer* population of Lake Kariba in Zimbabwe. *Ibis* **134**: 250–258.
— 1993. Effects of DDT applied for tsetse fly control on birds in Zimbabwe. Pp. 608–610 in R. Trevor Wilson, ed. *Proc. VIII Pan-Africa Ornith. Congr.* Bujumbura, Burundi (extended abstract).
FOGGIN, B.J.M. 1980. The bateleur survey. *Honeyguide* **101**: 5–9.
FRANCIS, J.T., P.J. MUNDY AND P.J. FEATHER. 1992. The status and distribution of fish eagles along the Middle Zambezi. *Honeyguide* **38**: 182–184.
GARGETT, V. 1990. *The black eagle.* Acorn Books and Russell Friedman Books, Randburg, South Africa.
GAU, W. 1990. A conservation technique for nesting martial eagles. *Honeyguide* **36**: 36–37.
HARTLEY, R.R. 1982a. Note on Ayres' eagle in the Mutare area. *Honeyguide* **110**: 23–27.
— 1982b. Notes from a falconer. *Honeyguide* **111/112**: 22–23.
— 1988. More on the Mutare bat hawks. *Honeyguide* **34**: 121–122.
— 1991a. Kori bustard and secretarybird electrocuted. *Honeyguide* **37**: 179.
— 1991b. The Zimbabwe Falconers Club — current research and conservation projects. *Endangered Wildlife* **8**: 9–15.
— 1993a. Falconry as an instrument of conservation. Pp. 105–110 in R. Trevor Wilson, ed. *Proc. VIII Pan-Africa Ornith. Congr.* Bujumbura, Burundi.
— 1993b. The Batoka Gorges, haven for birds of prey. *African Wildlife* **47**: 75–78.
—, G. BODINGTON, A.S. DUNKLEY AND A. GROENEWALD. 1993. Notes on the breeding biology, hunting behavior, and ecology of the Taita Falcon in Zimbabwe. *J. Raptor Res.* **27**: 133–142.
— 1994. Black eagle caught in gin trap. *Gabar* **8**: 36.
— In press a. DDT impact on raptors in the Zambezi Valley. In D. Solomon, ed., *Proc. of Our Endangered Environment, Birds of a feather-endangered wildlife trust*, Harare, Zimbabwe.
— In press b. Breeding and releasing the peregrine falcon in captivity—a Third World option for conservation. *Vision:* Endangered Wildlife Trust Annual 1995.
— AND R.J. DOUTHWAITE. In press. Effects of DDT treatments applied for tsetse fly control on the African goshawk in north west Zimbabwe. *Africa J. Ecol.*
HARWIN, R.M. 1972. Lanners in Central Salisbury. *Honeyguide* **72**: 14–15.
HOLTHUIJZEN, A.M.A., W.G. EASTLAND, A.R. ANSELL, M.N. KOCHERT, R.D. WILLIAMS AND L.S. YOUNG. 1990. Effects of blasting on behaviour and productivity of nesting prairie falcons. *Wildl. Soc. Bull.* **18**: 270–281.
HOWELLS, W.W. AND C.W. HUSTLER. 1984. The status and breeding success of eagles and vultures in the Hwange National Park, Zimbabwe. Pp. 99–107 in J.M. Mendelsohn and C.W. Sapsford, eds. *Proc. 2nd Symp. Africa Predatory Birds.* Natal Bird Club, Durban, South Africa.
HUNT, K.A., D.M. BIRD, P. MINEAU AND L. SHUTT. 1992. Selective predation of organophosphate-exposed prey by American kestrels. *Anim. Behav.* **43**: 971–976.
HUSTLER, K. AND W.W. HOWELLS. 1987. Breeding periodicity, productivity and conservation of the martial eagle. *Ostrich* **58**: 135–138.
IRWIN, M.P.S. 1981. *The birds of Zimbabwe.* Quest Publishing, Salisbury, Zimbabwe.
LAMBERT, M.R.K., I.F. GRANT, C.L. SMITH, C.C.D. TINGLE AND R.J. DOUTHWAITE. 1991. *Effects of deltamethrin ground-spraying on non-target wildlife.* Environmental impact of ground-spraying operations against tsetse fly in Zimbabwe (Tech. Rep. No 1). Nat. Resources Institute: Chatham.
LENDRUM, A.L. 1982. Ayres' hawk eagle in Bulawayo, Zimbabwe. *Honeyguide* **110**: 15–22.

MACLEAN, G.L. 1993. *Roberts' birds of southern Africa*. John Voelcker Bird Book Fund, Cape Town, South Africa.
MATTHIESSEN, P. 1985a. DDT insecticide residues in Zimbabwean wildlife and their potential environmental impact. *Zimbabwe Science News* 19: 3–8.
— 1985b. Contamination of wildlife with DDT insecticide residues in relation to tsetse fly control operations in Zimbabwe. *Envir. Pollut. B.* 10: 189–211.
MUNDY, P.J., K.I. GRANT, J. TANNOCK AND C.L. WESSELS. 1982. Pesticide residues and eggshell thickness of Griffon vulture eggs in southern Africa. *J. Wild. Manage.* 46: 769–773.
— 1988. Some recent bird strikes on aeroplanes in Zimbabwe. *Honeyguide* 34: 66.
— AND M.F.J. JARVIS, eds. 1989. *Africa's feathered locust*. Baobab Books, Harare, Zimbabwe.
— 1990. Bat hawks in the hand. *Honeyguide* 36: 89–90.
—, D. BUTCHART, J. LEDGER AND S. PIPER. 1992. *The vultures of Africa*. Acorn Books and Russel Friedman Books, Randburg, South Africa.
— 1993a. Large birds and powerlines — collisions and electrocution. *Honeyguide* 39: 27–30.
— 1993b. Mind the traffic dear vulture. *Honeyguide* 39: 139–146.
MUROMBEDZI, J. 1992. The communal areas management programme for indigenous resources (CAMPFIRE): a Zimbabwean initiative for natural resources conservation. *Zimbabwe Science News* 26: 77–81.
NEWTON, I. AND I. WYLLIE. 1992. Recovery of a sparrowhawk population in relation to declining pesticide contamination. *J. Appl. Ecol.* 29: 476–484.
PARKS AND WILD LIFE ACT. NO. 14. 1975. The Government Printer, Salisbury, Rhodesia.
PEAKALL, D.B. AND L.F. KIFF. 1979. Eggshell thinning and DDE residue levels among peregrine falcons *Falco peregrinus*: a global perspective. *Ibis* 121: 200–204.
RATCLIFFE, D.A. 1967. Decrease in eggshell weight in certain birds of prey. *Nature* 215: 208–210.
SAYCE, K. 1990. *Tabex encyclopedia of Zimbabwe*. Quest, Harare, Zimbabwe.
SILL, M. 1992. Cultivating a new system. *Geographical Magazine* 14: 45–50.
STEYN, P. 1982. *Birds of prey of southern Africa*. David Phillip, Cape Town, South Africa.
TANNOCK, J., W.W. HOWELLS AND R.J. PHELPS. 1983. Chlorinated hydrocarbon pesticide residues in eggs of some birds in Zimbabwe. *Envir. Pollut.* 5B: 147–155.
TARBOTON, W.R. AND D.G. ALLAN. 1984. *The status and conservation of birds of prey in the Transvaal*. Transvaal Museum, Pretoria, South Africa.
THOMSETT, S. 1987. Raptor deaths as a result of poisoning quelea in Kenya. *Gabar* 2: 33–38.
THOMSON, W.R. 1981. *A report on chemical contamination of the environment in Zimbabwe*. Mimeo. Dep. of Nat. Parks and Wild Life Manage., Harare, Zimbabwe.
— 1984a. DDT in Zimbabwe. Pp. 169–171 in J.M. Mendelsohn and C.W. Sapsford, eds. *Proc 2nd Symp. Africa Predatory Birds*. Natal Bird Club, Durban, South Africa.
— 1984b. Comparative notes on the ecology of peregrine, lanner, taita falcons in Zimbabwe. Pp. 15–18 in J.M. Mendelsohn and C.W. Sapsford, eds. *Proc. 2nd Symp. Africa Predatory Birds*. Natal Bird Club, Durban, South Africa.
WHITE, C.M., W.B. EMISON AND W.M. BREN. 1988. Atypical nesting habitat of the peregrine falcon (*Falco peregrinus*) in Victoria, Australia. *J. Raptor Res.* 22: 37–43.
WHITLOW, R. 1988. *Land degradation in Zimbabwe*. A geographical study. Dep. Nat. Res., Harare, Zimbabwe.
WHITWELL, A.C., R.J. PHELPS AND W.R. THOMSON. 1974. Further records of chlorinated hydrocarbon pesticide residues in Rhodesia. *Arnoldia* (Rhod.) 6: 1–8.
YELD, J. 1993. The quelea problem. *African Wildlife* 47: 112–115.
ZINYAMA, L. AND R. WHITLOW. 1986. Changing patterns of population distribution in Zimbabwe. *GeoJournal* 13: 365–384.

33

Response of Common Black Hawks and Crested Caracaras to Human Activities in Mexico

Ricardo Rodríguez-Estrella

Abstract – Two cases of raptors that have benefited from human activity in Mexico are presented here: the common black hawk and the crested caracara. In 1987–88, the largest river of Sonora, the Rio Yaqui, supported a higher number of common black hawk breeding pairs than its tributary, the Rio Bavispe. Apparently, this raptor has directly benefited from a human-altered environment as its preferred habitat, the riparian woodland, has increased over 300% in extension in the Rio Bavispe since the construction of a dam at its upper part in the mid-1930s. In 1990–91 the crested caracara was apparently associated with human activity at the Cape Region, Baja California Sur. The caracara nests were located in both natural vegetation and close to human activity, but more frequently in areas with low human disturbance. Nests were located near crop fields and small ranches. The productivity of the nests in areas with low and high human disturbance was not different. Young, juvenile and adult caracaras intensively foraged in henhouses, garbage sites and slaughterhouses. When waste disposal practices changed in the area, caracaras showed a rapid response to the changes in carrion availability, switching to the most available human refuse source. These examples show that some raptors are capable of adapting and benefiting from moderate human alterations in habitat that produce suitable nest sites, roosting places and/or increased food availability.

Key words: common black hawk; crested caracara; human activity; Mexico.

Like other plants and animals, raptors have been affected by human activity. The overall decline in the world raptor population has been attributed to a number of these activities including predator control by "gamekeepers" and ranchers, falconry, changes in cattle carcass management, hunting, electrocution, egg collectors, furtive commerce, use of pesticides, and especially, habitat loss (Newton 1979, Olendorff et al. 1981, Enderson et al. 1982, Schmutz 1984, Henny et al. 1985, Phillips 1986; Kochert et al. 1988, Holthuijzen et al. 1990). Nevertheless, some raptors have also benefited from human activities by increasing their distribution and abundance (Pruett-Jones et al. 1980, Friedman and Mundy 1983, Vannini 1989). In this paper I present the case studies of two raptors that have partially benefited from human activities in Mexico: the common black hawk and the crested caracara.

The common black hawk ranges from southern United States to western Peru, chiefly coastal in South America. This raptor is associated with streams, river bottoms and lagoons (Brown and Amadon 1968). The common black hawk feeds mainly on fish and amphibians (Brown and Amadon 1968, Millsap 1981, Hiraldo et al. 1991a). Black hawks seem to be disappearing in the US following the increasing changes of the species' preferred habitat for agricultural purposes (Johnsgard 1990). In Mexico this raptor is still common in riparian habitats.

The distribution of the crested caracara ranges from Florida, Texas and southern Arizona south to Tierra del Fuego (Brown and Amadon 1968). The crested caracara is found mainly in open and semi-open country, where it feeds mostly on carrion but also on invertebrates and vertebrates (Brown and Amadon 1968, Johnsgard 1990, Rivera and Rodríguez-Estrella (RR-E) 1993). Caracaras in southwestern United States are apparently declining as a result of habitat loss (Johnsgard 1990, Morrison 1993). The species is listed as threatened in US. In Mexico the crested caracara has disappeared from historical areas, i.e. Durango; J. Herrera pers. comm., by unknown causes, although it seems to be a common species in tropical and subtropical areas (RR-E unpubl. data).

STUDY AREA AND METHODS

The study of the common black hawk was carried out in March and April of 1987 and 1988 along the northwestern Mexican river, Rio Yaqui, and its tributaries, Rio Bavispe and Rio Aros, located in Sonora, Mexico (Fig. 1). Descriptions of the area can be found in White (1948) and Rodríguez-Estrella and Brown (1990). The riparian vegetation of the rivers corresponds to a gallery forest that contains pithecellobium, honey mesquite, goodding willow, fig tree, canyon ragweed, soapberry, as principal species, and a few scattered cottonwoods and palms. In the mid-1930s a dam, the Angostura Reservoir, was constructed in the upper part of the Rio Bavispe. The woodland distribution changed above and below the dam as a result of this modification. The dam eliminated the large annual floods which had scoured away all vegetation below the pre-dam high water mark, allowing a dense new zone of woody vegetation to develop at the post-dam river's edge. Our observations suggest that mesquite and pithecellobium-dominated riparian woodland along the Rio Bavispe increased from an estimated pre-dam 50 ha to a post-dam 200 ha. A continuous dense strip of even-aged riparian woodland dominated by mesquite and pithecellobium averaging 12 m in height has been present in the Rio Bavispe since the 1980s. However, riparian vegetation was less well developed along the Rio Yaqui due to the recurrence of annual floods from the undammed Rio Aros. In addition, the riparian zone of the Rio Yaqui was greatly disturbed by human activities (agriculture, grazing, mining, roads, several ranches and a village) while the Rio Bavispe was relatively undisturbed except for grazing and two small ranches. I intended to determine the effect of the dam on the

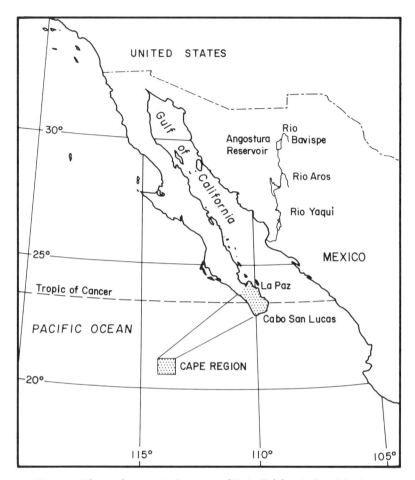

Figure 1. The study areas in Sonora and Baja California Sur, Mexico.

abundance and distribution of raptors in both rivers, especially on the common black hawk.

The raptor populations were surveyed to determine their density and habitat use along each river. Canoes were used to survey these populations as we floated downstream. All raptors observed within 0.5 km of the river, and their nesting and foraging habitat, were recorded.

The second case is related to a survey of several raptor species in the Cape Region of Baja California (Fig. 1) focusing on breeding ecology and foraging patterns. The species discussed here is the crested caracara. From February to November of 1990–91, a road count survey method (Fuller and Mosher 1981) was used to estimate the abundance of caracaras along the roads from La Paz to Cabo San Lucas. At the same time, the birds were also counted in four henhouses, one slaughterhouse and several garbage sites located around the city of La Paz. In 1991 the owners of the henhouses decided to incinerate the chicken

Table 1. Pairs of common black hawks and their density river-km^{-1} on the Rio Bavispe (65 km) and Rio Yaqui (80 km), Sonora, Mexico in 1987 and 1988 (from Rodríguez-Estrella and Brown 1990).

Year	Rio Bavispe		Rio Yaqui		Total	
	No.	Density[a]	No.	Density	No.	Density
1987	23	0.35	8	0.10	33	0.22
1988	21	0.32	8	0.10	30	0.20

[a]Density expressed as number of pairs km^{-1}

carcasses as a sanitation action. Thus, I recorded the variations in caracara numbers in all these places to denote their response to the mentioned human-action.

To estimate the association between nesting caracaras and human activity, I plotted their nests on 1:50 000 maps of the study area in 1991–92. The distance to ranches and field crops was then calculated. I developed a qualitative scale to associate nests with habitat disturbance by man:

(1) nests surrounded by undisturbed natural vegetation;
(2) some human activity present, but natural vegetation well preserved;
(3) strong human activity, natural vegetation remaining in patches; and
(4) extreme human development, only some residual natural vegetation remaining.

Chi-square goodness of fit tests were used to compare the association of nests to habitat disturbance.

RESULTS AND DISCUSSION

Common Black Hawk

This species was the third most abundant bird of prey along the rivers (after the turkey vulture and black vulture), with a mean density of 0.22 pairs km^{-1}, and was largely found in riparian areas dominated by mesquite and pithecellobium (Rodríguez-Estrella and Brown 1990). However, significant differences were found in the density (pairs river-km^{-1}) of common black hawk pairs between rivers in both years ($\chi^2_{1987} = 10.8$; df = 1; $P < 0.01$; $\chi^2_{1988} = 8.9$; df = 1; $P < 0.01$; chi-square goodness of fit tests; Table 1). This raptor has directly benefited from a human-altered environment, as the riparian woodland has increased over 300% in extension in the Rio Bavispe since the construction of the dam. In response to increasing riparian woodland, the common black hawk seems to have increased its numbers below the dam since it prefers this kind of habitat for nesting, roosting and to get fish, its main food (Rodríguez-Estrella and Brown 1990, Hiraldo *et al.* 1991a). It is estimated that at least 50 yr is required for a dense band of riparian woodland of mesquite and pithecellobium (suitable nesting/roosting trees for common black hawk) to be developed and reach 12 m in height (J. L. León and M. Cota pers. comm.).

Differences in the abundance of other raptors between rivers were also detected (Rodríguez-Estrella and Brown 1990). Turkey vultures were more common in the Rio Bavispe and they used more frequently than mesquite-pithecellobium riparian woodland, whereas black vultures were more associated with highly human-inhabited areas of the Rio Yaqui. In addition, bald eagle nests were observed only in the Rio Yaqui. Other authors (Jackson 1988, Hiraldo *et al.* 1991b) reported that the black vulture is densest in human adjacent-regions and the turkey vulture in wilderness regions, which agrees with my observations. To understand why the bald eagle was present only in the Rio Yaqui, studies on fish abundance between rivers are needed to determine the effect of the dam on the food availability for eagles (see Brown 1994, Bryan *et al.* 1994). Furthermore, studies on food availability and suitable nesting/roosting places above and below the dam are recommended to understand better the effects of the dam on the raptor population dynamics of the Rio Yaqui and tributaries.

Crested Caracara

One of the most abundant raptors in the Cape region is the crested caracara (RR-E unpubl. data), ranging from 0.24 to 0.97 birds road-km^{-1} throughout the year. The caracara nests mainly on 8 m high columnar cacti, the cardon (Rivera and Rodríguez-Estrella 1993). Nests were located in both natural vegetation and close to human activity, but they were more frequently (73% of nests, $N = 22$) located in areas with low human disturbance (Fig. 2). Nests were also located near field crops ($\chi^2 = 45.0$; df = 1; $P < 0.001$; Fig. 3) and small ranches ($\chi^2 = 7.80$; df = 1; $P < 0.01$; Fig. 4). No significant differences were found in the productivity of the nests located in areas with low and high human disturbance ($t = 2.01$; df = 15; $P > 0.05$), but some nests failed in high human impact areas because people took the nestlings to rear them as pets.

The caracara is an opportunistic species with a high component of carrion in its diet (55% carrion, 45% live prey; RR-E unpubl. data). We have observed that small prey (insects, mice) were commonly obtained from crop fields and ranches, and carrion from roadkills, henhouses and garbage sites. In 1990–91 young, juvenile and adult caracaras intensively foraged in henhouses and garbage sites (Fig. 5; RR-E *et al.* unpubl. data). In 1990 the caracara was observed more frequently feeding in henhouses than in garbage sites. In 1991 the proportions of caracaras in each feeding place was reversed as a result of the incineration of chicken carcasses and remains in the henhouses. At the present, the birds, especially juveniles, feed mainly in the new open slaughterhouse of La Paz (Fig. 5). Since the immature birds seem to depend mainly upon carrion and insects during the post-fledging period (RR-E unpubl. data), caracaras showed

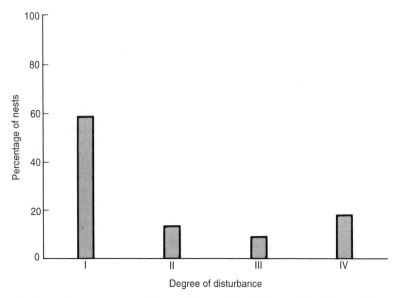

Figure 2. Crested caracara nests ($N = 22$) located in habitats with different degrees of human alteration. I = undisturbed natural vegetation; II = natural vegetation is well preserved, but evident habitat alterations; III = natural vegetation remains in patches; IV = only residual natural vegetation remaining.

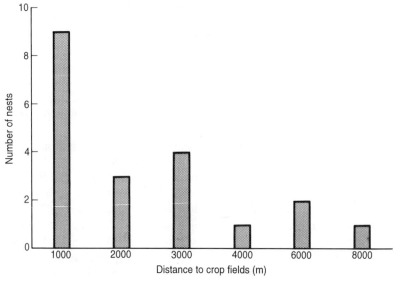

Figure 3. Distance of crested caracara nests ($N = 20$) to crop fields.

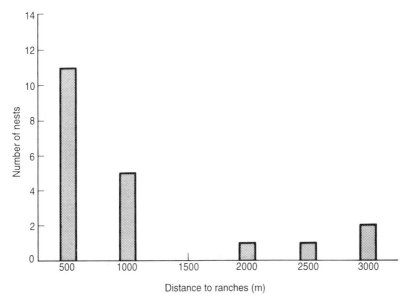

Figure 4. Distance of crested caracara nests ($N = 20$) to ranches.

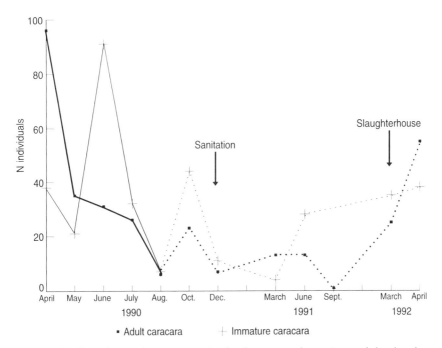

Figure 5. Number of crested caracaras using henhouses, garbage sites and the slaughterhouse in La Paz, B.C.S. during 1990–91. The arrows indicate the dates when chicken carcass incineration (sanitation) began and the opening of the slaughterhouse in the region. Continuous lines join monthly survey data. Dot lines were used when data for some months were lacking.

a rapid response to the changes in carrion availability, switching to the most available human refuse source according to the waste disposal practices.

These two examples show that some raptors are capable of adapting and benefiting from moderate human alterations in habitat that produce suitable nest sites, roosting places and/or increased food availability. Thus, some raptors may be more opportunistic and flexible in their behaviour than commonly believed. However, it is important to determine the direct effect of human activities on the breeding success of each raptor adapted to a man-made environment. For example, although caracaras seem to have learned to obtain food from human-altered environments, i.e. crop fields, I have not measured the effects of the intensive agricultural applications of malathion (a common practice in the Cape region) on the productivity of the caracara. These studies will help to develop new raptor management techniques in an increasingly man-made world.

ACKNOWLEDGMENTS

Special thanks are due to L. Rivera for field assistance at the Cape region. F. Hiraldo, an anonymous reviewer and J. J. Negro made helpful comments to an early draft. R. Bowers made the English revision. Centro de Investigaciones Biológicas del Noroeste and Consejo Nacional de Ciencia y Tecnología gave financial support.

REFERENCES

BROWN, L. AND D. AMADON. 1968. Eagles, hawks and falcons of the world. McGraw-Hill Book Co., New York.

BROWN, R.D. 1996. Attraction of Bald Eagles to habitats just below dams in Piedmont North and South Carolina. Pp. 299–306 in D. Bird, D.E. Varland and J.J. Negro, eds. *Raptors in Human Landscapes*. Academic Press, London.

BRYAN, A.L., JR., T.M. MURPHY, K.L. BILDSTEIN, I.L. BRISBIN AND J.J. MAYER. 1994. Use of reservoirs and other artificial impoundments by Bald Eagles in South Carolina. Pp. 285–298 in D. Bird, D.E. Varland and J.J. Negro, eds. *Raptors in Human Landscapes*. Academic Press, London.

ENDERSON, J.H., G.R. CRAIG, W.A. BURNHAM AND D.D. BERGER. 1982. Eggshell thinning and organochlorine residues in Rocky Mountain Peregrines, *Falco peregrinus*, and their prey. *Can. Field-Nat.* **96**: 255–264.

FRIEDMAN, R. AND P.J. MUNDY. 1983. The use of "Restaurants" for the survival of vultures in South Africa. Pp. 345–355 in S.R. Wilbur and J.A. Jackson, eds. *Vulture Biology and Management*. Univ. Calif. Press, Berkeley, California.

FULLER, M.R. AND J.A. MOSHER. 1981. Methods of detecting and counting raptors: a review. *Studies in Avian Biology* **6**: 235–246.

HENNY, C.J., L.J. BLUS, E.J. KOLBE AND R.E. FITZNER. 1985. Organophosphate insecticide (Famphur) topically applied to cattle kills magpies and hawks. *J. Wildl. Manage.* **49**: 648–658.

HIRALDO, F., M. DELIBES, J. BUSTAMANTE AND R. RODRÍGUEZ-ESTRELLA. 1991a. Overlap in the diets of diurnal raptors breeding at the Michilía Biosphere Reserve, Durango, Mexico. *J. Raptor Res.* **25**: 25–29.
—, — AND J.A. DONAZAR. 1991b. Comparison of diets of Turkey Vultures in three regions of northern Mexico. *J. Field Ornithol.* **62**: 319–324.
HOLTHUIJZEN, A.M.A., W.G. EASTLAND, A.R. ANSELL, M.N. KOCHERT, R.D. WILLIAMS AND L.S. YOUNG. 1990. Effects of blasting on behavior and productivity of nesting Prairie falcons. *Wildl. Soc. Bull.* **18**: 270–281.
JACKSON, J.A. 1988. Black vulture and Turkey vulture. Pp. 11–42 in R.S. Palmer, ed. *Handbook of North American birds*, Vol. 4. Varl-Ballou Press, Binghamton, New York.
JOHNSGARD, P.A. 1990. *Hawks, eagles and falcons of North American*. Smithsonian Institution Press, Washington, USA.
KOCHERT, M.N., B.A. MILLSAP AND K. STEENHOF. 1988. Effects of livestock grazing on raptors with emphasis on the southwestern U.S. Pp. 325–334 in R.L. Glinski, B.G. Pendleton, M.B. Moss, M.N. LeFranc, B.A. Millsap and S.W. Hoffman, eds. *Proc. Southwest Raptor Manage. Symp. and Workshop.* Nat. Wildl. Fed., Washington, D.C.
MILLSAP, B.A. 1981. *Distribution of Arizona Falconiformes*. US Dept. Interior, Bur. Land Manage., Tech. Note **355**: 1–102.
MORRISON, J.L. 1993. The elusive Caracara: preliminary information from South Central Florida. Abstracts Raptor Research Foundation Meeting, Bellevue, Washington. *J. Raptor Res.* **27**: 77–78.
NEWTON, I. 1979. *Population ecology of raptors*. Buteo Books, Vermillion, South Dakota.
OLENDORFF, R.R., A.D. MILLER AND R.N. LEHMAN. 1981. *Suggested practices for raptor protection on powerlines*. Raptor. Res. Rep. 4. Raptor Res. Found.
PHILLIPS, R.L. 1986. Current issues concerning the management of Golden Eagles in western USA. *Birds of Prey Bull.* **3**: 149–156.
PRUETT-JONES, S.G., M.A. PRUETT-JONES AND R.L. KNIGHT. 1980. Status of Black-shouldered kites in North America. *Amer. Birds* **34**: 682–688.
RIVERA, L.R. AND R. RODRÍGUEZ-ESTRELLA. 1993. Breeding ecology of the crested caracara (*Polyborus plancus*) in the Cape region, B.C.S., Mexico. Abstracts Raptor Research Foundation Meeting, Bellevue, Washington. *J. Raptor Res.* **27**: 91–92.
RODRÍGUEZ-ESTRELLA, R. AND B.T. BROWN. 1990. Density and habitat use of raptors along the Rio Bavispe and Rio Yaqui, Sonora, Mexico. *J. Raptor Res.* **24**: 47–51.
SCHMUTZ, J.K. 1984. Ferruginous and Swainson's hawk abundance in relation to land use in southeastern Alberta. *J. Wildl. Manage.* **48**: 1180–1187.
VANNINI, J.P. 1989. Neotropical raptors and deforestation: notes on diurnal raptors at Finca El Faro, Quetzaltenango, Guatemala. *J. Raptor Res.* **23**: 27–38.
WHITE, S. 1948. The vegetation and flora of the region of the Rio Bavispe in northeastern Sonora, Mexico. *Lloydia* **11**: 229–302.

34

Occurrence and Distribution of Diurnal Raptors in Relation to Human Activity and Other Factors at Rocky Mountain Arsenal, Colorado

Charles R. Preston and Ronald D. Beane

Abstract – Rocky Mountain Arsenal (RMA) is a federal superfund site surrounded by urban and agricultural lands near Denver, Colorado. The site has recently been designated a National Wildlife Refuge, pending extensive contamination cleanup. As part of a wide-ranging study of the effects of human activity on wildlife at RMA, we began conducting roadside surveys of diurnal raptors in 1991. We recorded 12 species during summer and winter surveys. American kestrel, Swainson's hawk, and red-tailed hawk were the most frequently encountered species in summer, and ferruginous hawk, red-tailed hawk, rough-legged hawk and bald eagle were the species most frequently encountered during the overwintering period. Using logit analysis, we identified statistical associations between species' distributions and three environmental factors: human activity, presence of black-tailed prairie dog colonies, and presence of woodland. The distribution of all species other than Swainson's and rough-legged hawks were linked with one or more of these factors, but there was no indication that any species avoided areas with high human activity. Attractive ecological features of the area may boost the tolerance of some species to human activity.

Key words: American kestrel; bald eagle; *Buteo* spp; logit analysis; human activity; Rocky Mountain Arsenal; Colorado.

Central to developing raptor conservation and management strategies in urban environments is an understanding of how raptors adapt to human activity. Increased human presence and activity have been found to alter raptor behavior and distribution in a variety of situations, e.g. Stalmaster and Newman 1978, Knight and Knight 1984, Andersen *et al.* 1986, 1989, 1990. The response of birds to human disturbance in a particular situation, however, is dependent on many factors, including the type, level and duration of disturbance, species and experience of the birds, and environmental setting (Skagen 1980, White and Thurow 1985, Knight and Skagen 1988, Grubb and King 1991, Skagen *et al.* 1991, Gonzalez *et al.* 1992, Klein 1993, Holmes *et al.* 1993). Interspecific differences in tolerance to humans might have a profound long-term influence on raptor community composition, whereby species more tolerant of humans become more abundant and widespread in a human-altered site (Voous 1977, Craighead and Mindell 1981).

Rocky Mountain Arsenal National Wildlife Area (RMA), near Denver, Colorado, is a federal superfund site designated to be a National Wildlife Refuge pending extensive (though not yet specified) contamination cleanup. This site, surrounded by urban development, provides important foraging and nesting opportunities for many raptor species (Morrison-Knudsen Environmental Services Inc. 1989a). These species are subjected to increasing human disturbance as contamination containment and cleanup proceed. As part of a larger study, we have monitored raptor occurrence and distribution at RMA for the past two years. Our objectives in this paper are to report seasonal variation in raptor species occurrence at RMA and to assess simultaneously the distribution of selected species with respect to human activity and other environmental factors.

STUDY AREA

RMA National Wildlife Area occupies nearly 70 km^2 in Adams County, Colorado, approximately 16 km northeast of downtown Denver. The property was purchased by the US Government in 1942 to produce chemical weapons and conventional munitions for World War II. In the early 1950s, Julius Hyman and Company (later acquired by Shell Oil Company) leased some facilities at RMA for the purpose of producing chemical pesticides. The US Army ceased munitions production at RMA in 1969, and Shell Oil Company stopped producing pesticides in 1982. For more than 40 yr, toxic chemicals were placed into waste basins and trenches on the property, in accordance with prevailing protocol. RMA was recently named a superfund site by the US Government, and plans for a massive cleanup project were begun.

Because RMA property has been restricted to human use since 1942, it has escaped the massive urbanization that has profoundly altered the surrounding region. As a result, isolated patches of native prairie habitat are intermingled with section roads, abandoned buildings, and other human-altered landscape (Morrison-Knudsen Environmental Services Inc. 1989b). Approximately 65% of the property is covered by "disturbance" vegetation, including cheatgrass, crested wheatgrass, prickly lettuce and Russian-thistle. The remaining vegetation includes grassland patches dominated by blue grama and buffalo grass, shrubland dominated by yucca, or sand sagebrush, and woodland dominated by plains cottonwood. Several large colonies of black-tailed prairie dogs are located on the property.

METHODS

Thirteen "summer" roadside surveys (Fuller and Mosher 1981, 1987) were conducted between 15 June and 25 August 1991 and 6 June and 18 August

1992. Similarly, 19 "winter" roadside surveys were conducted between 16 November 1991 and 22 February 1992 and 16 November 1992 and 23 February 1993. All surveys were conducted between 0600 h and 1100 h on days when visibility exceeded 1.5 km and wind velocity ≤ 16 km h^{-1}. Observations were recorded from a vehicle moving 15–25 km h^{-1}. All raptors identified within 100 m of either side of the road were recorded. The transect route included 24 sample segments of equal length (approximately 1.6 km). Segments were separated by at least 400 m.

Features varying most among survey segments were the amount of regular human traffic and activity, the presence of woodland and the presence of prairie dog colonies. Each segment was characterized with respect to these factors, i.e. low or high human activity, absence or presence of woodland and absence or presence of active prairie dog colony. Segments classified as having high human activity were located on improved, heavily traveled primary roads, generally providing access to interim contamination response action or office buildings. If a segment included at least one patch of 10 trees (≥ 10 m in height)/0.04 ha, it was classified as having woodland present. Logit analysis (CATMOD, SAS Institute Inc. 1988, Preston *et al.* 1989, Tabachnick and Fidell 1989) was used to test the hypothesis that the occurrence of each raptor species (that was encountered frequently enough to be included) was independent of human activity, the presence of woodland, and the presence of an active prairie dog colony. To test the hypothesis, a 2 (low, high human activity) × 2 (woodland, no woodland) × 2 (prairie dogs, no prairie dogs) × 2 (presence, absence of raptor species) multiway frequency table was constructed, with presence of raptor species used as the dependent variable. Total number of cases was 312 (24 sample segments × 13 surveys) in summer, and 456 (24 × 19) in winter.

RESULTS

Twelve diurnal raptor species were identified during our surveys (Table 1). The species most frequently encountered during summer surveys were American kestrels and Swainson's and red-tailed hawks (Fig. 1). Encounters with both Swainson's and red-tailed hawks increased from June through August, reflecting the production of fledglings through this period. Kestrel numbers peaked in July and decreased markedly in August, presumably indicating earlier dispersal from nest sites by this species.

Statistical associations between occurrence of each of these species and each of the independent factors and their interactions were determined (Table 2). Sample sizes for other species encountered in summer were inadequate for analysis. The occurrence of Swainson's hawks was independent of the factors considered. Although no one factor was associated with American kestrel distribution, a statistically significant relationship between kestrel occurrence and the interaction of woodland and human activity was identified. Kestrels were more frequently encountered in wooded areas with high human activity

Table 1. Diurnal raptor species encountered during summer and winter roadside surveys conducted at Rocky Mountain Arsenal.

Species	Season of Occurrence
Bald eagle	Winter[a]
Northern harrier	Both
Cooper's hawk	Summer[b]
Swainson's hawk	Summer
Red-tailed hawk	Both
Rough-legged hawk	Winter
Ferruginous hawk	Both
Golden eagle	Both
American kestrel	Both
Merlin	Winter
Prairie falcon	Both
Peregrine falcon	Winter

[a] mid-November–February
[b] June–August

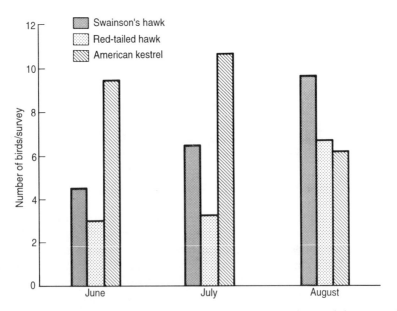

Figure 1. Relative frequencies of species most commonly encountered during roadside surveys in summer months at Rocky Mountain Arsenal, Colorado.

Table 2. Results of logit analysis. Dependent variable is occurrence of raptor species; test statistic is Wald's χ^2 (SAS Institute Inc. 1988).

	AK[a] (Summer)			SH (Summer)			RTH (Summer)			RTH (Winter)			RLH (Winter)			FH (Winter)			BE (Winter)		
	χ^2	df	P	χ^2	df	P	χ^2	df	P	χ^2	df	P	χ^2	df	P	χ^2	df	P	χ^2	df	P
HA[b]	0.29	1	0.590	2.70	1	0.102	6.78	1	0.009	0.15	1	0.701	3.02	1	0.082	1.17	1	0.280	1.49	1	0.222
W	0.09	1	0.764	0.01	1	0.940	3.08	1	0.079	28.62	1	<0.001	0.02	1	0.978	2.70	1	0.102	13.88	1	<0.001
PD	1.91	1	0.168	0.96	1	0.327	7.76	1	0.005	19.92	1	<0.001	1.06	1	0.303	40.96	1	<0.001	0.17	1	0.684
HA*W	5.48	1	0.019	0.08	1	0.782	1.32	1	0.250	0.84	1	0.360	2.41	1	0.121	2.69	1	0.101	0.08	1	0.778
HA*PD	2.63	1	0.105	0.03	1	0.860	9.16	1	0.002	1.20	1	0.274	3.00	1	0.083	0.48	1	0.489	1.50	1	0.222
W*PD	0.18	1	0.674	0.46	1	0.497	2.69	1	0.103	21.53	1	<0.001	1.98	1	0.159	0.12	1	0.729	0.82	1	0.364
HA*W*PD	1.59	1	0.208	1.17	1	0.280	1.90	1	0.167	0.04	1	0.836	0.00	1	0.000	0.75	1	0.387	0.08	1	0.778

[a] AK = American kestrel; SH = Swainson's hawk; RTH = red-tailed hawk; RLH = rough-legged hawk; FH = ferruginous hawk; BE = bald eagle.
[b] HA = level of human activity; W = presence of woodland; PD = presence of prairie dog colony.

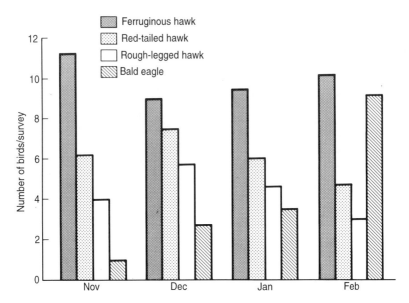

Figure 2. Relative frequencies of species most commonly encountered during roadside surveys in overwintering period at Rocky Mountain Arsenal, Colorado.

than expected by chance. Red-tailed hawk distribution was statistically linked with human activity, prairie dogs, and the interaction of these factors. These hawks favored areas where prairie dogs were present and/or human activity was high.

Species most frequently encountered during late fall/early winter surveys were bald eagle and ferruginous, rough-legged, and red-tailed hawks (Fig. 2). Ferruginous hawks outnumbered other species in each month of the overwintering period. Of the species encountered during winter, only these four were encountered frequently enough for logit analysis. Ferruginous hawks were strongly associated with prairie dogs (Table 2); where prairie dogs occurred, so did ferruginous hawks. Conversely, ferruginous hawks were not frequently encountered in areas where prairie dogs were absent. Like Swainson's hawks in summer, the distribution of rough-legged hawks in winter was independent of the factors considered. Red-tailed hawk distribution in winter was related to prairie dogs, woodland, and the interaction of these factors, such that they frequented areas containing either or both of these features. Bald eagles were rarely encountered away from woodland, and their statistical association with this variable was strong (Table 2).

DISCUSSION

Despite more than 40 yr of widespread chemical contamination at Rocky Mountain Arsenal, the site supports a wide diversity and high density of raptors

(Morrison-Knudsen Environmental Services Inc. 1989a). Concentrated use of this site by raptors is due to its resources and location in the midst of a relatively inhospitable urban environment. Two obvious, potential threats to the continued presence and welfare of raptors at RMA are lingering chemical contamination and increased human activity related to contamination cleanup. An examination of the direct effects of chemical contamination on raptors at RMA was beyond the scope of this study, but we were able to examine the relationship of diurnal raptor distribution to current spatial variation in human activity on the property. Because human activity does not occur in a vacuum, we included two other key, site-specific factors in our analysis. The null hypothesis of independence between raptor distribution and the three factors considered was rejected for most species in both seasons, but there was no indication that any species avoided areas with high human activity. To the contrary, our analyses indicate that the species studied may have habituated to the relatively high human activity found in some areas of RMA. In a similar study, Knight and Knight (1984) found that overwintering bald eagles occurring along popular boating areas were more tolerant of human activity than those frequenting areas with less disturbance. Habituation to human activity may be adaptive in the absence of persecution (Keller 1989). Certain changes, i.e. grass mowing, accompanying increased human activity at RMA may attract and benefit some raptors by increasing availability of prey. The associations identified by our analyses probably reflect direct raptor responses to either foraging or nesting opportunities.

The distribution of American kestrels coincided with the distribution of nest boxes and was probably not a response to the combination of human activity and woodland *per se*. These boxes had been placed in wooded areas along heavily traveled roadways prior to this study.

The distribution of ferruginous hawks was predictably associated closely with their primary food source in the study area, i.e. prairie dogs. There was no evidence that either human activity or presence of woodland modified this relationship.

The interpretation of bald eagle distribution is equally straightforward. The association of these birds with woodland reflects their propensity to perch in tall trees, apparently irrespective of other factors we examined. Bald eagles avoided perching on utility poles, even though these structures were abundant throughout the study area.

The distribution pattern of red-tailed hawks differed slightly between seasons. In summer, red-tailed hawks favored prairie dog colonies and heavily traveled roadways. Grassy margins approximately 30 m either side of heavily traveled roadways were mowed regularly during summer. Thus, prairie dog colonies and heavily traveled roadways shared the characteristic of sparse ground cover and, presumably, high prey vulnerability (see Preston 1990). Red-tailed hawks were again associated with prairie dog colonies in winter, but were also associated with woodland during this period. The affinity of red-tailed hawks for open woodland is well documented (see Preston and Beane 1993 for review), but it is unclear why birds in our study area were associated significantly with woodland only in winter. They readily perched on utility poles and trees in both seasons. One possibility is that open areas attracted red-tailed hawks away from some wooded areas during summer. For example, prairie dogs may have been more vulnerable in early to mid-summer, when they were more consistently active and when young were abundant (D. Serri pers. comm.). It is possible that hawks in woodland were simply more easily observed when trees were defoliated in winter.

Distribution patterns of Swainson's and rough-legged hawks were statistically independent of the factors considered here. This may indicate that these birds were more catholic in their use of available resources, or they may have responded to other factors uncorrelated with the ones we considered. One potential factor not explicitly considered was interspecific competition for space. Interspecific territoriality has been reported between Swainson's and red-tailed hawks in the breeding season (Rothfels and Lein 1983, Janes 1984), and we observed both red-tailed and ferruginous hawks supplant rough-legged hawks at RMA during winter. Interference competition among these birds might exert a particularly strong influence on species distribution in a small, relatively isolated habitat patch such as RMA.

One of the major weaknesses of roadside surveys is that roads do not always traverse landscape representative of the study area. In our study area, the routes we traversed provided a good representation of the available environment in our small, well-defined study area. However, due to possible differences in detectability, comparisons between seasons and between species must be interpreted cautiously (Fuller and Mosher 1981, 1987). Roadside survey techniques generally do not provide adequate information to detail behavior, habitat use, home range, reproductive success, or other crucial aspects of raptor biology. Therefore, results presented here cannot address the effects of human activity on these important parameters at RMA. Neither does our study distinguish among various types of human activity, and other workers have demonstrated that bird species are more sensitive to humans on foot than to humans in vehicles, e.g. Skagen 1980, Holmes et al. 1993, Klein 1993.

Effects of human activity on raptors must be viewed in a broad ecological context. Given attractive foraging, nesting, and/or roosting opportunities, raptors may be able to adapt to many types of human activity. Where feasible, adverse effects of human activities may be mitigated by changing the ecological context, i.e. by enhancing foraging or other opportunities in the area. Studies should be continued at RMA to determine how raptors respond to escalating levels and new kinds, i.e. more pedestrian traffic, of human disturbance associated with large-scale cleanup efforts.

Multiway frequency analysis offers a robust nonparametric technique for examining complex relationships. It is ideal for scrutinizing the relationships among easily categorized factors, but has rarely been used thus far in studies of raptor distribution or response to habitat alteration (but see Stinson et al. 1981, Preston et al. 1989). Logit analysis is a special case of multiway frequency analysis applied when one or more dependent variables are identified a priori. For potential users, it is recommended that sample size should be at least five times the number of cells (in our case 16 × 5 = 80), that all expected cell frequencies for two-way associations exceed one, and that no more than 20% are less than five (Tabachnick and Fidell 1989). Interpretation becomes difficult with the addition of too many variables, and the power of the analysis is severely reduced as expected cell frequencies drop.

ACKNOWLEDGMENTS

Funding for this project was provided by the US Army, under a cooperative agreement among the Army, US Fish and Wildlife Service (USFWS), National Fish and Wildlife Foundation, and Denver Museum of Natural History. J. J. Negro, W. Russell and two anonymous reviewers provided helpful suggestions on an earlier version of the manuscript. Personnel from USFWS, especially D. Gober, L. Malone, M. Lockhart, J. Griess, D. Matiatos and G. Langer provided invaluable logistical support. We are also grateful to C. Mackey, Morrison-Knudsen Environmental Services, Incorporated, for helping us understand the current composition and distribution of plant communities at RMA. R. DeBaca and F. Hein assisted with fieldwork.

REFERENCES

ANDERSEN, D.E., O.J. RONGSTAD AND W.R. MYTTON. 1986. The behavioral response of a Red-tailed Hawk to military training activity. *Raptor Res.* **20**: 65–68.
—, — AND —. 1989. Response of nesting Red-tailed Hawks to helicopter overflights. *Condor* **91**: 296–299.
—, — AND —. 1990. Home-range changes in raptors exposed to increased human activity levels in southeastern Colorado. *Wildl. Soc. Bull.* **18**: 134–142.
CRAIGHEAD, F.C., JR AND D.P. MINDELL. 1981. Nesting raptors in western Wyoming, 1947 and 1975. *J. Wildl. Manage.* **45**: 865–872.
FULLER, M.R. AND J.A. MOSHER, 1981. Methods of detecting and counting raptors: a review. Pp. 235–246 in C.J. Ralph and J.M. Scott, eds. *Estimating numbers of terrestrial birds.* Stud. Avian Biol. 6.
— AND —. 1987. Raptor survey techniques. Pp. 37–65 in B.A. Giron Pendleton, B.A. Millsap, K.W. Cline and D.M. Bird, eds. *Raptor management techniques manual.* Sci. Tech. Ser. No.10. National Wildlife Federation, Washington, DC, USA.
GONZALEZ, L.M., J. BUSTAMANTE AND F. HIRALDO. 1992. Nesting habitat selection by the Spanish Imperial Eagle *Aquila adalberti. Biol. Conserv.* **59**: 45–50.
GRUBB, T.G. AND R.M. KING. 1991. Assessing human disturbance of breeding Bald Eagles with classification tree models. *J. Wildl. Manage.* **55**: 500–511.
HOLMES, T.L., R.L. KNIGHT, L. STEGALL AND G.R. CRAIG. 1993. Responses of wintering grassland raptors to human disturbance. *Wildl. Soc. Bull.* **21**: 461–468.
JANES, S.W. 1984. Influences of territory composition and interspecific competition on Red-tailed Hawk reproductive success. *Ecology* **65**: 862–868.
KELLER, V. 1989. Variations in the response of Great Crested Grebes *Podiceps cristatus* to human disturbance – a sign of adaptation? *Biol. Conserv.* **49**: 31–45.

KLEIN, M.L. 1993. Waterbird behavioral response to human disturbances. *Wildl. Soc. Bull.* **21**: 31–39.

KNIGHT, R.L. AND S.K. KNIGHT. 1984. Responses of wintering Bald Eagles to boating activity. *J. Wildl. Manage.* **48**: 999–1004.

— AND S.K. SKAGEN. 1988. Effects of recreational disturbance on birds of prey: a review. Pp. 355–359 in R.L. Glinski, B.G. Pendleton, M.B. Moss, M.N. LeFranc, Jr., B.A. Millsap and S.W. Hoffman, eds. *Proceedings of the Southwest Raptor Management Symposium and Workshop.* Nat. Wildl. Fed., Washington, DC, USA.

MORRISON-KNUDSEN ENVIRONMENTAL SERVICES INC. 1989a. *Wildlife Resources of Rocky Mountain Arsenal, Adams County, Colorado.* Final Rep., Shell Oil Company, Holme Roberts and Owen, Denver, CO USA.

—. 1989b. *Vegetation Resources of Rocky Mountain Arsenal, Adams County, Colorado.* Final Rep., Shell Oil Company, Holme Roberts and Owen, Denver, CO USA.

PRESTON, C.R. 1990. Distribution of raptor foraging in relation to prey biomass and habitat structure. *Condor* **92**: 107–112.

—, C.S. HARGER AND H.E. HARGER. 1989. Habitat use and nest-site selection by Red-shouldered Hawks in Arkansas. *Southw. Nat.* **34**: 72–78.

— AND R.D. BEANE. 1993. Red-tailed Hawk (*Buteo jamaicensis*). In A. Poole and F. Gill, eds. *The Birds of North America, No. 52.* The Academy of Natural Sciences, Philadelphia, Pennsylvania; The American Ornithologists' Union, Washington, D.C. USA.

ROTHFELS, M. AND M.R. LEIN. 1983. Territoriality in sympatric populations of Red-tailed and Swainson's Hawks. *Can. J. Zool.* **58**: 1075–1089.

SAS INSTITUTE, INC. 1988. SAS/STAT user's guide, release 6.03 edition. SAS Institute, Inc. Cary, North Carolina, USA.

SKAGEN, S.K. 1980. Behavioral response of wintering Bald Eagles to human activity on the Skagit River, Washington. Pp. 231–241 in R.L. Knight, G.T. Allen, M.V. Stalmaster and C.W. Servheen, eds. *Proceedings of the Washington Bald Eagle Symposium.* The Nature Conservancy, Seattle, Washington USA.

—, R.L. KNIGHT AND G.H. ORIANS. 1991. Human disturbance of an avian scavenging guild. *Ecol. Appl.* **1**: 215–225.

STALMASTER, M.V. AND J.R. NEWMAN. 1978. Behavioral responses of wintering Bald Eagles to human activity. *J. Wildl. Manage.* **42**: 506–513.

STINSON, C.H., D.L. CRAWFORD AND J. LAUTHNER. 1981. Sex differences in winter habitat of American Kestrels in Georgia. *J. Field Ornithol.* **52**: 29–35.

TABACHNICK, B.G. AND L.S. FIDELL. 1989. *Using multivariate statistics.* 2nd Ed. Harper Collins Publ., New York, New York, USA.

VOOUS, K.H. 1977. Three lines of thought for consideration and eventual action. Pp. 343–347 in R.D. Chancellor, ed. *World conference on birds of prey; report of proceedings.* Int. Counc. for Bird Preserv., Cambridge UK.

WHITE, C.M. AND T.L. THUROW. 1985. Reproduction of Ferruginous Hawks exposed to controlled disturbance. *Condor* **87**: 14–22.

Appendix: List of species mentioned in the text

Common name | Scientific name

Birds
Common loon | *Gavia immer*
Horned grebe | *Podiceps auritus*
Pied-billed grebe | *Podilymbus podiceps*
Double-crested cormorant | *Phalacrocorax auritus*
Anhinga | *Anhinga anhinga*
Great blue heron | *Ardea herodias*
Great egret | *Casmerodius albus*
Snowy egret | *Egretta thula*
Wood stork | *Mycteria americana*
Little blue heron | *Egretta caerulea*
Tricolored heron | *Egretta tricolor*
Green-backed heron | *Butorides striatus*
Black-crowned night heron | *Nycticorax nycticorax*
American bittern | *Botaurus lentiginosus*
Least bittern | *Ixobrychus exilis*
Mallard | *Anas platyrhynchos*
American black duck | *Anas rubripes*
Gadwall | *Anas strepera*
Green-winged teal | *Anas crecca*
Blue-winged teal | *Anas discors*
Ruddy duck | *Oxyura jamaicensis*
Bufflehead | *Bucephala albeola*
Wood duck | *Aix sponsa*
Ring-necked duck | *Aythya collaris*
Lesser scaup | *Aythya affinis*
Northern pintail | *Anas acuta*
Gadwall | *Anas strepera*
American widgeon | *Anas americana*
Red-breasted merganser | *Mergus serrator*
Hooded merganser | *Lophodytes cucullatus*
Secretary bird | *Sagittarius serpentarius*

Common name	Scientific name
Turkey vulture	*Cathartes aura*
Lesser yellow-headed vulture	*Cathartes burrovianus*
White-headed vulture	*Trigonoceps occipitalis*
Greater yellow-headed vulture	*Cathartes melambrotus*
Black vulture	*Coragyps atratus*
King vulture	*Sarcoramphus papa*
California condor	*Gymnogyps californianus*
Andean condor	*Vultur gryphus*
Osprey	*Pandion halietus*
Jerdon's baza	*Aviceda jerdoni*
Cuckoo hawk	*Aviceda cucculoides*
Grey-headed kite	*Leptodon cayanensis*
Hook-billed kite	*Chondrohierax uncinatus*
Honey buzzard	*Pernis apivorus*
Oriental honey buzzard	*Pernis ptilorynchus*
Lizard buzzard	*Kaupifalco monogrammicus*
Swallow-tailed kite	*Elanoides forficatus*
Bat hawk	*Macheiramphus alcinus*
Pearl kite	*Gampsonyx swainsonii*
White-tailed kite	*Elanus leucurus*
Black-shouldered kite	*Elanus caeruleus*
Snail kite	*Rostrhamus sociabilis*
Slender-billed kite	*Rostrhamus hamatus*
Double-toothed kite	*Harpagus bidentatus*
Plumbeous kite	*Ictinia plumbea*
Mississippi kite	*Ictinia mississipiensis*
Black kite	*Milvus migrans migrans*
Pariah kite	*Milvus migrans govinda*
Black-eared kite	*Milvus migrans lineatus*
Yellow-billed kite	*Milvus migrans parasitus*
Red kite	*Milvus milvus*
Brahminy kite	*Haliastur indus*
White-bellied sea-eagle	*Haliaeetus leucogaster*
Pallas's fish-eagle	*Haliaeetus leucoryphus*
Bald eagle	*Haliaeetus leucocephalus*
African fish eagle	*Haliaeetus vocifer*
Lesser fish-eagle	*Icthyophaga humilis*
Cinereous vulture	*Aegypius monachus*
Lappet-faced vulture	*Torgos tracheliotus* (*Aegypius tracheliotus*)
Long-billed vulture	*Gyps indicus*
European griffon	*Gyps fulvus*
Cape griffon	*Gyps coprotheres*
Indian white-backed vulture	*Gyps bengalensis*
White-backed vulture	*Gyps africanus*

Common name	Scientific name
Egyptian vulture	*Neophron percnopterus*
Lammergeier	*Gypaetus barbatus*
Brown snake eagle	*Circaetus cinereus*
Short-toed eagle	*Circaetus gallicus*
Black-breasted snake eagle	*Circaetus pectoralis*
Southern banded snake eagle	*Circaetus fasciolatus*
Western banded snake eagle	*Circaetus cinerascens*
Bateleur	*Terathopius ecaudatus*
Crested serpent-eagle	*Spilornis cheela*
White-eyed buzzard	*Butastur teesa*
African marsh harrier	*Circus ranivorus*
Marsh harrier	*Circus aeruoginosus*
Hen harrier/Northern harrier	*Circus cyaneus*
Pale/Pallid harrier	*Circus macrourus*
Montagu's harrier	*Circus pygargus*
Pied harrier	*Circus melanoleucos*
Pale chanting goshawk	*Melierax canorus*
Dark chanting goshawk	*Melierax metabates*
Gabar goshawk	*Micronisus gabar*
African goshawk	*Accipiter tachiro*
Crested goshawk	*Accipiter trivirgatus*
Goshawk	*Accipiter gentilis*
Shikra	*Accipiter badius*
Little banded goshawk	*Accipiter badius*
Tiny hawk	*Accipiter superciliosus*
Besra	*Accipiter virgatus (Arran bemmeli)*
Red-breasted sparrowhawk	*Accipiter rufiventris*
Black sparrowhawk	*Accipiter melanoleucus*
Sparrowhawk	*Accipiter nisus*
Little sparrowhawk	*Accipiter minullus*
Cooper's hawk	*Accipiter cooperi*
Bicolored hawk	*Accipiter bicolor*
Ovambo sparrowhawk	*Accipiter ovampensis*
Crane hawk	*Geranospiza caerulescens*
Black-face hawk	*Leucopternis melanops*
White hawk	*Leucopternis albicollis*
Common black hawk	*Buteogallus anthracinus*
Great black hawk	*Buteogallus urubitinga*
Savanna hawk	*Heterospizias meriodionalis (Buteogallus meridionalis)*
Black-collared hawk	*Busarellus nigricollis*
Harris hawk	*Parabuteo unicinctus*
Roadside hawk	*Buteo magnirostris*
Red-shouldered hawk	*Buteo lineatus*
Red-shouldered hawk	*Buteo lineatus elegans*

Common name	Scientific name
Broad-winged hawk	*Buteo platypterus*
Short-tailed hawk	*Buteo brachyurus*
Swainson's hawk	*Buteo swainsonii*
White-tailed hawk	*Buteo albicaudatus*
Zone-tailed hawk	*Buteo albonotatus*
Red-tailed hawk	*Buteo jamaicensis*
Augur buzzard	*Buteo augur*
Common buzzard	*Buteo buteo*
Long-legged buzzard	*Buteo rufinus*
Ferruginous hawk	*Buteo regalis*
Grey hawk	*Buteo nitidus (Asturina nitida)*
Crowned eagle	*Stephanoaetus coronatus*
Crested eagle	*Morphnus guianensis*
Long-crested eagle	*Lophaetus occipitalis*
Harpy eagle	*Harpia harpyja*
Indian black eagle	*Ictinaetus malayensis*
Black eagle	*Aquila verreauxii*
Tawny eagle	*Aquila rapax vindhiana*
Steppe eagle	*Aquila rapax nipalensis*
Lesser spotted eagle	*Aquila pomarina*
Golden eagle	*Aquila chrysaetos*
Wedge-tailed eagle	*Aquila audax fleayi*
Wahlberg's eagle	*Aquila wahlbergi*
African hawk eagle	*Hieraaetus spilogaster*
Booted eagle	*Hieraaetus pennatus*
Martial eagle	*Polemaetus bellicosus*
Rufous-bellied eagle	*Hieraaetus kienerii*
Ayres' hawk eagle	*Hieraaetus ayresii*
Black and white hawk-eagle	*Spizastur melanoleucus*
Changeable hawk-eagle	*Spizaetus cirrhatus*
Blyth's hawk-eagle	*Spizaetus alboniger*
Wallace's hawk-eagle	*Spizaetus nanus*
Black hawk-eagle	*Spizaetus tyrannus*
Ornate hawk-eagle	*Spizaetus ornatus*
Black caracara	*Daptrius ater*
Red-throated caracara	*Daptrius americanus*
Crested caracara	*Polyborus plancus*
Yellow-headed caracara	*Milvago chimachima*
Laughing falcon	*Herpetotheres cachinnans*
Collared forest falcon	*Micrastur semitorquatus*
Black-thighed falconet	*Microhierax fringillarius*
Western red-footed kestrel	*Falco vespertinus*
Dickinson's kestrel	*Falco dickinsoni*
Lesser kestrel	*Falco naumanni*
Eastern red-footed kestrel	*Falco amurensis*

Common name	Scientific name
American kestrel	*Falco sparverius*
Greater kestrel	*Falco rupicoloides*
Common kestrel	*Falco tinnunculus*
Rock kestrel	*Falco tinnunculus*
Red-headed falcon	*Falco chicquera chicquera*
African hobby falcon	*Falco cuvierii*
Hobby falcon	*Falco subbuteo*
Bat falcon	*Falco rufigularis*
Sooty falcon	*Falco concolor*
Aplomado falcon	*Falco femoralis*
Prairie falcon	*Falco mexicanus*
Red-necked falcon	*Falco chicquera*
Merlin	*Falco columbarius*
Taita falcon	*Falco fasciinucha*
Lagger falcon	*Falco biarmicus jugger*
African peregrine falcon	*Falco peregrinus minor*
Peregrine falcon	*Falco peregrinus*
Siberian peregrine falcon	*Falco peregrinus calidus*
Shaheen falcon	*Falco peregrinus peregrinator*
Gymnogene	*Polyboroides typus*
Red-legged partridge	*Alectoris rufa*
Grey partridge	*Perdix perdix*
Pheasant	*Phasianus colchicus*
Red grouse	*Lagopus lagopus*
Black grouse	*Tetrao tetrix*
Virginia rail	*Rallus limicola*
Sora	*Porzana carolina*
Yellow rail	*Coturnicops noveboracensis*
American coot	*Fulica americana*
Common moorhen	*Gallinula chloropus*
Purple gallinule	*Porphyrula martinica*
Black-bellied plover	*Pluvialis squatarola*
Killdeer	*Charadrius vociferus*
Marbled godwit	*Limosa fedoa*
Whimbrel	*Numenius phaeopus*
Willet	*Catoptrophorus semipalmatus*
Greater yellowlegs	*Tringa melanoleuca*
Lesser yellowlegs	*Tringa flavipes*
Wandering tattler	*Heteroscelus incanus*
Solitary sandpiper	*Tringa solitaria*
Spotted sandpiper	*Actitis macularia*
Ruddy turnstone	*Arenaria interpres*
Phalarope, spp.	*Phalaropus spp.*
Woodcock	*Scolopax minor*
Common snipe	*Gallinago gallinago*

Common name	Scientific name
Dowitcher, spp.	*Limnodromus spp.*
Red knot	*Calidris canutus*
Dunlin	*Calidris alpina*
Sanderling	*Calidris alba*
Western sandpiper	*Calidris mauri*
Least sandpiper	*Calidris minutilla*
Bonaparte's gull	*Larus philadelphia*
Ring-billed gull	*Larus delawarensis*
Herring gull	*Larus argentatus*
Caspian tern	*Sterna caspia*
Forster's tern	*Sterna forsteri*
Common tern	*Sterna hirundo*
Least tern	*Sterna antillarum*
Black tern	*Chlidonias niger*
Western gull	*Larus occidentalis*
Passenger pigeon	*Ectopistes migratorius*
Woodpigeon	*Columba palumbus*
Rock dove	*Columba livia*
Cape turtle dove	*Streptopelia capicola*
White-winged dove	*Zenaida asiatica*
Mourning dove	*Zenaida macroura*
Cockatiel	*Nymphicus hollandicus*
Monk parakeet	*Mylopsitta monachus*
Budgerigar	*Melopsittacus undulatus*
Amazon parrot, spp.	*Amazona spp.*
Red-billed quelea	*Quelea quelea*
Yellow-billed cuckoo	*Coccyzus americanus*
Black-billed cuckoo	*Coccyzus erythropthalmus*
Grass owl	*Tyto capensis*
Barn owl	*Tyto alba*
Scops owl	*Otus senegalensis*
White-faced owl	*Otus leucotis*
Eastern screech owl	*Otus asio*
Cape eagle owl	*Bubo capensis*
Eagle owl	*Bubo bubo*
Giant eagle owl	*Bubo lacteus*
Great horned owl	*Bubo virginianus*
Spotted eagle owl	*Bubo africanus*
Snowy owl	*Nyctea scandiaca*
Spotted owl	*Strix occidentalis*
Tawny owl	*Strix aluco*
Wood owl	*Strix woodfordii*
Barred owl	*Strix varia*
Barred owl	*Glaucidium capense*
Pearl-spotted owl	*Glaucidium perlatum*

Common name	Scientific name
Long-eared owl	*Asio otus*
Short-eared owl	*Asio flammeus*
Marsh owl	*Asio capensis*
Little owl	*Athene noctua*
Spotted owlet	*Athene brama*
Brown hawk owl	*Ninox scutulata*
Western burrowing owl	*Speotyto cunicularia hypugaea*
Florida burrowing owl	*Speotyto cunicaluria floridana*
Northern saw-whet owl	*Aegolius acadicus*
Pel's fishing owl	*Scotopelia peli*
Whip-poor-will	*Caprimulgus vociferus*
Common nighthawk	*Chordeiles minor*
Chimney swift	*Chaetura pelagica*
White-throated swift	*Aeronatus saxatalis*
Belted kingfisher	*Ceryle alcyon*
Hairy woodpecker	*Picoides villosus*
Downy woodpecker	*Picoides pubescens*
Yellow-bellied sapsucker	*Sphyrapicus varius*
Red-headed woodpecker	*Melanerpes erythrocephalus*
Red-bellied woodpecker	*Melanerpes carolinus*
Acorn woodpecker	*Melanerpes formicivorus*
Northern flicker	*Colaptes auratus*
Eastern kingbird	*Tyrannus tyrannus*
Western kingbird	*Tyrannus verticalis*
Olive-sided flycatcher	*Contopus borealis*
Purple martin	*Progne subis*
Cliff swallow	*Hirundo pyrrhonota*
Tree swallow	*Tachycineta bicolor*
Violet-green swallow	*Tachycienta thalassina*
Blue jay	*Cyanocitta cristata*
American crow	*Corvus brachrhynchos*
Jackdaw	*Corvus monedula*
Raven	*Corvus corax*
Black-capped chickadee	*Parus atricapillus*
White-breasted nuthatch	*Sitta carolinensis*
Brown creeper	*Certhia americana*
Golden-crowned kinglet	*Regulus satrapa*
Eastern bluebird	*Sialia sialis*
Wood thrush	*Hylocichla mustelina*
Gray-cheeked thrush	*Catharus minimus*
Swainson's thrush	*Catharus ustulatus*
American robin	*Turdus migratorius*
Varied thrush	*Ixoreus naevius*
Northern mockingbird	*Nimus polyglottos*
Gray catbird	*Dumetella carolinensis*

Common name	Scientific name
Brown thrasher	*Toxostoma rufum*
Cedar waxwing	*Bombycilla cedrorum*
European starling	*Sturnus vulgaris*
Red-eye vireo	*Vireo olivaceus*
Yellow-rumped warbler	*Dendroica coronata*
Bay-breasted warbler	*Dendroica castanea*
Blackburnian warbler	*Dendroica fusca*
Common yellowthroat	*Geothlypia trichas*
American redstart	*Setophaga ruticilla*
Scarlet tanager	*Piranga olivacea*
Dickcissel	*Spiza americana*
Northern cardinal	*Cardinalis cardinalis*
Rose-breasted grosbeak	*Pheucticus ludovicianus*
Rufous-sided towhee	*Pipilo erthryophthalmus*
Vesper sparrow	*Pooecete gramineus*
White-throated sparrow	*Zonotrichia albicollis*
Clay-colored sparrow	*Spizella pallida*
Dark-eyed junco	*Junco hyemalis*
Song sparrow	*Melospiza melodia*
Fox sparrow	*Passerella iliaca*
Snow bunting	*Plectrophenax nivalis*
Bobolink	*Dolichonyx oryzivorus*
Brown-headed cowbird	*Molothrus ater*
Yellow-headed blackbird	*Xanthocephalus xanthocephalus*
Red-winged blackbird	*Agelaius phoeniceus*
Brewer's blackbird	*Ephagus cyanocephalus*
Western meadowlark	*Sturnella neglecta*
Eastern meadowlark	*Sturnella magna*
Northern oriole	*Icterus galbula*
Common grackle	*Quiscalus quiscula*
Great-tailed grackle	*Quiscalus mexicanus*
Evening grosbeak	*Coccothraustes vespertinus*
Purple finch	*Carpodacus purpureus*
House finch	*Carpodacus mexicanus*
Common redpoll	*Carduelis flammea*
American goldfinch	*Carduelis tristis*
Pine siskin	*Carduelis pinus*
House sparrow	*Passer domesticus*

Mammals

Brazilian free-tailed bat	*Tadarido brasiliensis*
Little brown bat	*Myotis lucifugus*
Hoary bat	*Lasiurus cinereus*
Domestic dog	*Canis familiaris*

Common name	Scientific name
Black-backed jackal	*Canis mesomelas*
Coyote	*Canis latrans*
Fox	*Vulpes vulpes*
Raccoon	*Procyon lotor*
Beech marten	*Martes foina*
Spotted hyena	*Crocuta crocuta*
Domestic cat	*Felis catus d.*
Tapir	*Tapirus terrestris*
Wild pig	*Sus scrofa*
Red deer	*Cerphus elephas*
North American grey squirrel	*Sciurus carolinensis*
European red squirrel	*Sciurus vulgaris*
American pine squirrel	*Tamiasciurus hudsonicus*
Red squirrel	*Tamiasciurus hudsonicus*
California ground squirrel	*Spermophilus beecheyi*
Rock squirrel	*Spermophilus variegatus*
Northern flying squirrel	*Glaucomys sabrinus*
Pocket gopher	*Geomys busarius*
Beaver	*Castor canadensis*
Black-tailed prairie dog	*Cynomys ludovicianus*
Deer mouse	*Peromyscus maniculatus*
California vole	*Microtus californicus*
Field vole	*Microtus agrestis*
Vole (Montane)	*Microtus montanus*
Black rat	*Rattus rattus*
Brown rat	*Rattus norvegicus*
Garden dormouse	*Eliomys quercinus*
Snowshoe hare	*Lepus americanus*
Audubon's cottontail rabbit	*Silvilagus audobonii*
Eastern cottontail rabbit	*Silvilagus floridanus*
Hare	*Lepus europaeus*
Varying hare	*Lepus timidus*
Rabbit	*Oryctolagus cunniculus*

Reptile
Ocellated lizard	*Lacerta lepida*
Horned lizard	*Phrynosoma platyrhinos*
Ladder snake	*Elaphe scalaris*
Montpelier snake	*Malpolon monspessulanus*

Fish
Black crappie	*Pomoxis nigromaculatus*
Carp	*Cyprinus carpio*
Goldfish	*Carassius auratus*
White perch	*Roccus americanus*

Alewife *Alosa pseudoharengus*
Largescale sucker *Catostomus macrocheilus*
Largemouth bass *Micropterus salmoides*

Plants
Ash, Green *Fraxinus pennsylvanica*
Ash, White *Fraxinus americana*
Aspen, Quaking *Populus tremuloides*
Basswood *Tilia americana*
Beech, American *Fagus grandifolia*
Birch, Red *Betula nigra*
Brome, Ripgut *Bromus diandrus*
Cardon *Pachycereus pringlei*
Catalpa *Catalpa bignonioides*
Cedar, Deodora *Cedrus deodora*
Cheatgrass *Bromus tectorum*
Cherry, Black *Prunus serotina*
Chess, Foxtail *Bromus rubens*
Clove *Eugenia aromatica*
Cottonwood, Black *Populus trichocarpa*
Cottonwood, Plains *Populus deltoides*
Cottonwood *Populus fremontii*
Creosote bush *Larrea tridentata*
Damar *Shorea javanica*
Durian *Durian zibethinus*
Elm, American *Ulmus Americana*
Elm, Slippery *Ulmus fulva*
Fig *Ficus petiolaris*
Fir, Douglas *Pseudotsuga menziesii*
Grama, Blue *Bouteloua gracilis*
Grass, Buffalo *Buchloe dactyloides*
Grass, Bunch *Agropyron spicatum, Elymus cinereus*
Greasewood *Sarcobatus vermiculatus*
Gum, Blue *Eucalyptus globulus*
Hemlock, Western *Tsuga heterophylla*
Hickory, Shagbark *Carya ovata*
Lettuce, Prickly *Lactura serriola*
Locust, Black *Robinia pseudoacacia*
Maple, Bigleaf *Acer macrophylum*
Maple, Sugar *Acer saccharum*
Maple, Silver *Acer saccharinum*
Mesquite, Honey *Prosopsis glandulosa*
Oak, Coast live *Quercus agrifolia*
Oak, White *Quercus alba*
Oak, Burr *Quercus macrocarpa*
Oak, Red *Quercus borealis*

Palm, California Fan	*Washingtonia filifera*
Palm	*Erythea roezlii*
Pine, Jack	*Pinus banksiana*
Pine, Red	*Pinus resinosa*
Pine, White	*Pinus strobus*
Pine, Austrian	*Pinus nigra*
Pine, Scots	*Pinus sylvestris*
Pine, Lodgepole	*Pinus contorta*
Pithecellobium	*Pithecellobium mexicana*
Rabbitbrush	*Chrysothamnus nauseosus*
Ragweed, Canyon	*Ambrosia ambrosioides*
Rubber	*Hevea brasiliensis*
Sagebrush, Big	*Artemisia tridentata*
Sagebrush, Sand	*Artemisia filifolia*
Spruce, Sitka	*Picea sitchensis*
Spruce, Norway	*Picea abies*
Sycamore, Western	*Platanus racemosa*
Thistle, Russian	*Salsola iberica*
Wheatgrass, Crested	*Agropyron cristatum*
Willow, Gooding	*Salix gooddingii*
Willow, Black	*Salix gooddingii*
Yucca	*Yucca glauca*

Index

Accipitridae, *see* hawks
African goshawk, in Zimbabwe, 350
aggression
  by Mississippi kite, 45, 50
  by peregrine falcon, interspecific, 11
  by red-shouldered hawk, 34
  by white-tailed kite, 171, 173, 174
agroforests, xvi
  Indonesian, 245–6, 249
    species composition, 253
  see also rain forest communities, and agroforestry
airports/aircraft strikes, xvi
  in India, 315–23
    airport associations, 320–2
    mitigation/management, 322
    strike analysis, 316–20
    study areas, 315–16
    study methods, 316
  in Zimbabwe, 348
American kestrel, xiv
  habitat, 155–6
    essential features, 161
  Rocky Mountain Arsenal, Colorado, 367–9, 371
  starling competition, nest boxes, 156–61
    box occupancy pattern, 158–9, 160
    breeding success, 159
    kestrel dominance, 161
    prey items, 159
    study area, Idaho, 156–8
    study methods, 158
  utility structure nesting, 93, 94
  Venezuelan rain forest, 270
Andean condor, conservation effort, 327
aplomado falcon, Venezuelan rain forest, 270
artificial nesting structures, 77
  for ferruginous hawks, 138–43
    breeding success, 140–2
    construction, 139
    introduction, 138
    management benefits, 143
    protection from nest failure, 142
    study areas, 138–9
    study methods, 139–40
  for ospreys, Great Lakes, 110, 112, 113–19
  for red-tailed hawks, *vs* natural sites, 78–84
  see also nest boxes; nesting platforms; utility structures, raptor use
association analysis, *see* landscape/raptor association analysis, Sacramento Valley
automobile collisions, *see* vehicle collisions
Ayres' eagle, in Zimbabwe, 350

bald eagle, xvi
  adaptability, *vs* osprey, 119
  at Rocky Mountain Arsenal, Colorado, 370, 371
  and dams, 299–306
    census results, 301–3
    features in dam attractiveness, 304
    fish availability, 304
    management recommendations, 304–5
    and season, 303
    study area, 300–1
    survey methods, 301
  hacking projects, 91, 92
  in Mexico, and human activity, 359
  powerline nesting, 89–91
  reservoir use, South Carolina, 285–98
    and avian use of reservoirs, 291, 292, 293, 294
    and fish availability, 293–5
    limiting factors, 295
    nest numbers, 289–90
    population recovery, 286, 292–3, 295
    productivity assessment methods, 287
    reproductive success, 291
    study area, 286–7, 288, 289
    survey methods, 287–9
banding
  of peregrine falcons, 16
  of urban red-shouldered hawks, 35
barn owl
  as burrowing owl predator, 66
  and pesticides in Zimbabwe, 345
  in spruce forests, lack of nest sites, 209
  utility structure nesting, 93, 94

bat falcon, Venezuelan rain forest, 270
bat hawk, habitat selection, 254
bateleur eagle, in Zimbabwe, 350
  decline, 344
  poisoning, 347
  superstition about, 349
bats, as burrowing owl prey, 66
besra, habitat selection, 255
bicolored hawk, Venezuelan rain forest, 267
black caracara, Venezuelan rain forest, 270
black eagle, in Zimbabwe, habitat destruction, 344
  see also Indian black eagle
black grouse, increase, and afforestation, 207
black hawk, see common black hawk; great black hawk
black sparrowhawk, in Zimbabwe, 350
black vulture
  in Mexico, and human activity, 359
  Venezuelan rain forest, 266
black-collared hawk, Venezuelan rain forest, 268
black-face hawk, Venezuelan rain forest, 268
black-shouldered kite, in Zimbabwe, increasing habitat, 343
black-thighed falconet, in rain forest, and agroforestry, 253, 257
Blyth's hawk-eagle, habitat selection, 254, 257
breeding programmes, vultures, 333–4
bridge nesting, peregrine falcons, 11, 15–23
  breeding success, 21, 22
  bridge sites, 16
  eggshell thinning, 21–2
  management options, 22
  methods, 16
  prey base, 21, 24
  study results, 17–20
  territory use, 21
Britain
  afforestation, 201, 204–5
  deforestation, historical, 202
  raptor species, 202–4
  see also conifer plantations, sparrowhawks in; spruce forests, in Britain
broad-winged hawk, Venezuelan rain forest, 267
buildings for nesting, peregrine falcons, 11
burrowing owls, xii
  Florida, 61
  human association, and deaths, 62
  see also western burrowing owl
buzzard, see common buzzard; oriental honey buzzard

California condor
  conservation effort, 327
  eggshell thickness, 330
  poisoning
    lead, 331
    strychnine, 330

California condor programme, 333
California voles
  as Swainson's hawk prey, and white-tailed kites, 174
  as white-tailed kite prey, 173
  see also voles
Cape griffon
  adaptation, 328
  conservation programme, 327
  in Zimbabwe, 350
captive breeding
  peregrine falcons, in Zimbabwe, 349
  vultures, 333–4
  see also hacking
caracara, Venezuelan rain forest, 270
  see also crested caracara
cattle troughs, vulture drowning, 331–2
cemeteries, red-tailed hawk habitat, 83
changeable hawk-eagle, habitat, and agroforestry, 256–7
chemical contamination
  Great Lakes basin, 25
  Rocky Mountain Arsenal, Colorado, 366
    cleanup, and raptor occurrence, 371
  see also organochlorine pesticides; poisoning of raptors
chimneys, peregrine falcon nesting, 11
  nest boxes, 92, 145, 146
cicadas, Mississippi kite prey, 47
cinereous vultures, population, and livestock farming, 329
collapse of nests, ospreys, 116, 119
  see also nest failure; nest loss
collared forest falcon, Venezuelan rain forest, 269
collisions, as raptor hazard
  for urban peregrine falcons, 11–12
  with vehicles, urban screech owl, 73
  in Zimbabwe, 340–1, 348–9
common black hawk
  and human activity in Mexico, 355
  response to dam construction, 358–9
  study area, 356–7
  study methods, 357
  range, 356
  see also great black hawk
common buzzard, and goshawks, 237
competition
  American kestrel/starling, see under American kestrel
  goshawks in America
    food, 237–8
    nesting, 236–7
  white-tailed kite/Swainson's hawk, 171, 173, 174
condors, conservation effort, 327
  see also California condor
conifer forests, in reafforestation of Britain, 204–5
  see also pine trees; spruce forests, in Britain

conifer plantations, sparrowhawks in, xv
conifer *vs* broadleaved preference, 199
habitat, 192
spatial variation, 193–7
temporal variation, 197
young plantation preference, 197, 198
conservation
raptors in Zimbabwe, 350–1
vultures, 327, 332–4
captive breeding, 333–4
feeding stations, 332–3
public education, 334
Cooper's hawk, xii
and goshawks, 237
urban
nesting biology, 41–4
territories, 36–7
corvids, as goshawk prey, 237
cottontail rabbits, as goshawk prey, 239
crane hawk, Venezuelan rain forest, 269
Crane Prairie Reservoir, osprey management area, 98, 104–5
crested caracara
and human activity in Mexico, 355, 359–62
range, 356
Venezuelan rain forest, 270
crested eagle, Venezuelan rain forest, 268–9
crested goshawk
habitat selection, 255, 257
in tropical agroforest, 254
crested serpent eagle, in rain forest, and agroforestry, 253, 255–6
crowned eagle, in Zimbabwe, 350
crows, forest colonization, raptor nest sites, 208
cuckoo, as peregrine falcon prey, 10

Damar agroforests, 248
dams
Angostura, Mexico, black hawk study, 356, 358–9
bald eagle use, xvi, 299–306
census results, 301–3
features in dam attractiveness, 304
and fish availability, 304
management recommendations, 304–5
and season, 303
study area, 300–1
survey methods, 301
in Zimbabwe, 343
see also reservoirs
DDE, and ospreys, in western Oregon, 104
DDT
and egg thinning, in peregrine falcons, 26–7, 28
Great Lakes contamination, 25
and ospreys
in Central Europe, 129
in western Oregon, 104
in Zimbabwe, 344–5, 346

deforestation
Britain, historical, 202
rain forest, Venezuela, 263–4
raptor adaptation, xv–xvi
deltamethrin, use in Zimbabwe, 345
desertion of nest site, and forestry operations, 275
see also nest failure
dieldrin, 345
double-toothed kite, Venezuelan rain forest, 267
drowning of raptors
vultures, in cattle troughs, 331–2
in Zimbabwe, 340–1, 348
ducks, reservoir use, and bald eagles, 291, 292, 293, 294
durian agroforests, 248–9

eagle owl
and goshawks in Europe/North America, 237
in Zimbabwe, vehicle collisions, 348
eagles
Venezuelan rain forest, 268–9
in Zimbabwe
exotic tree use, 339, 342
habitat destruction, 344
increasing habitat, 343
see also by particular species
eastern screech owl
powerline nesting, 89, 90
urbanization, 69–74
life history, 70, 71
model, 72–4
model assumptions, 71
population ecology, 70, 71–2
rural-urban environmental continuum, 70, 71
study populations, 69–70
education, in vulture conservation, 334
eggs
collecting, in Zimbabwe, 349–50
DDT concentrations, in Zimbabwe, 345, 346
eggshell thinning
peregrine falcons, 21–2
urban, Wisconsin, 26–9
vultures, 330
Egyptian vulture
adaptation to human activity, 328
population, and farming, 329
electric utility structures, *see* utility structures, raptor use
see also powerline nesting
electrocution of raptors
vultures, 331
in Zimbabwe, 340–1, 348
enhancement programs
ospreys, Crane Prairie Reservoir, 98, 104–5
by utility companies, 91–3

ethno-medicine, in Zimbabwe, raptors in, 340–1, 349
eucalyptus, raptor use, Zimbabwe, 339, 342
Eurasian hobby, 204
Eurasian sparrowhawk, forest colonization, 208
European griffon, conservation programme, 327
exotic trees, Zimbabwe, raptor use, 339, 342

falconry, in Zimbabwe, 340–1, 349
falcons
  population dynamics, Sacramento Valley, 180, 181–3, 184, 186, 187
  Venezuelan rain forest, 269, 270
  see also by particular species
farming, see agroforests; livestock farming
ferruginous hawk, 32
  artificial nesting structures, see under artificial nesting structures
  at Rocky Mountain Arsenal, Colorado, 370, 371
  population decline, 137–8
  population status, 137
field voles, on moorland, and afforestation, 207
  see also California voles; voles
fish
  Great Lakes changes, osprey adaptation, 119–20
  as peregrine falcon prey, 5
  in reservoirs, and bald eagles, 293–5
    in discharge, 304
  in Willamette River, Oregon, and ospreys, 105
fish eagle, in Zimbabwe
  and dams, 343
  DDT in eggs, 345
  habitat destruction, 344
  nest destruction, 347
fish-eating birds, reservoir use, 291, 292, 293, 294, 298
Florida burrowing owl, 61
flying squirrels, spotted owl prey, 220, 222, 223
Forestry Commission, in Britain, 204
forestry operations
  and nest desertion, 275
  and wedge-tailed eagle, see wedge-tailed eagle, Tasmania, disturbance
forests, Washington/Oregon
  management, 217
    for owls, 226–8
  natural succession, 216–17
  and spotted owl
    adaptation, 216, 217–18, 225–6
    habitat selection, 218–22
    home ranges, 222–3
    population, and habitat, 223–4

  see also conifer forests; conifer plantations; deforestation; rain forests; spruce forests, in Britain
freeway intersections, red-tailed hawk habitat, 83

garbage/garbage dumps
  and caracara in Mexico, adaptability, 359, 361, 362
  management recommendations, 322
  and raptors at airports, 321
gin traps, 349
golden eagle
  and afforestation, 210–11
  hacking projects, 91, 92
  powerline nesting, 90
  in spruce forests, 209
goshawk, xv
  crested
    habitat selection, 255, 257
    in tropical agroforest, 254
  in North America, vs Europe, 233, 236–9, 239–40
    and food competition, 237–8
    future research questions, 240
    and nesting competition, 236–7
    and winter food shortage, 238–9
  in northern Europe, adaptation, 233, 234–6
  ospreys as prey, 131
  sparrowhawks as prey, and sparrowhawk habitat, 198
  in Zimbabwe, 350
    and DDT use, 345
    exotic tree use, 339, 342
    increasing habitat, 343
Gotland, goshawk study, 234–5
grassland, see reclaimed surface mines, for northern harrier
gray hawk, Venezuelan rain forest, 268
gray-headed kite, Venezuelan rain forest survey, 266
great black hawk, Venezuelan rain forest survey, 268
  see also common black hawk
great horned owl
  electric utility nesting sites, 93, 94
  as goshawk nesting competitor, 237
  as spotted owl predator, 220
Great Lakes basin
  chemical contamination, 25
  ospreys, see under osprey
grey squirrels, as goshawk prey, 239
griffon vultures
  Cape, adaptation to disturbance, 328
  conservation effort, 327
  population, and livestock farming, 328–9
  in Zimbabwe, 350
grouse, increase, and afforestation, 207
gulls, peregrine falcon interactions, 18–19, 22

hacking projects, by utility companies, 91–2
  see also captive breeding
hares, as goshawk prey, 234
  Europe vs America, 238
harpy eagle, Venezuelan rain forest, 269
harriers
  aircraft strikes, 317
  at airports, 321
  hen, forest colonization, 207
  marsh, 204
  Montagu's, 204
  northern, see northern harrier
Harris' hawk, powerline nesting, 89, 90
hawk-eagle
  Venezuelan rain forest, 269
  in Zimbabwe, shooting, 347
  see also changeable hawk-eagle
hawks
  population dynamics, Sacramento Valley, 180, 181, 182, 184, 186, 187
  urbanization model, 72–4
    assumptions, 71
  Venezuelan rain forest, 267–9
  in Zimbabwe, exotic tree use, 339, 342
  see also by particular species
hen harrier, forest colonization, 207
hevea agroforests, 248
hook-billed kite, Venezuelan rain forest survey, 266
horned owl, see great horned owl

Imataca Forest Reserve, see Venezuela, rain forest fragmentation
Indian black eagle, habitat selection, 255
insecticides, for mosquitoes, and ospreys, 128
  see also organochlorine pesticides
insects, Mississippi kite prey, 47

Jerdon's baza, habitat selection, 254

K-selection, in stable population, 73
kestrels
  aircraft strikes, 317
  nest boxes, electric utility company projects, 92–3
  powerline nesting, 77
  in Zimbabwe, man-made structure use, 342
  see also American kestrel; lesser kestrel, in Spain
king vulture, Venezuelan rain forest survey, 265
kites
  aircraft strikes, 316–20
  at airports, 321–2
  black-shouldered, in Zimbabwe, 343
  collisions in Zimbabwe, 348
  Venezuelan rain forest, 266–7
  see also Mississippi kite; white-tailed kite

lagomorphs, as goshawk prey, 238
landscape ecology, 178
landscape engineering, Sacramento Valley, 178
landscape/raptor association analysis, Sacramento Valley, 178–88
  association identification, 180
  falcons, 180, 181–3, 184, 186, 187
  hawks, 180, 181, 182, 184, 186, 187
  landscapes preferred/avoided, 185
  limitations, 183
  perch use patterns, 186, 188
  and season, 184, 187
  survey methods, 179–80
  validity, 183, 185
  turkey vulture, 180, 183, 184, 186, 187
  see also white-tailed kite, and land-use, Sacramento Valley
lanner falcon
  man-made structure use, 342
  in Zimbabwe
    habitat increase, 343
    nest destruction, 348
    population increase, 350
    use by falconers, 349
lappet-faced vulture, and livestock farming, 329
largescale sucker, as osprey prey, 105
laughing falcon, Venezuelan rain forest, 269
lead poisoning, of vultures, 330–1
least terns, 22
lesser kestrel, in Spain, xii, 53–60
  breeding parameters, 58
  food supply, 53
  foraging distances, 59
  predation pressure, 53, 55–6
  prey items/delivery, 55, 56–8, 58–9
  study areas, 54
  study methods, 54–5
Levins formula, in rain forest habitat analysis, 251–2
lighting, artificial, and burrowing owl success, 66
little owl, 204
livestock farming
  predator poisoning
    and raptors in Zimbabwe, 347
    strychnine, and vulture decline, 330
  vulture adaptation, 328–9
logging, rain forest, Venezuela, 263–4
logit analysis, for raptor analysis, 367, 369, 370, 372–3
long-eared owl, and urbanization, 31

man-made nest sites, see artificial nest sites
marsh birds, reservoir use, 291, 292, 293, 294, 298
marsh harrier, 204
martial eagle, in Zimbabwe, 350–1
  drowning, 348
medicine, traditional, raptors in, 340–1, 349

mercury
  egg levels, in peregrine falcons, 28
  Great Lakes contamination, 25
  in reservoirs, and bald eagle population, 295–6
merlin, forest colonization, 207
Mississippi kite, xii, 45–52
  demography, 48–9
  future of populations, 51–2
  history of urbanization, 46–7
  initial response to humans, 49
  natural history, 47–8
  natural selection pressure, 49
  nest defense diving, 45
  population shift, 49–50
  study methods, 46
  urban nesting benefits, 50–1
Montagu's harrier, 204
mosquito insecticide control
  DDT, in Zimbabwe, 344–5
  and osprey decline, 128

nature reserves, and agriculture, 177
nest boxes
  competition, *see* American kestrel, starling competition
  for kestrels, 92–3
  for peregrine falcons, 11
    on power plant chimneys, 92
    on shoreline power plants, 145–52
  see also artificial nesting structures
nest defense, by Mississippi kites
  against humans, 45
  against predators, 50
  see also aggression
nest destruction, in Zimbabwe, 347–8
nest failure
  desertion, and forestry operations, 275
  white-tailed kite, 171, 172
nest loss, kites, and urbanization, 50
  see also collapse of nests
nest sites
  airports, 321
  protection, *see* wedge-tailed eagle, Tasmania
  spotted owl, 219, 220–1
nesting platforms
  installation by utility companies, 92
  for ospreys
    on electricity pylons, 127
    Great Lakes basin, 113–14, 116, 118–19
  see also artificial nesting structures; utility structures, raptor use
northern flying squirrels, owl prey, 220, 222, 223
northern goshawk, in spruce forests, 208
northern harrier, xvi
  decline, 307–8
  reclaimed habitat, 308–13
    breeding attempts, 308–11
    management guidelines development, 312
    plant species, 308

organochlorine pesticides
  eggshell analysis, urban peregrine falcons, 27–9
  and ospreys
    in Central Europe, 129–30
    in Great Lakes basin, 110
  reducing levels, and raptor increases in Britain, 204
  and vulture decline, 330
  in Zimbabwe, 340–1, 344–7
  see also chemical contamination; poisoning
oriental honey buzzard, 254, 255–6
osprey, xiii–xiv
  artificial nest sites, 77, 98
    utility structures, 89, 90, 93, 94, 101–3, 104–5
  in Central Europe, 126–35
    breeding habitat, 127
    breeding success, German study, 130–3
    distribution, 126
    management, 127
    pesticides, and breeding success, 129–30
    population density, 127–8
    population trends, 128–9
  Great Lakes basin, 109–20
    adaptability, 117–20
    historical view, 110
    nest sites, artificial *vs* natural, 113–17
    and organochlorine pesticides, 110
    study area, 110–12
    study methods, 112
  hacking projects, 91, 92
  historical abundance, in Oregon, 97–8
  in Western Europe, 125
  Willamette River, Oregon, 98–106
    historical population, 103–4
    population increase, 101–3, 104–6
    study area, 98–100
    survey methods, 100
Ovambo, in Zimbabwe
  eucalyptus nesting, 339, 342
  increasing habitat, 343
owls, xii
  aircraft strikes, 317
  at airports, 321
  Mississippi kite nest predation, 50
  urbanization model, 72–4
    assumptions, 71
  in Zimbabwe
    exotic tree use, 339, 342
    increasing habitat, 343
    superstition about, 349
  see also by particular species

parks as raptor territory
  Cooper's hawk, 37
  red-shouldered hawk, 34
PCBs
  and egg thinning, in peregrine falcons, 27–9
  Great Lakes contamination, 25

pearl kite, Venezuelan rain forest, 266
perches
  falcons, 183
  hawks, 181
  Sacramento Valley raptors, 186, 188
  white-tailed kite, 170
peregrine falcon, xi–xii, xiv
  artificial nest sites, 78
  bridge nesting, see bridge nesting, peregrine falcons
  forest colonization, 207
  hacking projects, 91–2
  power plant nest boxes, 145–52
  urban nesting, eggshell contaminant analysis, 25–30
  urban North America, 3–13
    eggshell analysis, Wisconsin, 26–9
    nest sites, 11
    numbers/distribution, 3–5, 6–7
    prey species, 5, 8–11
    urban hazards, 11–12
  in Zimbabwe
    collisions, 349
    DDT in eggs, 345
    habitat destruction, 344
    increasing habitat, 343
    man-made structure use, 342
    nest destruction, 348
    population threats, 350
    use by falconers, 349
persecution of raptors
  in Britain
    current, 205
    nineteenth century, 202
  ospreys, shooting, in Oregon, 105
  vultures, shooting, 330
  in Zimbabwe, 340–1, 347–8
pesticides, see organochlorine pesticides
  see also chemical contamination; poisoning of raptors
pheasants, as goshawk prey, 234
pigeons, as goshawk prey, 235, 239
pine squirrel, as goshawk prey, 239
pine trees, Zimbabwe, raptor use, 339, 342
  see also conifer forests; conifer plantations
plumbeous kite, Venezuelan rain forest, 267
poisoning of raptors
  urban screech owls, 73
  vultures, 330–1
  in Zimbabwe, 340–1, 347
  see also chemical contamination; organochlorine pesticides
Poland, osprey population, 128, 133
power plants, for peregrine falcon nest boxes, 92, 145–52
powerline nesting, 77, 87
  mail survey, 88–94
    cooperating agencies, 94
    methods, 88
    raptor enhancement programs, 91–3
    response rate, 88
    species reported, 88–91
  red-tailed hawks, 79, 80, 89, 90
  see also red-tailed hawks, nesting study; utility structures, raptor use
prairie dogs, and raptor occurrence, 367, 370, 371–2
prairie falcon, powerline nesting, 89, 90
prey
  of American kestrel, 159
  at airports, 321
  of bald eagle, fish in reservoirs, 293–5
  of burrowing owl, bats, 66
  of goshawk
    Europe vs America, 234–5, 237–9
    ospreys, 131
    sparrowhawks, 198
  of lesser kestrel, 55, 56–8, 58–9
  of Mississippi kite, 47
  of osprey, 105
  of peregrine falcon, 21, 24
    urban, 5, 8–11
  of spotted owl, 220, 222, 223
  of Swainson's hawk, 174
  of white-tailed kite, 173
public education, in vulture conservation, 334
pylons, see powerline nesting; utility structures, raptor use

queleas, pesticide control, and raptor deaths, 345
queletox, 345

r-selection, in young communities, 73
rabbits, as goshawk prey, 234
  cottontail, 239
raccoons, as osprey egg predators, 118
radio-tagging, urban red-shouldered hawks, 35
rain forest communities, and agroforestry, 246–61
  community composition/structure, 252–4
  conservation implications, 259
  habitat categories, 250–1
  habitat requirements, 261
  habitat selection, 254–7
  habitat types, 246–9
    agroforests, 246–9
    cultivated areas, 246
    primary forest, 249
  pooling of primary forest data, 257–8
  species adaptation, 258–9
  species as disturbance indicators, 258
  statistical methods, 251–2
  study area, 246, 247
  survey method, 249–50
rain forests, human disturbance, 245
  see also Venezuela, rain forest fragmentation
Ratcliffe egg thickness index, 26–7

reclaimed surface mines, for northern harrier, 308–13
  breeding attempts data analysis, 308–11
  management guidelines development, 312
  plant species, 308
red squirrels, as goshawk prey, 239
red-shouldered hawk, xii
  behavior, 34–5
  territory types, 33–4
  urban, southern California, 31–8
    food habits, 36
    habitat, 32
    nest trees, 35
    reproductive success, 35
    space use, 36
    study area, 32
    study methods, 32–3
red-tailed hawk, xiv, 32
  as goshawk nesting competitor, 236–7
  nesting study, man-made *vs* natural sites, 78–84
    advantages of man-made substrates, 83
    habitat comparisons, 81–2
    macrohabitat comparisons, 82–3
    man-made substrates, 79, 80
    natural substrates, 79–80
    nearest neighbor distances, 84
    site comparisons, 81
    study area, 78
    study methods, 78–9
  Rocky Mountain Arsenal, Colorado, 367, 368, 369, 370, 371–2
  urban, territories, 36–7
  urbanization pressure, 84
  utility structure nesting, 79, 80, 89, 90, 93, 94
red-throated caracara, Venezuelan rain forest, 270
reservoirs
  avian guild components, 298
  and bald eagle increase, xvii
  bald eagle use, South Carolina, 285–98
    and avian use, 291, 292, 293, 294
    and fish availability, 293–5
    limiting factors, 295
    nest numbers, 289–90
    population recovery, 286, 292–3, 295
    productivity assessment, 287
    reproductive success, 291
    study area, 286–7, 288, 289
    survey methods, 287–9
  concrete tank, raptor drowning, 348
  in Zimbabwe, additional habitat, 343
  see also dams
'restaurants' for vultures, 332–3
rice stubble, as white-tailed kite habitat, 170, 171, 173
roadside hawk, Venezuelan rain forest, 268
Rocky Mountain Arsenal, Colorado, 366–74
  raptor distribution patterns, 371–2

raptor tolerance of disturbance, 371, 372
species encountered
  summer survey, 367–70
  winter survey, 370
study area, 366
study methods, 366–7
  roadside survey weakness, 372
roosting sites
  at airports, 321
  spotted owl, 219, 221
rough-legged hawk, 370, 372
rufous-bellied eagle, habitat selection, 254, 257

savanna hawk, Venezuelan rain forest, 268
sciurids, as goshawk prey, 237
screech owl, suburban-nesting, xiii
  see also eastern screech owl
shooting of raptors
  ospreys, in Willamette Valley, Oregon, 105
  vultures, 330
  in Zimbabwe, 347
short-eared owl, and afforestation, 210
short-tailed hawk, Venezuelan rain forest, 268
sitka spruce, in British reafforestation, 205
  see also spruce forests, in Britain
slender-billed kite, Venezuelan rain forest, 267
smoke stacks, *see* chimneys, peregrine falcon nesting
snail kite, Venezuelan rain forest, 267
snowshoe hares, as goshawk prey, 238
songbirds
  and afforestation, 207
    second generation crops, 209
  densities, conifer *vs* broadleaved forests, 192
sparrowhawk
  Eurasian, forest colonization, 208
  in Zimbabwe, 350
    exotic tree use, 339, 342
    increasing habitat, 343
  see also conifer plantations, sparrowhawks in
spotted eagle owl, in Zimbabwe, vehicle collisions, 348
spotted owl
  adaptation to managed forests, 216, 218, 225–6
  forest management for, 226–8
  and forestry industry, xv
  habitat
    natural, 215–16
    relationships, 219–24
    selection, ultimate/proximate factors, 218–19, 225
  life cycle stages, 224
  population decline, 216
spruce forests, in Britain, and raptors, 210–11
  and clear-felling, 209–10
  and forest succession, 205–9

squirrels, 208
  as goshawk prey, 234
    Europe *vs* America, 239
  as spotted owl prey, 220, 222, 223
starling, *see* American kestrel: starling competition
starvation, lesser kestrel, urban, 56, 57
storms, and osprey nest collapse, 119
strychnine, for livestock predators
  raptor poisoning, 347
  and vulture decline, 330
sugarbeet, as white-tailed kite habitat, 170, 171, 173
superstition about raptors, in Zimbabwe, 349
Swainson's hawk, 32
  Rocky Mountain Arsenal, Colorado, 367, 368, 372
  white-tailed kite competition, 171, 173, 174
swallow-tailed kite, Venezuelan rain forest, 266

taita falcon, in Zimbabwe, 350
Tasmania, *see* wedge-tailed eagle, Tasmania
tawny owl, in spruce forests, 208
  and clear-felling, 210
telodrex, raptor poisoning in Zimbabwe, 347
terns, 22
territorial defense, *see* aggression
thrushes, as goshawk prey, 237
tiny hawk, Venezuelan rain forest, 267
toxaphene, raptor poisoning in Zimbabwe, 347
trapping of raptors
  burrowing owls, 64
  in Zimbabwe, 347
    carnivore gin traps, 349
tsetse fly, DDT control in Zimbabwe, 344–5
turkey vulture, xviii
  landscape association analysis, 180, 183, 184, 186, 187
  in Mexico, and human activity, 359
  Venezuelan rain forest, 266

urbanization
  Cooper's hawk
    nesting biology, 41–4
    territories, 36–7
  eastern screech owl, *see under* eastern screech owl
  kites, nest loss and, 50
  lesser kestrel, *see* lesser kestrel, Spain
  long-eared owl and, 31
  Mississippi kite
    history of, 46–7
    nesting benefits, 50–1
  model, 72–4
    assumptions, 71
  peregrine falcon, *see under* peregrine falcon
  see also bridge nesting, peregrine falcons

red-shouldered hawk, *see under* red-shouldered hawk
red-tailed hawk
  pressure, 84
  territories, 36–7
utility structures, raptor use, xiii
  American kestrel, 93, 94
  at airports, in India, 321
  and electrocution
    vultures, 331
    in Zimbabwe, 340–1, 348
  hawk perches, 181, 186, 188
  ospreys
    in Great Lakes basin, 112, 113, 114, 115, 116, 118
    pylons in Germany, 127, 130–3
    Willamette River, Oregon, 100, 101, 102–3, 104–5
  red-tailed hawks, 93, 94
  in Zimbabwe, 342, 343–4
  see also artificial nesting structures; powerline nesting

vehicle collisions
  raptors in Zimbabwe, 348
  and screech owl urbanization, 73
Venezuela, rain forest fragmentation, 263–73
  deforestation pressures, 263–4
  research
    future, 271–2
    present inadequacy, 264
    species observed, 265–70, 271
    study area, 264, 270
    survey methods
      road survey, 264–5
      walking survey, 271
voles, and afforestation, 207, 209
  see also California voles
vultures, xvii, 327–35
  adaptation to agriculture, 328–9
  aircraft strikes, 316–20
  at airports, 321
  conservation management, 327, 332–4
    captive breeding, 333–4
    feeding stations, 332–3
    public education, 334
  convergent evolution, 327
  in ethno-medicine in Zimbabwe, 349
  landscape association analysis, 180, 183, 184, 186, 187
  in Mexico, and human activity, 359
  mortality factors, 329–32
    drowning, 331–2
    electrocution, 331, 348
    pesticides, 330
    poisoning, 330–1
    shooting, 330
  reproductive rate, and mortality factors, 329
  Venezuelan rain forest, 265–6

wading birds, reservoir use, 291, 292, 293, 294, 298
Wallace's hawk eagle, habitat selection, 254
waterfowl, reservoir use, 291, 292, 293, 294, 298
wedge-tailed eagle, Tasmania, xvi, 276
  disturbance
    and breeding success, 277–9
    and protection measures, 279–81
    survey methods, 276–7
  protection measures, 276, 279–81
western burrowing owl, 61
  nest sites, 61–2
  nesting success, 62–8
    and artificial lighting, 66
    burrow abandonment, 64–5, 66
    burrow collapse, 65
    and burrow proximity, 65–6
    and neighbor competition, 66–7
    nestlings/fledglings produced, 64–5
    and predation, 66
    prey, 66
    study methods, 64
    study sites, 62–3
Western Hemlock Zone, see forests, Washington/Oregon
white hawk, Venezuelan rain forest survey, 268
white-backed vulture, Indian, adaptation, 328
white-tailed hawk, Venezuelan rain forest, 267
white-tailed kite, xiv
  and land-use, Sacramento Valley, 166–75
    habitat improvement, 174–5
    land-use association, 167
    nesting habitat study, 167–8, 169, 171–2, 173
    population changes, 172–3
    prey, 173
    road survey, 166–7, 168, 170–1
    study area, 166, 167
  and Swainson's hawk competition, 174
  and urban development, 174
  range, 165–6
  urban, territories, 37
  see also landscape/raptor association analysis, Sacramento Valley
wind, and osprey nest collapse, 119
woodpecker cavities, screech owl use, 89, 90
woodpigeon, as goshawk prey, 234
woodrats, spotted owl prey, 223

yellow-headed caracara, Venezuelan rain forest, 270
yellow-headed vulture, Venezuelan rain forest, 266

Zimbabwe raptor review, 337–53
  conservation priorities, 350–1
  exotic tree use, 339, 342
  habitat
    destruction, 344
    increasing, 343–4
  human impact summary, 340–1
  land-use categories, 339
  man-made structure use, 342–3
  mortality factors, 340–1
    collisions, 348–9
    drowning, 348
    electrocution, 348
    organochlorine pesticides, 344–7
    persecution, 347–8
    poisoning, 347
  species diversity/abundance, 339
  study area, 338–9
  study methods, 338
  utilization of raptors, 340–1, 349–50
zone-tailed hawk
  powerline nesting, 89, 90
  Venezuelan rain forest, 267